FRICTION AND WEAR OF POLYMERS

TRIBOLOGY SERIES

Vol. 1 Tribology – A Systems Approach to the Science and Technology of Friction, Lubrication and Wear (Czichos)
Vol. 2 Impact Wear of Materials (Engel)
Vol. 3 Tribology of Natural and Artificial Joints (Dumbleton)
Vol. 4 Tribology of Thin Layers (Iliuc)
Vol. 5 Surface Effects in Adhesion, Friction, Wear, and Lubrication (Buckley)
Vol. 6 Friction and Wear of Polymers (Bartenev and Lavrentev)

TRIBOLOGY SERIES, 6

FRICTION AND WEAR OF POLYMERS

G.M. BARTENEV
Institute of Physical Chemistry of the Academy of Science of the U.S.S.R, Moscow, U.S.S.R.

and

V.V. LAVRENTEV
Department of Physics, Second Moscow Medical Institute, Moscow, U.S.S.R.

Translated by D.B. PAYNE

Edited by

LIENG-HUANG LEE
Xerox Corporation, Rochester, New York

and

K.C. LUDEMA
University of Michigan, Ann Arbor, Michigan

ELSEVIER SCIENTIFIC PUBLISHING COMPANY
AMSTERDAM — OXFORD — NEW YORK 1981

ELSEVIER SCIENTIFIC PUBLISHING COMPANY
1, Molenwerf
P.O. Box 211, 1000 AE Amsterdam, The Netherlands

Distributors for the United States and Canada:

ELSEVIER/NORTH-HOLLAND INC.
52, Vanderbilt Avenue
New York, N.Y. 10017

Library of Congress Cataloging in Publication Data

Bartenev, G. M. (Georgiĭ Mikhaĭlovich), 1915-
Friction and wear of polymers.

(Tribology series, ISSN 0-444-41677-3 ; v. 6)
Translation of: Trenie i iznos polimerov.
Bibliography: p.
Includes index.
1. Polymers and polymerization. 2. Friction.
3. Mechanical wear. I. Lavrent'ev, V. V. (Valentin Vi͡acheslavovich) II. Lee, Lieng-Huang, 1924-
III. Ludema, K. C. IV. Title. V. Series.
TA455.P58B3413 621.8'9 81-9907
ISBN 0-444-42000-2 AACR2

ISBN: 0-444-42000-2 (Vol. 6)
ISBN: 0-444-41677-3 (Series)

Printed in The Netherlands

EDITORS' PREFACE

Polymers are becoming more widely used as bearing and slider materials with each year of availability of these materials. As polymers improve and as designers learn the advantages of polymers, public acceptance also increases. The friction and wear properties of polymers, therefore, becomes of greater interest and importance, which in turn encourages research and development efforts in this field. The use of rubber in tires and in power transmission belts dates back at least 80 years. The use of Bakelite and other thermosetting resins, particularly in automotive brakes, is about as old, whereas the use of nylon and Teflon has grown tremendously since their introduction only 40 years ago.

The understanding of the mechanisms of friction and wear of polymers has progressed very slowly, undoubtedly because of the great range of compositions of polymers and because of the wide range of influence of various operating conditions on friction and wear of plastics and rubbers. It is, therefore, commendable when authors such as Bartenev and Lavrentev take up the task of reviewing the state of knowledge in this subject. Their book is primarily an account of Russian work with some reference to Western and Japanese authors. The book has no rival, however, in the breadth of its coverage of the topic. It includes the friction and wear of polymers in every state ranging from the viscous state to the rigid and crystalline state.

The book gives Western readers a rare opportunity to read the breadth of Russian work and this was the reason for proceeding with the translation of the book. Papers of Russian authors appear in international journals but most Western readers have difficulty comprehending Russian papers. It is often difficult to determine whether lack of understanding is due to differences in basic approach or language difficulties. To some extent, these difficulties appear also in this translation but it is obvious that the Russian writers have a very different philosophy from that of Western writers. For example, the Russian writers are usually of a Mechanics orientation or are Physical Chemists and they are strongly oriented to writing mathematical models of the phenomena they are studying. They build their models with great dispatch, invoking all possible physical and chemical quantities, whether defined or not. By contrast, Western writers will construct a model using only those quantities that are known to be relevant or else those quantities that yield dimensional compatibility. For example, taking the model of Kragelsky, who is given credit for the "most advanced" theory, namely,

> "$F = f_{mol} \times S_{mol} + f_{mech} \times S_{mech}$, where F is the friction force, S is the area of contact, and $f_{mol} = A_1 + B_1 \cdot P_\theta = N/S$, A_1 is the normal molecular interaction stress, B_1 is the coefficient that characterizes molecular roughness (and N is the normal load). Likewise, $f_{mech} = A_2 + B_2 P_\theta$, where A_2 is the shear stress without normal load and B_2 is the coefficient that characterizes the material strengthening under a normal load".

This formulation is characteristic of their method. Western readers would be left with the impression that these models consist of a veritable morass of constants, factors with coefficients, subscripts and exponents that do not always describe known quantities. A second difficulty Western readers may have with Russian papers is characterized in a comment concerning another theory in the book developed by a physical chemist. The model includes a quantity known as the activation energy, and

> "It follows that despite great advances in the field of the polymer friction, physics at the present time is still unable to predict friction force. This is explained mainly by the fact that it is possible to determine one of the fundamental characteristics of friction, the activation energy, only by experiment".

A third difficulty Western readers have with Russian papers is in the nature of their presentation of data. Russian data points are almost always exactly on the plotted line, or alternately above and below the line. On the other hand, Western writers lament the fact that data from wear tests show considerable unexplained scatter.

In the translation of this book, it became apparent that in some areas Russian words are not exactly translatable to English. The word that caused the greatest difficulty is a word which could be translated either "viscoelasticity", or "highly elastic". In some instances, the word was seen from the text to mean material of high compliance or low stiffness, i.e., highly elastic. In other instances, the word described stress or strain cycles with some irreversibility or damping loss, i.e., visco-elasticity. Again, there seemed to be some ambiguity around the state which we refer to as the rubbery state of the polymer. In some instances, it seemed to refer to a state of full, though delayed, recoverability after some strain and other instances it seemed to mean high compliance with damping loss. The Russian seems to contain a word for rubbery morphology as well. All of these were simply translated as "rubbery".

A major difficulty in the book is the great confusion in the use of symbols or nomenclature. Some of the mix of nomenclature is understandable since in the area of polymer friction and wear references are taken from studies in visco-elasticity, mechanics, engineering, and physics. However, the authors could have separated their nomenclature in a more logical order. For example, letters A, B, C, α, β, γ, and one or two others have been used to symbolize up to 30 different quantities each. On the other hand, half of the English and Greek alphabet has gone unused. The consequence is a chaos in which it is impossible to trace, in many instances, whether the quantity referred to on one page is the same quantity referred to on the next page. The matter is not aided by several instances where a coefficient is totally unexplained, or where the authors offer confusing explanations. We have taken the liberty to change many of the symbols where it was possible, or where it was necessary to reduce confusion. We ask the authors' indulgence in making such changes. Finally, invoking editorial privilege we differ with the authors on one central point. They state that the coefficient of friction is the most valuable indicator of wear mechanisms. We cannot support that view from our own work. On the other hand, we commend the authors for their work in producing this book.

L. H. Lee
K. C. Ludema
February, 1981

PREFACE

Polymers have great importance among engineering materials. Without many polymer components, contemporary manufacturing of machines and instruments could not continue. Thus, not only all forms of vibration dampers - automobile and airplane tires and cylinders, but also friction blocks and brake devices as well as packing material employ polymers. Moreover, in important machine components and in friction blocks, the combinations of metal and polymer materials - plastics and vulcanized rubbers, especially when reinforced - occupy a prominent position.

Molecular interactions of polymers and also of these materials with metals on the boundary of their common surfaces determine the friction phenomena and adhesion (sticking), and, consequently, the friction properties of the corresponding friction pair, including wear resistance, that is, durability of connected components.

Therefore, the study of the regularities and molecular mechanisms of friction and adhesion of polymers on the boundary between them and with other materials (metals) has a special importance for physico-chemical mechanics. This science is one of the most important tasks for the production of engineering materials and their combination in machine components (friction blocks and braking) with their given properties, of high strength and durability under conditions of use.

In the Soviet Union during the last decade, significant progress has been made in the development of the physico-chemical mechanics of polymers. Of great significance in this area is the work of G. M. Bartenev and his school, which had added a large contribution to contemporary concepts on strength and failure regularities of polymer materials, and also their rheological and friction properties (especially for high elastic materials).

In 1954, G. M. Bartenev proposed the molecular kinetic theory of friction for high elastic materials with network structure (cross-linked rubbers). Experimental studies in this area and related further development of the theory are presented in this book by G. M. Bartenev and V. V. Lavrentev.

This is the first book in Soviet scientific literature devoted to the physico-chemical mechanics of the friction of polymers. It contains the results of the author's original studies and also analyzes and generalizes all the basic scientific literature in the field.

A valuable special feature of the present book is the authors' propensity to establish a close link between the friction properties of polymer materials and the special characteristics of their molecular structure and physical state, with due consideration given to changes in the real contact area, temperature, pressure, and sliding velocity.

This book is designed for a wide circle of specialists: engineer-technologists and mechanical engineers, designers and users, and also physicists and physical chemists, researchers, teachers, aspirants and university students - all who are interested in the special characteristics of polymeric materials and their production, processing and various uses.

For all these specialists, the present book will undoubtedly provide valuable assistance.

FOREWORD BY P.A. REBINDER

In this book, detailed consideration is given to the nature of friction and wear, the influence of temperature on friction, the effects of sliding velocity, the duration of contact and similar factors. Special attention is given to the question of relations between frictional properties and structure of polymers.

Methods are given for the calculation of the friction coefficients for a wide range of pressures, velocities and temperatures. Principles are illustrated for the design of new products, among which are anti-friction polymeric materials of easy flow and durability.

This book is intended for a wide circle of scientific workers, designers, engineers and technologists in the chemical and mechanical industries, and deals with selection, reprocessing and application of polymers for engineering materials. It will also be of use to teachers and undergraduate and graduate students at colleges and universities.

NOMENCLATURE

a	local constant, a(T) - visco-elastic transformation quantity
b	local constant
c	local constant, $\bar{c}$ - plastic contact flow stress constant
d	diameter of contact area
e	shear strain
f	specific friction force, f_e - elementary tangential force, f_{mol} specific molecular force
g	gravity
h	height of asperities, thickness
i	specific wear
j	mathematical operator
k	Boltzmann constant, local constant
m	strain hardening exponent, mass, strengthening coefficient, local constant
n	number of units or segments, local constant
p	pressure, p_r - real contact pressure, p_n - nominal contact pressure, $p_{m(ol)}$ - molecular attraction stress
q	temperature coefficient of U
r	radius, relaxation rate, local constant
s	shear stress, $\bar{s}$ - surface area of one irregularity
t	time, durability, $\bar{t}$ - mean bond life time
u	displacement, deformation, u_a - distance of approach
v	velocity, v_p - velocity of crack propagation, v_c - actual rate of crack growth
w	width
y	elevation or height
z	(wavelength), z_a - pitch of asperities, z_w - pitch of wear bands
A	local constant
B	local constant
C	local constant
D	rebound number
E	Young's modulus, E' - storage modulus, E" - loss modulus, E_1 - unrelaxed modulus, E_∞ - relaxed modulus, $\bar{E}$ - elasticity, E* - effective modulus
F	force
G	shear modulus
H	hardness, HB - Brinell hardness, H' - hardening coefficient
I	thermal power (energy), Bessel function
J	wear intensity
K	a coefficient, K' - stiffness
L	molecule length
N	intermolecular interaction
O	compressibility
P	load
Q	discharge volume
R	wear resistance, strain gage resistance
S	contact area
S_r	real contact area
S_n	apparent contact area

T	temperature
U	activation energy
V	volume of wear, $\overline{V}$ - specific wear
W	energy, work
Y	shear strength
Z	stress concentration
α	first visco-elastic loss peak, local constants, α_o - coefficient of heat flux
β	second visco-elastic loss peak, β_q, β_n, β_w - coefficients of wear resistance
γ	third visco-elastic loss peak
Γ	local constant
δ	loss angle, δ' - hysteresis loss coefficient
Δ	loss, Δ' - loss decrement
ε	normal strain
ζ	bearing surface, relevant angle
η	viscosity, fourth visco-elastic loss peak
θ_α	contact angle, θ_z - surface tension, $\theta_{v\ell}$ - surface free energy at the vapor-liquid interface, etc.
ι	local constant
κ	thermal conductivity coefficient
λ_a	interatomic distance, λ_j - jump distance, λ_p - wave length of phonon.
Λ	local constant
μ	coefficient of friction
ν	local constant, $\underline{\nu}$ - frequency
ξ	angle
ρ	density
σ	normal stress
Σ	normal strength
τ	relaxation time, period, τ_M - Maxwell relaxation time
ϕ	S_r/S_n
Φ	scaling factor
ω	time factor, frequency.
Ω	elastic after effect, $\dot{\Omega}$ - rate of elastic after effect
$\times$	Poisson's ratio, a coefficient
χ	function of ..

TABLE OF CONTENTS

Chapter 1

STRUCTURE AND PHYSICAL PROPERTIES OF POLYMERS

1.1 PHYSICAL STATES OF POLYMERS

Polymers can exist in four physical states--the crystalline and three amorphous states (glassy, rubbery, and viscous flow). Polymers that exist in the glassy or crystalline state are sometimes called rigid polymers. Each state has its own complex of mechanical properties and its own area of technical application.*

The physical state of a polymer is most often determined by studying its mechanical properties, for example, its compliance. One can judge the basic technological properties of a polymer on the basis of its compliance over a wide range of temperatures (10). One can determine the degree of compliance by use of thermo-mechanical method proposed by Alexandrov and Lazurkin (11) for cyclic deformation and by Kargin et al. (12,13) for static deformation.

Figure 1.1 depicts a thermo-mechanical curve for polymers. At temperatures lower than the glass transition temperature T_g, polymers deform in the manner of low-molecular-weight glass. A significant increase in reversible strain occurs at temperatures above this, indicating the rubbery state. In this range the elastic modulus changes little with temperature up to the flow temperature T_f. Above the flow temperature, a polymer is a liquid of high viscosity.

The rubbery regime for a crystalline polymer lies between T_m and T_f. Depending on the rate of cooling, a polymer can crystallize at a temperature below T_m (curve DCBF); however, with rapid cooling a polymer does not crystallize but passes into the glassy state (curve DCBA) (Fig. 1.2).

---

*The reader who is interested in the preparation of polymers can find considerable information in sources (1-6) and modern views on the structure and classification of polymers in (7-9).

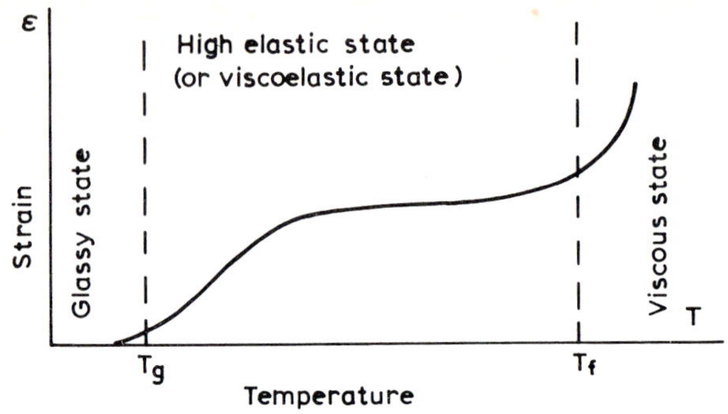

Fig. 1.1 Thermo-mechanical curve for polymers.

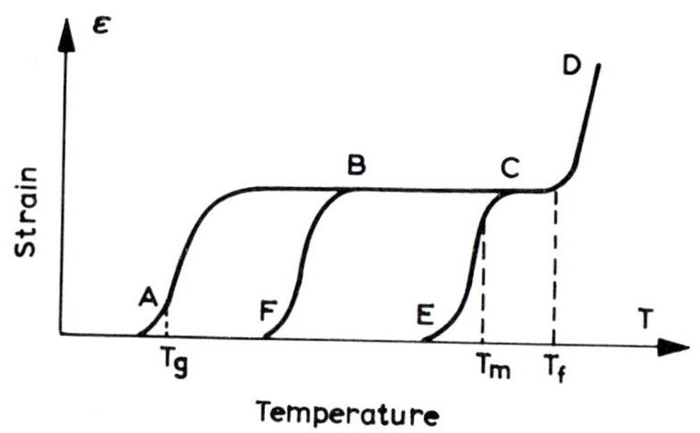

Fig. 1.2 Compliance strain vs. temperature for various polymers.

1.2 PHYSICAL PROPERTIES OF MACROMOLECULES AND POLYMER STRUCTURE

Flexibility is the basic property of chain molecules. A polymer molecule has a transverse dimension of several angstroms and a length of several thousand angstroms, which may form a fiber 10 microns in diameter and several millimeters in length (13). A macromolecule's flexibility is effected by the rotation of its links around ordinary C-C bonds. Through thermal motion and their own flexibility, macromolecular segments take on various spatial forms or conformations. If a macromolecule is stretched, its conformations will be elongated. If the external stretching force is removed, then through thermal motion the macromolecule folds and bends into a random spatial pattern as in Fig. 1.3.

In a real macromolecule, the motion of molecular segments is hindered by internal and intermolecular interactions. The hindrance leads to an increase in stiffness of the chain over that of a chain with free rotation. The potential energy curve for internal rotation has several

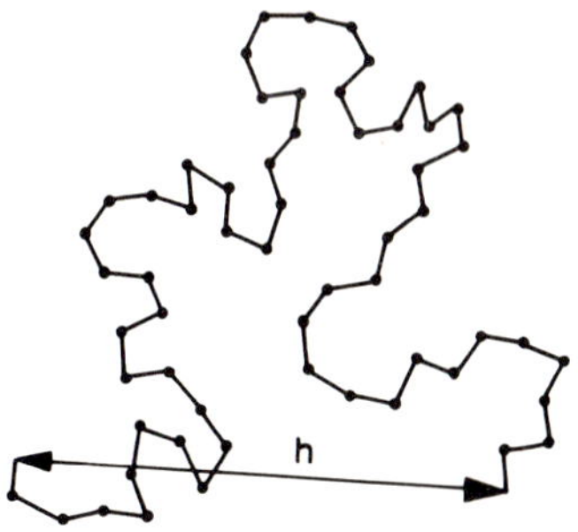

Fig. 1.3 A flexible macromolecule chain in a spatially folded state.

minima that differ in depth. Most often the molecular segments or links are found in places that correspond to the energy minima. The stable conformations of a chain that result from rotation of individual bonds are called rotational isomers (8,14). On the whole, a polymer can be viewed, with qualifications, as a complex of rotational isomers. In the rotational isomer approach, internal friction, leading to the hindrance of rotation in the network, represents couplings among different rotational isomers.

Polymer macromolecules always exist in a condensed phase and interact with the environment, which can either be a polymer as well (as in a pure polymer) or be a liquid (as in solutions of polymers).

At an earlier stage of polymer science, a polymer was described as a complex of random, tangled molecular chains (15). Modern theory, however, describes a polymer as an ordered molecular system. The simplest kinds of ordering in macromolecules (i.e., supermolecular structures) also occur in amorphous polymers. The supermolecular structures characteristically are of two main types: bundles and globules (13,16,17). Bundles form in polymers that have sufficiently rigid chains; they are very long clusters of more or less parallel macromolecules. The lateral dimensions of a bundle amount to several intermolecular distances; its length far exceeds the size of elongated macromolecules. Globules form in polymers whose macromolecules are very flexible and are folded into little "coils". Discrete segments of a macromolecule chain within a globule are randomly arranged. Polymeric materials in the globule state lose those properties which are associated with the great length of macromolecules, and behave like small molecules. Crystalline polymers form supermolecular structures more readily than do amorphous polymers.

1.3 SPECIAL CHARACTERISTICS OF THERMAL MOTION IN POLYMERS

Thermal motion of macromolecules or their segments is quite important in polymer friction processes. Let us examine the special characteristics of polymer thermal motion that bear a resemblance to thermal motion in liquids. Previously, under the influence of van der Waals' ideas, liquids were viewed as extremely dense gases. Thermal motion in liquids was reduced to the translational motion of particles. In 1926, Frenkel (18,19) proposed a new view of thermal motion. According to Frenkel, liquids, especially when they are near the crystallization temperature, are more like solids than like dense gases in structural arrangement and in the character of their thermal motion. In solids and liquids, a constant magnitude of kinetic energy can be attributed to the thermal motion of particles. Each particle (atom or molecule), from time to time, can acquire sufficient kinetic energy to surmount the potential barrier that separates two adjacent states. Usually, a particle is found in places that have a minimum of potential energy and it oscillates about an equilibrium point. As a result of such displacements, in crystalline solids vacancies or "holes" (free lattice points) and dislocated atoms, located between the lattice points, are formed. In liquids, such "holes" are microvoids of indefinite size and shape distributed among molecules. The lives of "holes" are short, but their number is significantly larger than in crystals. For the most part, free space in liquids consists of these "holes", which constantly disappear from one place and appear in other places. The existence of free space in liquids results in great mobility of the particles.

A molecule's duration in a transitional equilibrium position, called "the time of a particle at rest", is expressed as a statistical mean value and is dependent upon temperature. Frenkel demonstrated that the time of a particle at rest, that is, the molecular relaxation time τ, is equal to

$$\tau = \tau_o \exp (U/kT), \tag{1.1}$$

where τ_o is the period of particle oscillation around equilibrium points ($\approx 10^{-12}$ to 10^{-13} second) and U is the corresponding activation energy, equal to the minimum kinetic energy required for a particle to surmount the energy barrier at a given temperature.

The viscosity of a liquid is determined by the molecular relaxation time τ. The larger τ is, the mcre viscous is the liquid. According to the Maxwell relation, the viscosity of a liquid is given by $\eta = x\tau G_o$, where G_o is the "instantaneous" shear modulus of a liquid, observable during high shear rates and x is the coefficient that relates the Maxwell relaxation time τ_M and the molecular relaxation time τ. In general, from the activation mechanism, the viscosity of liquids is expressed in terms of the well-known Frenkel-Andrade equation for Newtonian flow:

$$\eta = A \exp (U/kT) .$$

Concepts of the mechanism of thermal motion and viscous flow in liquids have been further developed in Eyring's theory of anomalously viscous systems (20). Modern ideas about activation mechanisms of viscous flow and diffusion are based on concepts of thermal motion in liquids formulated by Frenkel and Eyring. According to Eyring, viscous flow occurs as a result of transfer from an equal-probability pattern of autodiffusing passage of kinetic units through a static liquid in all spatial directions, to an asymmetrical probability distribution of particle transfer in a viscous stream, where with the highest probability particles are transferred in the direction of a tangential force. Eyring is one of the scientists who have made a more precise description of viscosity (21). We have since learned that the passage of particles occurs in all spatial directions, not only in the direction of the operating tangential force. For small shear stresses the probability distribution appears as a linear function of the shear stress. Consequently, the shear rate is proportional to the shear stress; that is, one observes Newtonian flow of constant velocity. Under high stresses, which are realized in highly viscous liquids that have complex structures (polymers, dispersed systems and others) (22-26), the linear approximation is impossible because viscosity diminishes with increase in stress or shear rate.

Thermal motion of macromolecules is distinguished from that of submolecular matter in a number of ways. In an isolated macromolecule, thermal motion is characterized by forward Brownian motion of the macromolecule as a whole, by rotating motion of the macromolecule chains, by oscillation of the macromolecule segments about one another, and by rotating intermolecular Brownian motion of macromolecule segments about one another. Brownian motion of polymer chains or discrete macromolecules is usually called macro-Brownian motion. Brownian motion of discrete segments of a chain is called micro-Brownian.

As a result of intramolecular interactions in a polymer chain, (consisting of interactions of separate segments of the chain or its lateral groups), repulsive forces arise that reduce molecular mobility, particularly where there are stable valence angles. Restrictions on thermal motion in polymer chains are described in terms of potential barriers. A segment is the kinetic unit that characterizes the mobility of a polymer chain. The size of a segment characterizes the rigidity of a polymer chain. The closer the size of the kinetic unit (segment) is to the chemical unit (link), the more flexible the polymer chain is. Thus, for example, in elastomers (e.g., natural rubber) the segment is several times larger than the link. The rigidity of a chain also depends in large measure upon both the structure of lateral groups and interactions among them. The presence of double and triple bonds in a polymer chain also increases the chain's rigidity.

A special characteristic of thermal motion in crystalline polymers is related to the fact that such polymers consist of amorphous and crystalline regions. In crystalline polymers molecular motion is more difficult than in amorphous polymers. An increase in the degree of crystallinity decreases mobility not only in the kinetic units of the chains (segments), but also in the lateral groups (27).

The viscous-flow state of a polymer, which is observed above the glass temperature in amorphous polymers or above the melting point in crystalline polymers, is due to great molecular mobility. In linear polymers, the viscous-flow state is characterized by viscosity in the range 10^3 to 10^{12} poise. Viscous flow is accompanied by the development of viscoelastic deformation, and related to this there is rectification and orientation of the macromolecules. The basic mechanism of polymer viscous flow is unclear, but it may be described as a micro-Brownian transfer of macromolecular segments, leading to general transfer of the macromolecules. However, it is necessary to keep in mind that during viscous flow the macromolecule orientation influences viscosity (28-32).

1.4 THE RUBBERY STATE IN POLYMERS

For rubbery polymers, one observes a special kind of friction. The predominant aspect of deformation in this case is viscoelastic (13,15,33-37). The temperatures in which a polymer exists in the rubbery state is above the glass transition temperature T_g. However, the region where rubbery properties develop depends upon the strain frequency and the observation time. In other words, the position of the rubbery region is related to polymer relaxation phenomena; its lower limit occurs at the mechanical transition temperature, T_m, which usually lies above T_g.

The most distinct rubber elasticity develops in crosslinked rubbers (vulcanized rubbers). In linear amorphous polymers, during a long enough period of observation, viscous flow is superimposed upon rubbery deformation. Deformation properties of crystalline polymers depend upon both the rubbery properties of amorphous micro-regions and the more rigid nature of crystalline micro-regions, in proportion to the degree of crystallinity in the polymer as a whole.

Natural rubbers, vulcanized rubbers, some rubberlike polymers, and even swollen rigid-chained polymers are typical rubbery materials, each in its own temperature ranges. Polymers that exist in the rubbery state are widely used in engineering, chiefly in the form of various vulcanized rubber products (packings, valves, vibration dampers, etc.), automobile and airplane tires, and so forth, where their frictional properties are important. The useful technical properties of rubbery materials are: low moduli of elasticity, high friction coefficients, and good shockabsorption capacity. The need for stability of these properties compels the use of vulcanized rubbers in those thermal regions, and restricts strain rates and strains to the linear range of behavior.

Crosslinked polymers (vulcanized rubbers) can completely recover their shape after unloading, just like resilient solids. But in other characteristics they are more like liquids. Submolecular liquids and crosslinked polymers are more alike in their coefficients of thermal expansion and compressibility, which are much larger than those of solids. These and other properties conform to Pascal's law. Thus, the bulk thermal expansion coefficients for gases are 3.6×10^{-3} degree^{-1}; for metals, 6×10^{-4} degree^{-1}; but for organic liquids and polymers the coefficients are close together: 10×10^{-4} and 3.6×10^{-4} degree^{-1}. The isothermal compressibility coefficients are approximately equal for air, 1 cm^2/kg (at a pressure of 1 atm) and for metals, 10^{-6} cm^2/kg, but for organic liquids and polymers they are nearly the same and differ from the value for metals by two orders of ten (10^{-4} and 0.5×10^{-4} cm^2/kg).

By its nature, rubbery deformation differs from deformation of hard crystalline and glassy bodies, but is analogous to molecular-kinetic (entropic) elasticity in gases. For example, the equilibrium stress in a deformed vulcanized rubber, like the pressure in a compressed gas, is proportional at constant volume to the absolute temperature. Such a conjunction of physical properties, characteristic of all three states of aggregation, distinguishes rubbery materials from others.

Rubbery deformation occurs by the orientation and transfer of links in flexible chains, that is, the transition of chains from folded to elongated conformations. The classical statistical theory of rubbery deformation was developed by Wohl, Kuhn, James and Good, Treloar, and others (5,7-9,15).

1.5 RELAXATION PROPERTIES OF POLYMERS

Molecular-kinetic processes in polymers include diffusion and self-diffusion, crystallization and fusion, vaporization and dissolution, glass transition and softening, deformation and disintegration, mechanical and dielectric losses, viscous flow, and many others. Some of these processes (in particular, relaxation) are characteristic of polymers. Among them are the relaxation of mechanical stresses, mechanical losses during repeated strain, viscous flow, the mechanical and structural glass transition, and dielectric and magnetic relaxation. The formation of the contact area between a polymer and hard surfaces and the friction process are also included among relaxation phenomena.

The study of mechanical relaxation in polymers (creep, stress relaxation, and the temperature-frequency dependence of dynamic characteristics) allows not only the evaluation of exploitable properties in polymer materials but also the determination of the relationship between the chemical and physical structure of polymers, and the nature of their molecular mobility and macroscopic relaxation properties (13,36-38).

Polymer relaxation processes are divided into linear and nonlinear (27). Linear relaxation is described by the linear theory of visco-elasticity (39-41). It is characterized by the rearrangement of macromolecular segments during the formation of equilibrium stress without alteration in the polymer structure. For constant strain (ε = const), linear stress relaxation follows the relaxation law $\sigma = \sigma_0 f(t)$, where σ_0 is the initial stress at the moment $t = 0$.

Nonlinear relaxation processes are observed during large stresses and deformations and are generally related to structural changes of polymers in the process of deformation. The nonlinear stress relaxation for a given deformation corresponds to the relaxation law $\sigma = \sigma_0 f(t,\varepsilon)$.

Nonlinear relaxations in amorphous and crystalline polymers have much in common, which allows us to associate them with the presence of supermolecular formations in rigid polymers (13,42). The disintegration of supermolecular structures during deformation leads to the beginning of nonlinear relaxation properties.

Mechanical losses play an important role in friction of different materials. The maximum friction coefficients are often associated with maximum mechanical losses (which are examined briefly below). The modulus of elasticity in the case of cyclic deformations has a complex form, for example $E^* = E' + iE''$, where storage modulus of elasticity E' and the loss modulus E'' are expressed by the formulas

$$E' = \sum_{j=1}^{n} E_j [(\omega^2\tau_j^2)/(1 + \omega^2\tau_j^2)] ,$$

$$E'' = \sum_{j=1}^{n} E_j [(\omega\tau_j)/(1 + \omega^2\tau_j^2)] .$$

Here, n is the number of elementary relaxation mechanisms; E_j are the coefficients that indicate the contribution of each relaxation mechanism. For the mechanical loss coefficients we have

$$(2\pi/E')\left(E_1 \frac{\omega\tau_1}{1 + \omega^2\tau_1^2} + E_2 \frac{\omega\tau_2}{1 + \omega^2\tau_2^2} + \ldots + E_j \frac{\omega\tau_j}{1 + \omega^2\tau_j^2} + \ldots\right). \quad (1.2)$$

If we ascribe to each molecular relaxation process its own characteristic relaxation time τ_j, then as the conditions $\omega\tau_1 \approx 1$, $\omega\tau_2 \approx 1$ etc., are successively attained, the mechanical losses successively pass through maxima. These losses often are not characterized by the mechanical loss coefficient, but by a quantity proportional to it, namely the loss tangent, tan δ. Experimental data on changes of mechanical losses usually are plotted tan δ vs the cyclic frequency ω or tan δ vs temperature T.

The relaxation times τ_j have a temperature relation $\tau_j = \tau_0 \exp(U_j/kT)$, where U_j is the activation energy for the jth molecular process, and τ_0 is a coefficient whose value can vary from one relaxation mechanism to another.

Each well-defined relaxation process that is characterized by a fixed value of activation energy U_j and a relaxation time τ_j appears as a sharp or diffuse maximum in the temperature dependence of the mechanical loss. The temperature, height, and width of mechanical-loss maxima for polymers of different structures have been identified (42). Figures 1.4 and 1.5 show the temperature dependence of tan δ, based upon an analysis of experimental data for amorphous linear and crystalline polymers (43-45).

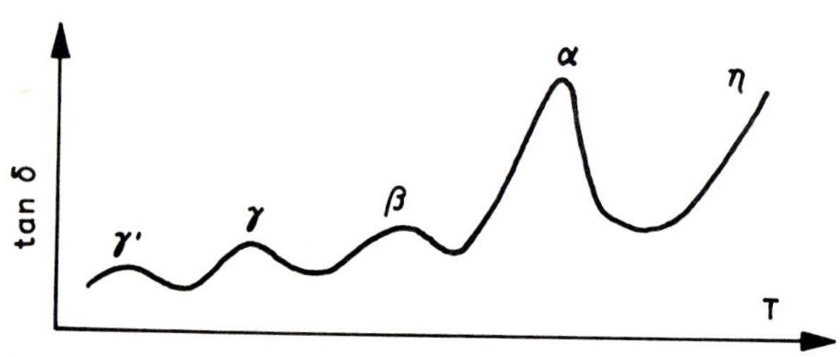

Fig. 1.4 Temperature dependence of loss tangent for an amorphous linear polymer: α is the maximum corresponding to the glass transition process; maxima γ', γ, and β are related to the molecular mobility in the glass state; η is the transition to viscous flow.

The dynamic modulus and mechanical losses of polymers depend upon the rate of deformation and the frequency of applied stress, because in dynamic systems work is done not only against large elastic forces, but also against internal friction, depending upon the rate of the process. For example, the modulus of elasticity in vulcanized rubbers consists of two parts: $E = E_\infty + E_1$, where E_∞ is the relaxed modulus, which corresponds to the state in which relaxation is almost complete and is generally dependent upon the extent of transverse crosslinking. The nonequilibrium part E_1, due to internal friction, is essentially dependent upon the number of polar groups in the polymer chain and the quantity of active filler, that is, upon the nature and intensity

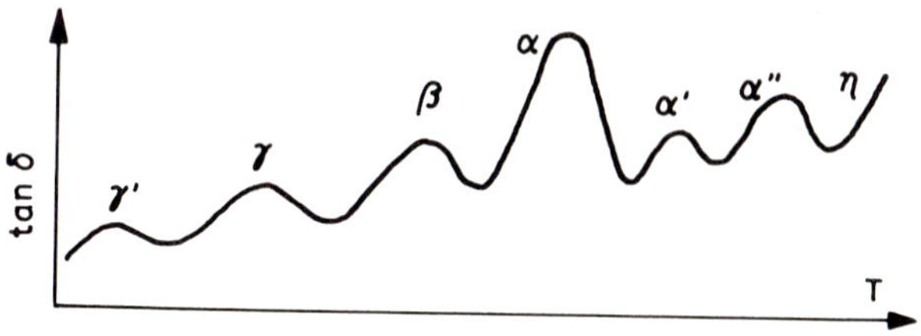

Fig. 1.5 Temperature dependence of the loss tangent for a crystalline polymer: α is the maximum corresponding to the glass transition for the amorphous component; γ', γ, β are related to molecular mobility in the glassy state; α and α' are related to the mobility of supermolecular structures; η is the melting transition.

of intermolecular interactions. E_1 and the mechanical losses increase with decreasing temperature and with increasing frequency and rate of deformation.

1.6 THE PRINCIPLE OF TIME-TEMPERATURE EQUIVALENCE

The principle of time-temperature equivalence has been applied in the calculation of many relaxation processes and several polymer friction properties.

It is known that all mechanical, electrical, and other polymer relaxation processes are characterized by corresponding relaxation times. The temperature dependence of polymer mechanical and electrical properties is directly related to the effect of temperature on relaxation times. With a decreasing temperature, relaxation times increase significantly and interaction times decrease. This indicates that relaxation processes are retarded so much at low temperatures that they almost disappear. By contrast, at high temperatures relaxation occurs so quickly that such a change as rubbery deformation instantaneously attains its limiting value.

The study of relaxation is often done by one of two equivalent methods: either by altering interaction time within broad limits or by varying the temperature. In the first method, the interaction time exceeds all relaxation times, and in the second method all relaxation times are altered and vary at a fixed interaction time. Experimentally the second method is simpler, and therefore it has been used widely.

Aleksandrov and Lazurkin (11) first applied the frequency method to the investigation of relaxation effects in rubberlike polymers during cyclid deformations. They demonstrated that an increase in interaction frequency exerts an effect similar to that produced by an equivalent decrease in temperature. Similar results in statistical tests have been obtained by Kobeko, Kuvshinsky, and Gurevich (46) and by Leadermann (47), and worked out in more detail by Tobolsky (33). These researchers have determined that curves of creep and stress relaxation, obtained at different temperatures, can be superposed by a translation along the logarithm-of-time axis.

Figure 1.6 schematically shows results for creep (or stress relaxation) obtained at different temperatures. As indicated, the curves can be brought into coincidence by translation to the right or left. This rule of translation is applicable to many polymer relaxation properties both static and dynamic. For dynamic properties, which include dynamic moduli and mechanical losses, in place of time (i.e., log t) axis we have the frequency (i.e., log ω) axis. For mechanical losses, instead of monotonic waves on the graph, there will be curves with maxima.

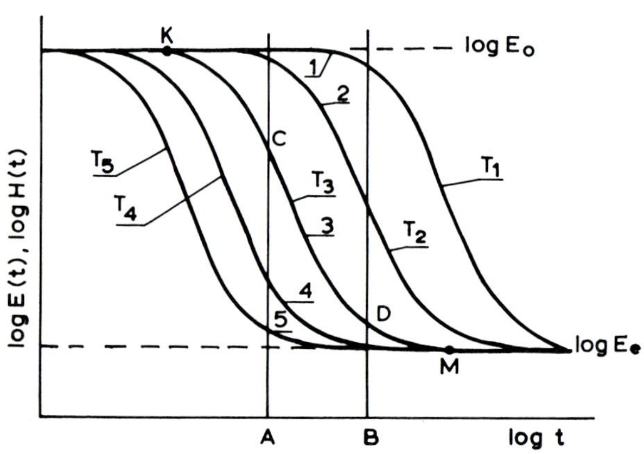

Fig. 1.6 Relation of the logarithm of the modulus of elasticity [or function H(t)] of a vulcanized rubber to the logarithm of time, for different temperatures ($T_1 < T_2 < T_3 < T_4 < T_5$).

This rule of curve shifting, which serves as a basis for formulation of the time-temperature equivalence principle, is equivalent to the assertion that the effect of temperature T on polymer stress-strain properties occurs independently of time t and the polymer's physical properties (for example, the modulus of elasticity) and can be expressed by the function

$$E(t,T) = E[t,a(T)] ,$$

where the coefficient a(T) is the function of temperature only and at a certain standard temperature T_{st} is chosen equal to one, i.e., $a(T_{st}) = 1$. The time dependence of the modulus at this temperature is written in the form $E_{st}(t)$. Knowing this relation from experiment, it is possible to calculate the dependence of the modulus upon time at any temperature with the aid of the function a(T). The determination of this function is the chief methodological task in predicting polymer relaxation properties at different temperatures.

To clarify the significance of the value a(T), let us examine the simplest case of stress relaxation according to Maxwell's law:

$$\sigma = E_0 \varepsilon \exp -(t/\tau_M) ,$$

where the Maxwellian relaxation time τ_M depends upon temperature (ε = const).

The elastic modulus for the standard temperature T_{st} and for any temperature T is equal to

$$E_{st}(t) = E_0 \exp -[t/(\tau_M(T_{st}))] ,$$

$$E(t) = E_0 \exp -[t/(\tau_M(T))] .$$

If it is supposed that

$$a(T) = \tau_M/\tau_M(T_{st}) , \qquad (1.3)$$

then, knowing $E_{st}(t)$, one can calculate E(t), having substituted t in t/a(T), that is, $E(t) = E_{st}(t/a)$.

Generally, any physical quantity $\Phi(t)$ that depends upon a relaxation process characterized by a relaxation time τ_M, is some function f of the ratio t/τ_M (for static processes) or of the product $\omega\tau_M$ (for cyclic processes):

$$\Phi(t) = f(t/\tau_M) ,$$

$$\Phi(\omega) = f(\omega\tau_M) \qquad (1.4)$$

If, as before, a(T) is determined by the ratio of relaxation times, then for static processes, by carrying out identical operations, we obtain

$$\Phi(t) = f(t/\tau_M) = f\left[\frac{t/\tau_M(T_{st})}{a(T)}\right] = \Phi_{st}(t/a) \quad , \tag{1.5}$$

and for cyclic modes we obtain

$$\Phi(\omega) = f(\omega\tau_M) = f[\omega a(T)\tau_M(T_{st})] = \Phi_{st}(\omega a) \quad , \tag{1.6}$$

where $\Phi_{st}(t)$ and $\Phi_{st}(\omega)$ are known relations at the temperature T_{st}. The following "reverse" relations are also valid

$$\Phi_{st}(t) = \Phi(ta) \quad , \tag{1.7}$$

$$\Phi_{st}(\omega) = \Phi(\omega/a) \quad , \tag{1.8}$$

where $\Phi(t)$ and $\Phi(\omega)$ are the relations for some temperature T. It is obvious that the coefficient a(T) has different values for various polymers and for different relaxation processes in the same polymer.

Let us apply the principle of time-temperature equivalence to experimental results such that the same relaxation times at a given temperature do not change with the passage of time. Because the relaxation time depends not only upon temperature but also upon the polymer's structure, the change in τ_M with time indicates a change in the polymer's structure during prolonged observation. For a given relaxation process, the equivalence principle is fulfilled if the relaxation time does not depend upon the observation time. In many cases this condition is fulfilled to a good approximation.

For practical application of the equivalence principle, in relation to polymer friction properties, it is necessary to select a standard temperature and to find the coefficient a(T) and the property $\Phi_{st}(t)$ or $\Phi_{st}(\omega)$ for the standard temperature. Then, in order to find the time and frequency dependences $\Phi(t)$ and $\Phi(\omega)$ for any other temperature, it is necessary to substitute t for t/a and ω for ω/a: $\Phi(t) = \Phi_{st}(t/a)$; $\Phi_{st}(\omega a)$, where Φ denotes each function that is determined by experiments at the standard temperature.

Let us examine some procedures for the calculation of $\Phi_{st}(t)$ and a(T). The function $\Phi_{st}(t)$ can be determined experimentally, with some difficulty, for a wide range of times. However, in practical experiments a rather narrow observation-time interval is realized (for example, from one second to one day). The limits of this interval are provisionally plotted as points A and B in Fig. 1.6. In such a case, an effort is made to obtain as many dependences as possible for different temperatures between points A and B. Suppose, as an example, in Fig. 1.6 for the standard temperature we take T_3. Segment 3 between points C and D, experimentally observable for this temperature, is chosen as a base. In order to obtain a complete curve beyond the limits of zone AB, segment 2 is shifted to the left parallel to the time axes so as to coincide with segment 3, and then segment 1 is shifted to coincide with the new location of segment 2. As a result, the left branch of the basic curve CK is obtained. Further, segment 4 is shifted to the right so as to coincide with segment 3, and then segment 5 is shifted to coincide with the new location of segment 4. As a result, the right branch of the basic curve DM is obtained. Thus, we determine the whole relation $\Phi_{cT}(t)$ graphically. The coefficient a(T) is determined for different temperatures by the magnitude of the necessary shifts and is expressed as a function of temperature in a graph or table.

For determining a(T), we propose the following analytic procedures. First of all, we look at the simplest, in which the relaxation time is expressed by the formula

$$\tau_M = b \exp (U/kT) , \tag{1.9}$$

where τ_M is the Maxwellian relaxation time, and b is the constant. From this formula it follows that

$$\log a(T) = -(U/2.3kT_{st})(T-T_{st}/T) .$$

Usually the glass temperature T_g is selected for T_{st}; therefore,

$$\log a(T) = -(U/2.3kT_g)(T-T_g/T) .$$

For many polymers this expression proves to be too imprecise. For this reason, Ferry (38) proposed the introduction of coefficient a(T) as in Eq. (1.3), and suggested a more precise empirical formula

$$\log a(T) = -A\{(T-T_g)/[B + (T-T_g)]\}. \tag{1.10}$$

where A = 17.4, B = 51.6°.

Ferry also indicated that $a(T) = \eta(T)/\eta(T_{st})$, where $\eta(T)$ and $\eta(T_{st})$ are the viscosities at temperatures T and $T_{st} > T_g$. This calculation follows from the Maxwell equation $\eta = \tau_M G_0$, where G_0 is the shear modulus and depends weakly upon temperature.

Fulcher (48) had already proposed the following equation for the viscosity

$$\eta = C \exp(a/T-T_0) , \tag{1.11}$$

which proved to be more precise than Frenkel and Andrade's equation. Because the viscosity is proportional to the relaxation time, it follows that there is a need for a more precise formula than (1.9) for the relaxation time

$$\tau_M = b \exp(a/T-T_0) . \tag{1.12}$$

This formula matches the usual formula (1.9), if we assume that the activation energy depends upon temperature by the law $U = k\alpha/(1 - (T_0/T))$, where $k\alpha$ is the activation energy U_∞ for $T \to \infty$. This agrees with the well-known experimental temperature dependence of a polymer's activation energy at and below the glass temperature. If we now assume that $T_{st} = T_g$, then we obtain

$$\log a(T) = -(a/2.3) \, [(T-T_g)/(T_g-T_0)(T-T_0)] \tag{1.13}$$

leading to the expression

$$A = a/2.3(T_g-T_0) ,$$

$$B = T_g - T_0 ,$$

from which we obtain Eq. (1.10).

From the above numerical values of A and B, it follows that $T_0 = T_g - 51.6°$ (that is, T_0 is approximately 50° below the glass temperature); $\alpha = 2.04 \times 10^3$ degrees; $U_\infty = 4$ kcal/mole.

The principle of time-temperature equivalence is applied in the processing of data on the friction of rubbery materials; therefore, it will be used frequently in the following sections.

1.7 POLYMER GLASS-TRANSITION PROCESSES

We commonly distinguish several polymer glass-transition processes. The most essential for polymer friction properties are mechanical and glass transitions (11,13,37,50-55). One of the authors has examined this problem in detail (52). Some data on glass transitions necessary for an understanding of polymer friction processes in different physical states are given below.

The glass transition is the change from the liquid state, which has constantly changing structure, to the solid state, which has fixed structure. The glass transition appears as a change in the temperature dependence of a substance's physical properties (thermal, expansion, heat capacity, viscosity, and similar changes) in the absence of cyclic mechanical interactions. The glass transition for a liquid or a fused solid may be defined as a loss of mobility in basic units of matter (ions and molecules) as a result of decrease in temperature or increase in pressure. The glassy state is thermodynamically unstable, but it is kinetically stable because of the extremely slow rearrangement of structure and crystallization at low temperatures. One characteristic of the glassy state is irreversibility of structure under changes of pressure even at a temperature far from T_g. Liquids, and even rubbery polymers, by contrast, are characterized by structural changes that occur quickly with a change of temperature or pressure. With a decrease in temperature, the structure of a liquid or amorphous polymer gradually and continuously changes because of the rearrangement of kinetic units, which leads directly to other changes, such as a degree of microstratification, and other special structural characteristics of the liquid. The rate of rearrangement decreases with decreasing temperature, so that in the range of the glass temperature a short-range-order equilibrium can no longer be maintained and the liquid's structure is immobilized. During heating up, softening--that is, a transition from glass to liquid--is observed. This softening process does not greatly differ from the glass transition process, if the rate of heating up is the same as the rate of cooling that produces the glass. Glasses that are created at different rates of cooling have different structures and different softening points.

The glass transition is not necessarily connected with any mechanical interactions. In contrast, the so-called "mechanical transition" develops under the action of external forces. At high strain rates, any liquid (or any amorphous polymer) in some range above T_g behaves as an elastic solid, because with reduction in operation time or period of forced oscillation, a liquid (polymer) gradually loses its flow capacity, so that it deforms viscoelastically and passes to the elastic state. This transition is observed when the molecular relaxation time, related to macromolecular segment mobility, is equal to the loading time or external force time.

For submolecular liquids, it is very difficult to realize a transition to the elastic state, inasmuch as the molecular relaxation time for all temperatures above the melting point is extremely short. An altogether different situation arises with viscoelastic supercooling of highly viscous liquids and especially with polymers, whose relaxation time is many orders of magnitude longer than that of simple liquids.

The transition-temperature range where rubbery deformation cannot occur, but where mechanical losses pass through a maximum, is characterized by the mechanical transition temperature T_M. This temperature is defined as the temperature at which the maximum mechanical losses are observed (Fig. 1.7). In actual experimental conditions rubbery deformation does not occur at all temperatures above T_g, but only in the range of temperatures above T_M. During an upward transition across T_M, a polymer's flexibility rises abruptly and the modulus of elasticity drops abruptly (Fig. 1.7). The mechanical transition temperature rises with decreasing duration of force or with increasing duration of force or with increasing deformation frequency.

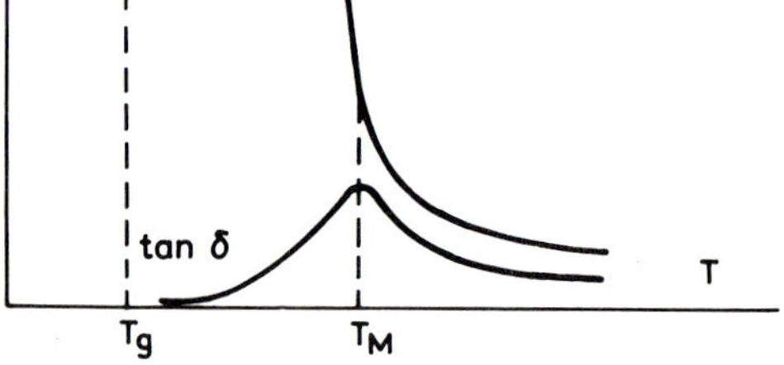

Fig. 1.7 Temperature dependence of the logarithm of the elastic modulus E and loss tangent tan δ. T_g = glass temperature; T_M = mechanical transition temperature.

It follows that a polymer's transition from the rubbery to the elastic-rigid state (as the temperature decreases) is of a molecular-kinetic nature, like the glass transition. However, the transition to the elastic-rigid state is not related to the congealing of structure but always takes place above T_g, though below T_M. Therefore, this transition is not associated with the glass transition. However, in practice it is often called by that name.

1.8 COLD FLOW OF POLYMERS

The mechanical transition usually is examined with small stresses and strains, so that the polymer's structure does not change. Qualitatively new phenomena normally arise with large stresses: cold flow and failure. Large stresses substantially affect the relaxation time, and the deformations that arise alter the polymer's structure (orientation, etc.). Rubbery deformation also can be observed in rigid polymers (amorphous and crystalline), but only for loads that exceed a certain limit, the so-called lower limit or strength of cold flow, σ_B. Below this limit, a rigid polymer deforms like an ordinary low-molecular-weight solid; above it, plastic deformation develops, in this case called cold flow. Cold flow appears in crystalline polymers below the melting point and in amorphous polymers below the glass transition temperature. It is characterized by reversibility in large forced deformations; however, deformation reversibility is observed only after heating to temperatures that are near the glass transition temperature or near the melting point. The phenomenon of cold flow is essential for understanding the friction properties of rigid polymer, because it explains the formation of the actual contact area of a rigid polymer under stress.

Cold flow in amorphous glassy polymers was first elucidated by Aleksandrov (56) and later investigated in detail by Lazurkin (57). These scientists continued to conduct experimental and theoretical investigation of cold flow (see also 58). They proposed the relaxation concept of this phenomenon, which is now generally accepted. Lazurkin, Bartenev, and their co-workers (59,60) carried out a thorough experimental investigation of cold flow in rubberlike polymers and rubbers at low temperatures. The cold flow of crystalline polymers was first described by Carothers and Hill, and was investigated in detail by Miklowit and others (61,62).

Contributions to the study of this phenomenon have also been made by Kargin and Sogolov (12,63) and others (64).

The lower limit of cold flow increases with increasing rate of deformation or decreasing loading time, indicating that it is a relaxation phenomenon (65). Hence, σ_f is a function of deformation rate

$$\sigma_f = A - \log \dot{\varepsilon}$$

where $\dot{\varepsilon}$ is the strain rate.

When the temperature rises to a certain value, σ_f becomes zero (Fig. 1.8). For slow deformations, this temperature practically coincides with the glass temperature; it has to be lower than the mechanical transition temperature because that temperature depends upon the rate of polymer deformation stretching.

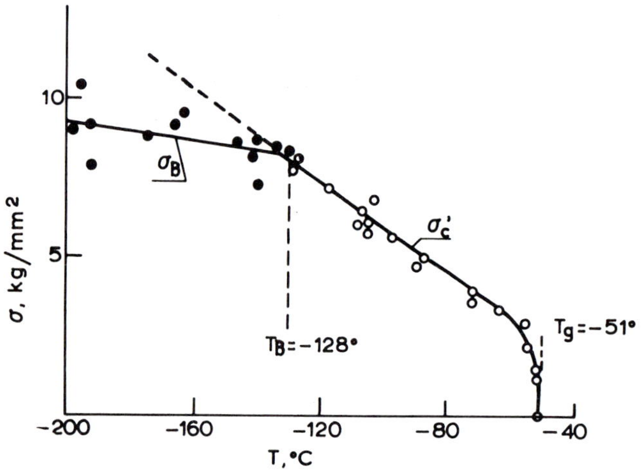

Fig. 1.8 Temperature dependence of the tensile cold flow stress σ_f (0) and strength for brittle fracture, σ_B (0), of butadiene-styrene rubber SBR-30.

The relaxation time τ, which characterizes the segment rearrangement rate, depends upon stress and temperature according to the equation

$$\tau = \tau_0 \exp\,[(U_0 - a\sigma)/kT], \qquad (1.14)$$

where U_0 is the activation energy for stress $\sigma = 0$, determined by potential barriers that segments overcome in transfer from one superposition site to another, and a is a constant that depends upon the size of a kinetic unit, i.e., a chain segment. From this equation, it follows that for low temperatures and small stresses, τ is longer than the testing time; therefore, rubbery deformations of chains appear "frozen". The relaxation time τ, comparable to the observation time (or to the inverse magnitude of the deformation rate) at which rubbery deformation "appears", may be attained either by a rise in temperature to the mechanical transition temperature T_M or by an increase in stress according to Eq. (1.14). Hence, it follows that σ_f depends upon the observation time or upon the deformation rate. From Eq. (1.14) it also follows that σ_f is the boundary condition of the stress state where rubbery deformation can develop slowly at smaller stresses. This is defined as the creep of polymers.

Einbinder and co-workers (66,67) have shown that the rheological and mechanical properties of polymers essentially depend upon the hydrostatic component of the pressure; therefore, the characteristics of polymeric materials even under simple stretching will differ from those under simple compression. Rheological properties differ considerably under high compression.

1.9 PROLONGED STRENGTH (FATIGUE) OF POLYMERS

Polymer wear under friction depends as much upon the friction force as upon the material's strength properties--in particular, it depends upon properties related to prolonged stressed conditions. Cyclic deformation on friction-surface asperities leads to typical fatigue effects. Therefore, fatigue properties can never be ignored in the investigation of friction properties in solids.

The present state of knowledge of polymer strength has been treated elsewhere (68). In this section, problems of strength will be pursued only to the degree that is necessary for further examination of polymer friction and wear properties.

There are two fundamental failure mechanisms for solids and polymers: athermal and thermal-fluctuating (52). Griffith studied and described athermal failure in solids under the operation of external force, but did not calculate the intermittent breaking of interatomic bonds during the thermal motion process (strictly speaking, he treated the problem at absolute zero).

For real solids containing defects, according to Griffith, failure occurs when the stress of the tips of the worst cracks attain critical value to to stress concentration.

The thermal-fluctuating mechanism presupposes the fact that random bond breaking occurs among atoms, i.e., by the increase in kinetic energy of atoms to a critical magnitude (the activation energy for bond breaking) in thermal motion. For this basic reason, bond breaking is fluctuating, but external force reduces the potential barrier, that is, increases the probability of bond breaking.

The conception of the failure process in solids and polymers as due to thermal fluctuations has been introduced elsewhere (56,69,70). Zhurkov and his co-workers have corroborated this conception experimentally in recent years. From it has developed the general form of the time-temperature dependence of strength. In the athermal failure mechanism in solids, time dependence of strength is practically absent.

The failure process must not be examined without taking into consideration the relaxation effects. One of the basic manifestations of these effects is mechanical losses. According to Bartenev's classification (52,68) it is possible to distinguish conditionally two types of losses: "surface" and "volume". The first type of loss, characteristic of any solid (theoretically including even brittle ones), is related to the necessary energy shift after the transfer of a breaking bond across a potential barrier and to the impossibility of fully quasistatic processes. In inelastic, plastic, and other materials, losses develop that are determined by local inelastic deformations, which occur even in externally induced brittle failure.

Relaxation processes determine the character of failure and alter its mechanism (71). Because temperature affects the relaxation properties of polymers, it is possible to distinguish three temperature zones (Fig. 1.9) corresponding to the different failure mechanisms. The general thermo-fluctuating mechanism is operative for all three zones, but under different conditions (69,72,73).

The lower brittle temperature T_B of a real polymer is determined by the development of one or several cracks initiated by the most catastropic defect. Micro-deformations in the elastic range and mechanical losses are comparatively small, local forced-elastic deformations can be realized in the rigid state below the glass

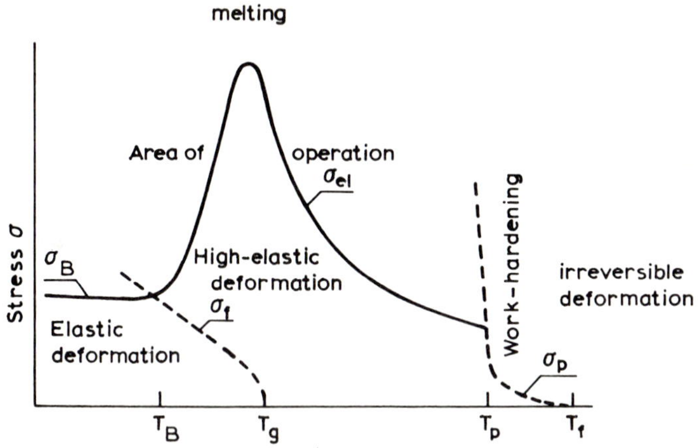

Fig. 1.9 Tensile strength of amorphous polymers over a wide temperature range. σ_B is the brittle failure stress; σ_f is the cold flow stress; σ_{el} is the strength in the rubbery state; σ_p is the plasticity flow stress. T_B, T_g, T_p, and T_f are the brittle, glass transition, plasticity, and fluid-flow temperatures.

temperature T_g, and local cold-flow deformations occur in proportion to the rise in temperature at points of stress concentration (on microcrack tips). At those tips, there appear areas in which fibrous material form and are elongated during cold flow.

Mechanical losses basically are determined by losses in local cold-flow micro-deformations. As this process develops, with rise in temperature, so-called "crazing" appears. After the "crazing" come failure cracks, in which the most elongated fibers are torn (74). With further rise in temperature, the polymer passes to the rubbery state. The failure mechanism is determined by the amount strain and its distribution throughout the material. These strains form as a result of micro-stratification in places of stress concentration, which are related to micro-viscous deformations. These deformations determine the major part of mechanical losses during failure of vulcanized rubbers.

Theories of polymer failure in all three basic zones have not been worked out to the same degree. Fluctuation theory has been applied to the brittle-failure zone (75,76). According to this theory, the breaking of polymer chemical and intermolecular bonds occurs as a result of thermal fluctuations, and the durability t of a material under stretching (stress σ = const) is expressed by the following approximate formula

$$t = \frac{LkT\exp(-q/k)}{\omega Z\phi^2 v_c}\;\frac{1}{\sigma-\sigma_o}\exp\left(\frac{U_o-\omega Z\phi\sigma}{kT}\right) + \frac{L}{v_c\sigma_c}\sigma\;, \qquad (1.15)$$

where L is the linear dimension of a transverse cross-section of the sample, ϕ is the scale factor, k is the Boltzmann constant, T is the absolute temperature, v_c is the critical rate of crack growth, ω is the fluctuation capacity, Z is the coefficient of stress concentration at a microfissure peak, σ is the tensile stress, σ_c is the critical stress, σ_o is the safe stress, U_o is the "zero" activation energy during failure, and q is the temperature coefficient of the activation energy. In concise form we obtain, keeping in mind that this formula is correct for tensile stresses from σ_o to σ_c,

$$t = \frac{A_o}{\sigma-\sigma_o}\exp\left(\frac{U_o-\gamma\sigma}{kT}\right) + B\pi\;, \qquad (1.16)$$

where A_o depends weakly upon temperature.

The curve in Fig. 1.10 fits Eq. (1.16). For $\sigma \to \sigma_o$ the durability tends to infinity; for $\sigma \to \sigma_c$ it is determined by the critical rate of crack growth. As can be seen, the equation in the significant zone of durability change approximates a linear relationship between the durability logarithm log t and the stress σ (Zhurkov's equation)

$$t = A\exp[(U_o-\gamma\sigma)/kT]\;, \qquad (1.17)$$

where $\gamma = \omega Z\phi$, $A = A_o/(\sigma-\sigma_o)$.

For small stresses, according to this theory, the zone of applicability of the durability Eq. (1.17) is limited to safe stresses σ_o (Fig. 1.10). If failure is determined by crack growth, then the safe stress is proportional to the surface free energy α of a rigid polymer in a given environment (68,77): $\sigma \simeq \alpha/\lambda_o$, where λ_o is the interatomic distance.

The zone of brittle (low-temperature) polymer failure has been studied by Peschanskaya and Stepanov (78) and others.

The "zero" energy barrier, U_o, according to Zhurkov and associates (70,79-82), is approximately equal to the chemical bond strength of a polymer. Sample imperfections, the scale factor,

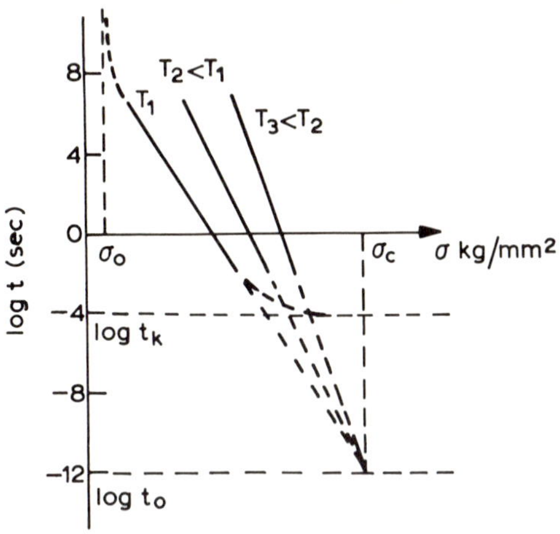

Fig. 1.10 Time dependence of durability for rigid polymer. Plot according to Eq. (1.17).

polymer orientation, and plasticization do not significantly change the value U_o. All these factors show up more strongly in value γ, the structure-sensitive coefficient.

In the rubbery state, the durability of polymers is characterized by a power law (83)

$$t = B\sigma^{-b} = B_o\sigma^{-b} \exp (U/kT) \qquad (1.18)$$

where B_o and b are constants that have no direct physical meaning but in a complex way take into account the way in which basic fluctuations add up during failure.

An analogous relation is observed for dynamic fatigue in vulcanized rubbers (Fig. 1.11). Static and dynamic fatigue (in cyclic deformations) are characterized by the same value of constant b in the durability formula, but by different values of B and B', so that it is possible to write for the dynamic durability

$$t' = B'\sigma^{-b} , \qquad (1.19)$$

where B' is a constant the depends exponentially upon temperature, and b is a constant that does not depend upon temperature and the normal stress operation of a vulcanized rubber--nor, and consequently, upon the deformation frequency ν.

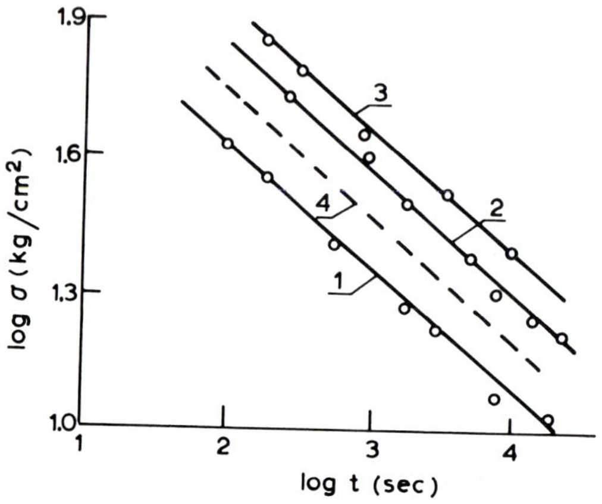

Fig. 1.11 Relations among maximum stresses in a cycle and durability for vulcanized SBR-30 for different modes of deformation stretching:

(1) Cyclic stretching with constant maximum deformation;
(2) The same, with a constant maximum load;
(3) The same, with constant maximum stress;
(4) Static deformation under a given load.

If, allowing for the number of cycles until failure ($n = \nu t'$), we assume $m = b$ and $C = \nu B'$, then it is easily seen that the formula $n\sigma^m = C$ and Eq. (1.19) express the same dynamic fatigue law for a vulcanized rubber. Because the constant C depends weakly upon frequency, the constant $B' = C/\nu$ is inversely proportional to frequency. During the transition to the static mode, when $\nu \to 0$, this correlation loses meaning, because the constant B ought not to approach infinity, but rather some constant B_o.

Because $b >> 1$, large changes in durability correlated with small changes in strength. Consequently, durability is a more sensitive characteristic of fatigue than strength. Hence, in engineering the number of cycles until failure is used as a characteristic of dynamic fatigue.

In particular, the kinetic nature of failure appears in the relation of polymer durability to time of experimentation (static loads, stretching with different rates, repeated and multiple deformations, etc.). This allows the recalculation of durability after each stress operation. Methods for such recalculation are described elsewhere (68).

REFERENCES

1 V.V. Korshak, Khimiya Vysokomolekulyarnykh Soyedinenii (Chemistry of High-Molecular Compounds), AN SSSR, Moscow, 1950.
2 I.P. Losev and Ye.B. Trostyanskaya, Khimiya Sintyetioheskikh Polimerov (Chemistry of Synthetic Polymers), Khimiya, Moscow, 1964.
3 F.W. Billmeyer, Jr., Textbook of Polymer Science, 2nd edn., Interscience, New York, 1971
4 B. Colding, Polymers and Resins, Their Chemistry and Chemical Engineering, Van Nostrand-Reinhold, New York, 1959.
5 S.Ye. Bresler and B.L. Yerusalimsky, Khimiya i Fizika Makromolekul (Macromolecule Chemistry and Physics), Nauka, Moscow, 1965.
6 S.Ya. Frenkel, Vvedeniye v Statisticheskuyu Teoriyu Polimerizatsii (Introduction to the Statistical Theory of Polymerization), Nauka, Moscow, 1965.
7 B.A. Dogadkin, Khimiya i Fizika Kauchukov (Chemistry and Physics of Natural Rubbers), Goskhimizdat, 1947.
8 M.V. Volkenshtein, Konfiguratsionnaya Statistika Polimernykh Tsepei (Configuration Statistics of Polymer Chains), AN SSSR, Moscow, 1959.
9 T.M. Birshtein and O.B. Ptsitsyn, Konformatsii Makromolekul (Macromolecule Conformations), Nauka, Moscow, 1964.
10 T.I. Sogolova and G.L. Slonimsky, Zh. Vses. Khim. Ova., 6(1961) 389.
11 A.P. Aleksandrov and Yu.S. Lazurkin, Zh. Tekh. Fiz., 9(1939)1249.
12 V.A. Kargin and T.I. Sogolova, Zh. Tekh. Fiz., 23(1949)538.
13 V.A. Kargin and G.L. Slonimskii, Kratkiye Ocherki po Fizko-Khimii Polimerov (Short Essays on Polymer Physical Chemistry), Khimiya, 1967.
14 M.V. Volkenshtein, Dokl. Akad, Nauk SSSR, 78(1951)879; 125(1959)523.
15 L.R.G. Treloar, The Physics of Rubber Elasticity, 2nd edn., Clarendon Press, Oxford, 1967.
16 V.A. Kargin, A.I. Kitaigorodsky and G.L. Slonimsky, Kolloidn. Zh., 19(1957)131.
17 C. Tenford, Physical Chemistry of Macromolecules, Wiley, New York, 1961.
18 Ya.I. Frenkel, Z. Phys., 35(1926)664.
19 Ya.I. Frenkel, Kineticheskaya Teoriya Zhidkosti (Liquid Kinetic Theory), AN SSSR, 1945; Vvedeniye v Teoriyu Metallov (Introduction to Metal Theory), Gostekhizdat, 1950.
20 H. Eyring, J. Chem. Phys., 4(1936)283.
21 G.M. Bartenev, Zh. Fiz. Khim., 29(1955)2007.
22 P.A. Rebinder, in Fiziko-Khimicheskaya Mekhanika Dispersnykh Struktur (Physico-Chemical Mechanics of Dispersed Structures), Nauka, Moscow, 1966, p. 3.
23 P.A. Rebinder and I.N. Vlodavets, in Problyemy Fiziko-Khimicheskoi Mekhaniki Voloknistykh i Poristykh Dispersynykh Struktur i Materialov (Problems of Physico-Chemical Mechanics of Fibrous and Porous Dispersed Structures and Materials), Zinatnye, Riga, 1967, p. 3.
24 V.A. Fedotova, Kh.Kh. Khadzhayev and P.A. Rebinder, Dokl. Akad. Nauk SSSR, 170(1966)1133.
25 L.A. Adburagimova, P.A. Rebinder and N.N. Serb-Serbina, Kolloidn. Zh., 17(1955)184.

26 G.M. Bartenev and N.V. Ermilova, in Fiziko-Khimicheskaya Mekhanika Dispersynkh Struktur (Physico-Chemical Mechanics of Dispersed Structures), Nauka, Moscow, 1966, pp. 371, 378.
27 G.M. Bartenev and Yu.V. Zelenev, Mekh. Polim., (no. 1)(1969)30.
28 G.V. Vinogradov, A.Ya. Malkin, N.V. Prozorovskaya and V.A. Kargin, Dokl. Akad. Nauk SSSR, 150(1963)574; 155(1964)406.
29 G.V. Vinogradov and A.Ya. Malkin, J. Polym. Sci., A2(1964)2357.
30 G.V. Vinogradov, B.A. Dogadkin, N.V. Prozorovskaya and A.P. Neverov, Kolloidn. Zh., 26(1964)567.
31 A.I. Leonov and G.V. Vinogradov, Dokl. Akad. Nauk SSSR, 155(1964) 406.
32 E. Mustafaev, A.Ya. Malkin, Ye.P. Plotnikova and G.V. Vinogradov, Vysokomol. Soedin., 6(1964)1515.
33 A.V. Tobolsky, Structure and Properties of Polymers, Wiley, New York, 1960.
34 B.A. Dogadkin and T.N. Tarasova, in Vulkanizatsiya Rezin (The Vulcanization of Rubbers), Goskhimizdat, 1954, p. 113.
35 V.A. Kargin and T.I. Sogolova, Dokl. Akad, Nauk SSSR, 108(1956)662; Problyemy Fizicheskoi Khimii, Vol. 1, Goskhimizdat, 1958, p. 18.
36 N.I. Shishkin and M.F. Milagin, Fiz. Tverd. Tela, 4(1962)2681.
37 P.P. Kobeko, Amorfnyye Veshchestva (Amorphous Substances), AN SSSR, Moscow, 1952.
38 J.D. Ferry, Viscoelastic Properties of Polymers, 2nd edn., Wiley, New York, 1970.
39 G.L. Slonimsky and L.B. Rogovina, Vysokomol Soedin., 6(1964)315, 620.
40 G.L. Slonimsky, Zh. Tekh. Fiz., 9(1939)1791; Dokf. Akad. Nauk SSSR, 140(1961)343.
41 A.P. Bronsky, Prikl. Mat. Mekh., 5(1941)132.
42 G.M. Bartenev and Yu.V. Zyelenev, Vysokomol. Soedin., 14A(no. 5)(1972)912.
43 R.F. Boyer, J. Polym. Sci., C, Polymer Symposia "Transitions and Relaxations in Polymers", No. 4, 1966.
44 G.M. Bartenev and Yu.V. Zelenev, J. Mater. Sci. Eng., 2(1967)137.
45 G.M. Bartenev and Yu.V. Zelenev, Vysokomol. Soedin., 4(1962)66.
46 P.P. Kobeko, Ye.V. Kuvshinsky and G.I. Gurevich, Izv. Akad. Nauk SSSR, Ser. Fiz., (No. 3)(1937)329.
47 H. Leadermann, J. Text. Res., 11(1941)171.
48 G. Fulcher, J. Am. Ceram. Soc., 8(1925)339.
49 E.I. Frenkin and Yu.G. Yanovsky, in Uspekhi Reologii Polimerov (Advances in Polymer Rheology), Khimiya, Moscow, 1970, p. 269.
50 G. Tamman, Stekloobraznoye Sostoyaniye (The Glassy State), ONTI, 1935.
51 N.I. Shishnkin, Zh. Tekh. Fiz., 25(1955)188; 26(1956)1461; Fiz. Tverd. Tela., 2(1960)350, 358.
52 G.M. Bartenev, Stroyeniye i Mekhanicheskiye Svoistva Neorganicheskik Stekol (The Structure and Mechanical Properties of Inorganic Glasses), Groningen, 1970.
53 M.V. Volkenshtein and O.B. Ptitsyn, Zh. Tekh. Fiz., 26(1956)2204.
54 S.N. Zhurkov and B.Ya. Levin, Khimiya i Fiziko-Khimiya Vysokomolekulyarnykh Soyedinenii (Chemistry and Physical Chemistry of High Polymers), AN SSSR, Moscow, 1952, pp. 280-289.
55 G.M. Bartenev, Dokl. Akad. Nauk SSSR, 110(1956)805; in Stekloobraznoye Sostoyaniye (The Glassy State), AN SSSR, Moscow, p. 147.
56 A.P. Aleksandrov, Trudy I and II i Konferentsii po Vysokomelekulyarnym Soyedineniyam (Transactions of the First and Second Conferences on High Polymers), AN SSSR, Moscow, 1945, p. 49.
57 Yu.S. Lazurkin and R.L. Fogelson, Zh. Tekh. Fiz., 21(1951)267.

58 G.M. Bartenev, Usp. Khim. (Advances in Chemistry), 24(1955)815.
59 Yu.S. Lazurkin, G.M. Bartenev et al., Vysokomol. Soedin., 6(1964)504.
60 G.M. Bartenev and M.V. Voyevodskaya, Kauch. Rezina N, 12(1964)14.
61 W.H. Carothers and J.W. Hill, J. Am. Chem. Soc., 54(1932)1579.
62 J. Miklowitz, J. Coll. Sci., 2(1947)193.
63 V.A. Kargin and T.I. Sogolova, Zh. Fiz. Khim., 27(1953)1039, 1208, 1213, 1325.
64 G.I. Barenblatt, Prikl. Mat. Melch., 28(1964)1048.
65 G.M. Bartenev, Yu.V. Zelenev and I.S. Lyakhovich, Mekh. Polim., (No. 5)(1971)885.
66 S.B. Einbinder et al., Dokl. Akad. Nauk SSSR, 159(No. 6)(1964)1244.
67 S.B. Einbinder et al., Mek. Polim., (No. 1(1965)65.
68 G.M. Bartenev and Yu.S. Zuyev, Prochnost' i Razrusheniye Vysokoelasticheskikh Materialov, Khimiya, Moscow, 1964; or Strength and Failure of Visco-elastic Materials, Oxford, 1968.
69 G.M. Bartenev, Mek. Polim., (No. 5)(1966)700.
70 S.N. Zhurkov, Neoganicheskiye Materialy (Inorganic Materials), 3(No. 10)(1967)1767.
71 G.M. Bartenev, in Trudy Mezhdunarodnoi Konferentsii po Kauchuku i Rezine (Transactions of the International Conference on Natural and Vulcanized Rubber), Khimiya, Moscow, 1971, p. 13.
72 G.M. Bartenev, Vysokomol. Soedin., 11A(1969)2341.
73 S.B. Ratner and Yu.I. Brokhin, Dokl. Akad. Nauk SSSR, 188(1969)807.
74 M.I. Bessonov and Ye.V. Kuvshinsky, Fiz. Tverd. Tela., 1(1959)1441; Vysokomol. Soedin., 2(1960)397; Fiz. Tverd. Tela., 3(1961)1314; Fiz. Tverd. Tela., 3(1961)607.
75 G.M. Bartenev, Izv. Akad. Nauk SSSR, Otd. Tekh. Nauk, (No. 9) (1955)53.
76 G.M. Bartenev and V.N. Kondratyev, Fiz.-Khim. Mekh. Mater., 4(1968) 201.
77 G.M. Bartenev and I.V. Razumoyskaya, Dokl. Akad. Nauk SSSR, 150(1963)784).
78 N.I. Peschanskaya and V.A. Stepanov, Fiz. Tverd. Tela., 4(1962)2926.
79 S.N. Zhurkov and B.N. Narxullayev, Zh. Tekh. Fiz., 23(1953)1677.
80 V.R. Regel, A.I. Slutsker and E.Ye. Tomashevsky, Usp. Fiz. Nauk, 106(1972)193.
81 S.N. Zhurkov and S.A. Abasov, Zh. Vses. Khim. Ova., (No. 3)(1961) 441.
82 S.N. Zhurkov and S.A. Abasov, Fiz. Tverd. Tela., (1962)2184.
83 G.M. Bartenev and L.S. Bryukhanova, Zh. Tekh. Fiz., 28(1958)287.

Chapter 2

FRICTION OF METALS

With the development of polymer materials, questions associated with polymer friction on hard surfaces were pursued on the basis of results obtained earlier from the study of metal friction. There is confirmation (1-3) that in the contact zone between a polymer and a metal, plastic deformation is observed in the polymer and adhesional "setting junctions" are produced (in analogy with such junctions for metals). The friction process is also associated with the failure and rebonding of setting junctions. Moreover, it is believed that the nature of static friction is the same for both metals and polymers (4-8). Other examples can be given to indicate that the concepts and methods used for investigating metal friction can also be used for polymer friction, despite the fact that the physico-mechanical properties of metals and polymers are essentially different. Therefore it is necessary to examine briefly the fundamental theoretical aspects of friction in ductile metals.

2.1 BASIC CHARACTERISTICS OF FRICTION

The friction force F, which is the fundamental characteristic of friction of two surfaces, is determined by the interaction of the two surfaces, whose real contact area will be denoted by S. Usually, the friction force is a function of the pressure p, the sliding velocity v, the temperature T, the contact time, and other external friction parameters. In practice, the specific nominal friction force, f_n, is often used, which is determined by the ratio of the friction force to the nominal geometric contact area: $f_n = F_f/S_N$. We also define the friction coefficient μ as the ratio of the friction force to the normal load P:

$$\mu = F_f/P. \tag{2.1}$$

Together with the specific friction force f, the concept of pressure or specific normal load, is employed: $p = P/S_N$. The most important characteristics of friction undoubtedly are the specific real friction force $f_r = F/S_r$ and the actual pressure $p_r = P/S_r$ (9). However, these characteristics are used rather rarely because of the difficulty of determining the real contact area.

With solid friction, stick-slip is sometimes observed; it has been studied in detail by Kragelsky, Deryagin, Push, Tolstoy, and others (10-13). This stick-slip motion is usually characterized by maximum (F_{max}), minimum (F_{min}), and intermediate (F_{cp}) friction forces.

According to prevailing ideas, the friction force can be divided into two terms (14-17):

$$F_f = F_a + F_d, \tag{2.2}$$

where F_a is the adhesion component, and F_d is the deformation component, which is associated in particular with the deformation asperities induced in the counterbody by the harder member of the friction pair.

Furthermore, it is assumed that the friction force obeys a monomial law and is proportional to the real contact area:

$$F_f = \text{const } S_r. \tag{2.3}$$

Despite the fact that so far there is no coherent understanding of metal friction, and that a general friction theory has yet to be formulated, Eqs. (2.2) and (2.3) are accepted by most researchers.

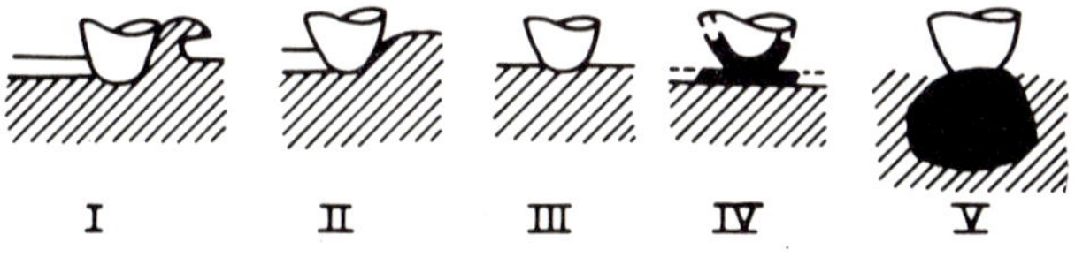

Fig. 2.1 Basic forms of friction-bond breaking (according to Kragelsky (14)): (I) shear; (II) plastic displacement; (III) elastic deformation; (IV) setting with fracture of layers that cover solid surfaces; (V) surface setting with ploughing.

According to Kragelsky's proposed classification of external-friction conditions (14), basically external friction is associated with aspects of the breaking of friction bonds (Fig. 2.1). Depending upon testing conditions, external friction usually is classified as dry friction, semi-dry friction, etc. From our point of view, the most useful classification of external friction modes is Akhmatov's

(16). According to him, the basic friction modes are determined by the states of the hard surfaces (Fig. 2.2):

Mode I. This is friction on virgin surfaces, with no interfacial phase. As a rule, the friction coefficient for virgin surfaces has the maximum value, and friction in this case if often accompanied by adhesion bonding of the surfaces.

Mode II. This is friction on "clean" surfaces. The friction coefficient is fairly large and depends upon the type and properties of oxide film.

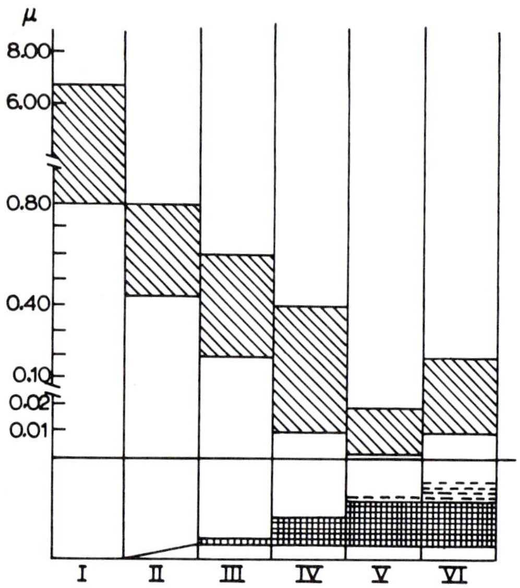

Fig. 2.2 Friction coefficient for different friction modes: (I) virgin surfaces; (II) clean oxidized surfaces; (III) boundary-friction transition zone; (IV) pure boundary friction; (V) hydrodynamic-friction transition zone; (VI) pure hydrodynamic friction. The lower part of the figure illustrates the extent of boundary-layer formation according to Akhmatov (16).

Mode III. This is transitional boundary friction, characterized by the presence between the surfaces of a very thin layer of interstitial, or tertiary, phase. This friction mode is transitional between clean-surface friction and boundary friction.

Mode IV. This is pure boundary friction, observed "in all cases when an extremely thin layer of tertiary phase lubricant occurs that separates hard surfaces within the range of their effect as lubricant." Boundary friction depends upon the physical nature, states, and properties of the hard surfaces, the structure and properties of oil molecules, and the experimental conditions (pressure, sliding velocity, temperature, surface roughness, etc.). Boundary friction thus is the most complex external friction process and also the most interesting technically and physically.

Mode V. This is boundary hydrodynamic friction, which represents a transition from boundary to hydrodynamic friction.

Mode VI. This is hydrodynamic friction. It is observed in the presence of an interposing layer of a lubricant. Hydrodynamic friction conforms to the laws of liquid-flow hydrodynamics and is largely dependent upon the liquid viscosity.

Static and kinetic friction forces are basic external-friction characteristics. The static friction force is determined by the boundary friction force that is necessary to set a body in motion.* When the transition from rest to motion cannot be determined exactly, the concept of initial friction force is used, which is determined in the initial shearing period by a given preliminary displacement. The kinetic friction force characterizes a steady friction mode with a constant sliding velocity.

Deryagin (17) believes that when the sliding velocity tends to zero, the friction force converges to the static friction force; that is, as $v \to 0$, $F_f \to F_s$. Experimentally, the static friction force is measured as the minimal force necessary to move a solid from rest. In establishing the static friction force for plastic solids, a series of difficulties arises, particularly the necessity of determining the start of motion. Here, with increase in the accuracy of determining the start of motion, the measure of the static friction force varies. The determination of the static friction force for plastic materials can be compared with the ascertaining of the plasticity limit. Experiments by Obreimov, Shubnikov (18), and Klassen-Neklyudov (19) have shown that the more precise the methods for measuring displacement, the smaller the force at which plastic flow is found to begin. The authors draw the conclusion that the plasticity limit is conditional in magnitude and tends to zero with increasing measurement precision. Burwell and Rabinowicz (20) have shown experimentally that in metal friction (indium on steel) the magnitude of static friction depends upon the rate of load application, observation time, and preliminary contact time.

*According to Kragelsky, the static friction force is the maximum value of the friction force in preliminary displacement.

Furthermore, these researchers have demonstrated that when $v \to 0$, the friction force tends to zero. An analogous pattern appears during the determination of polymer static friction. There is no doubt that the concept of static friction has practical value for a large class of materials.

2.2 THE REAL CONTACT AREA

The contact of real surfaces is always discrete. This means, that the contacting of surfaces occurs on a real contact area that is always less than the nominal geometric contact area.

The size of the real contact area and the kinetics of its formation are determined by the mechanical properties of the materials and by the geometry of the surfaces (21-52). Bowden and Tabor (15, 44) have stated that under plastic contact conditions the real contact area S is determined by the shear strength of the material. It follows that $S_r = P/Y$, where P is the normal load and Y is the shear strength of the softer member of the friction pair. It is often claimed (15, 53, 54) that $S_r = P/HB$, where HB is the material's hardness according to Brinell. Later, analogous conceptions of the contact area were utilized not only in the field of metal friction, but also for polymers (44-47, 55).

The roughness of a hard surface is usually presented in the form of profilograms, scaled lengthwise and crosswise in direction (40, 41). Conforming to the profilogram is a bearing-surface curve within the coordinates $1 - L$, ϕ, where $L = u_a/h_{max}$ is the relative approach distance, defined as the ratio of the approach distance (u_a) to the maximum height of micro-asperities (h_{max}), and ϕ is the relative contact area, equal to S_r/S_N. The initial segment of the bearing surface curve can be represented, according to Demkin (37), by the relation

$$\phi = \alpha_\phi b L^\nu \tag{2.4}$$

in which b and ν are constants that characterize the bearing curve (usually ν is between 2 and 3, b is between 1 and 16) and α_ϕ is the coefficient that takes into account the difference between the real contact area and the cross-sectional area of the asperities for relative approach ε. This method for the calculation of the real contact area has been described in detail elsewhere (37, 40).

Besides the bearing-surface technique, the real contact area can also be computed on the basis of micro-asperities on real surfaces. The most common models of surface roughness are spherical, elliptical, and conical. Theoretical calculations of the real contact area by the surface-asperity model have been made by several researchers (14, 37, 40, 48, 50).

Zhuravlev (50) has examined the problem of computing the real contact area for elastic-sphere models with a linear distribution of height. The contact area equals

$$S_r = k[(1 - x^2)/\pi E]^{10/11} P^{10/11} \qquad (2.5)$$

where x is Poisson's ratio and k is the proportionality coefficient.

Archard (45) considered an analogous problem, giving due consideration of surface waviness and nonlinear dispersion of asperities by height. In the absence of waviness the relation he obtained was

$$S_r = k_1 P^{8/9} \qquad (2.6)$$

but with waviness,

$$S_r = k_2 P^{26/27} \qquad (2.7)$$

where k_1 and k_2 are coefficients.

Fundamental contributions to the calculation of the actual contact area were made by Kragelsky (14, 56, 57), Dyachenko (40), and Demkin (37, 41). Kragelsky (56) calculated the real contact area for elastic, plastic, and elastic-plastic deformations of asperities, taking into consideration regular surface waviness as simulated by an assembly of spheres and cylinders. In this calculation, it is better to describe the properties of real rough surfaces by a cylindrical model rather than by a spherical model. For the cylindrical model, the relation obtained was

$$S_r = 1.76[(S_H \gamma_1 \gamma_2)/k_3^2]^{1/3} P^{2/3} \qquad (2.8)$$

where γ_1 and γ_2 are bearing-curve constants and k_3 is the rigidity coefficient.

For the spherical roughness model, by using a curved bearing surface, Kragelsky (14) and Demkin (37) obtained the relation of the relative real contact area to the pressure:

$$\phi = \left[\frac{(ab)^{\frac{\omega}{\nu}} r_1^{\omega} p_c}{h_{max}^{\omega} k_4 B}\right]^{\nu/(\nu + a)} \tag{2.9}$$

where p_c is contour pressure, ω is a coefficient that depends upon the nature of the deformation, B is a coefficient that characterizes the material's properties, and k_4 is a coefficient that depends upon ν and ω.

Equation (2.9) takes the following form for elastic-contact boundary conditions:

$$\phi = k_4 \left(\frac{\gamma_1}{h_{max}}\right)^{\frac{\nu}{\nu + 1}} \left(\frac{p_c}{E}\right)^{\frac{2\nu}{2\nu + 1}} \tag{2.10}$$

and for plastic contact:

$$\phi = p_c/\bar{c}Y, \tag{2.11}$$

where Y is the shear strength of the softer of the contacting bodies and $\bar{c}$ is a constant, equal to 3.

Calculations of the real contact area on the basis of models have also been made by Ilchenko (58), Rubinstein (59), Dyachenko (40), Holm (60), Bowden (61), and others.

An important circumstance, derived from experimental study, is that the real contact area increases with an increase in load, mainly through an increase in the number of contacts (with an insignificant increase of their size) in the initial contact period. As an example, Fig. 2.3 shows the dependence of the real contact area upon pressure for some metals. From the figure, one can see that in general this dependence is nonlinear. The rate of growth of the contact area as a function of pressure is determined by the properties of the solid. For example, with an increase in hardness of the material, the rate of change of S(N) decreases. Such a relation for the actual contact area is typical for a number of metals and, as we shall see later, for polymers too.

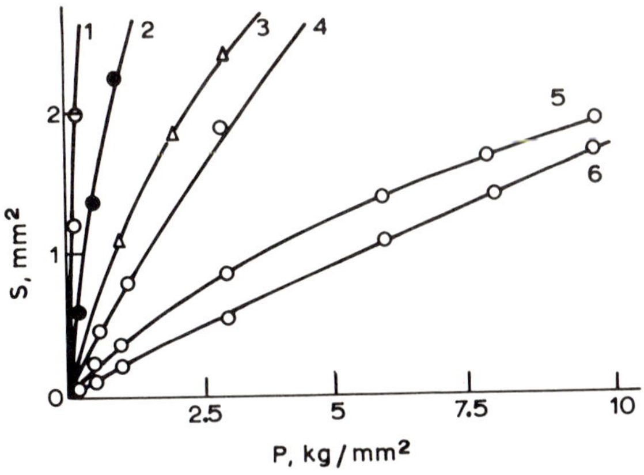

Fig. 2.3 Relation of the real contact area to pressure for various metals [fourth class grinding (14)]: (1) lead, (2) cadmium, (3) magnesium, (4) aluminum, (5) copper, and (6) steel-10.

A series of Soviet (36, 52) and foreign (45) studies record changes in the real contact area during shear, viz., a slight increase has been observed. Figure 2.4 shows findings of Demkin (52), and obtained by Mekhau's optical method. As shown, the real contact area increases about 10 to 15 per cent during shear. This increase in the contact area is associated with the process of surface conditioning or "run-in" treatment. For well "run-in" surfaces, shear does not produce significant change in the contact area.

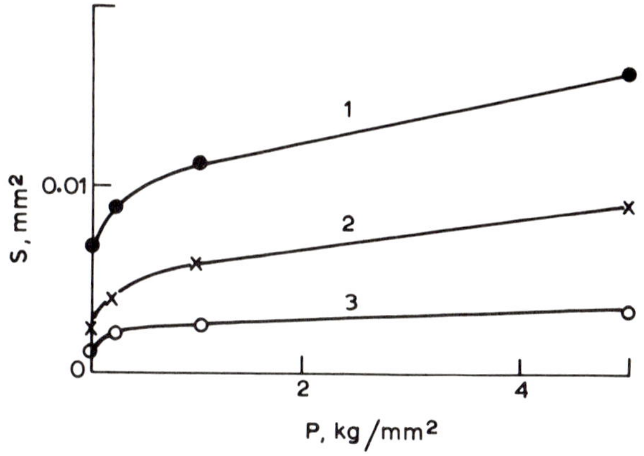

Fig. 2.4 Change in the plastic contact area relative to track length (52): (1) aluminum, (2) copper, and (3) steel a-3.

According to Kragelsky's theory, the real contact area [using the Maxwell-Ishlinsky equation (10)] depends upon the contact time:

$$S = S_c b_0 \left[\varepsilon_\infty^{\nu+1} - (\varepsilon_\infty^{\nu+1} - \varepsilon_0^{\nu+1}) \exp\left(-t \frac{\nu + 1}{\frac{r_1}{\dot{\Omega}} + \nu} \right) \right]^{\nu/(\nu+1)} \tag{2.12}$$

where ε_o is the strain in asperities at time t = 0, and ε_∞ is the strain for $t \to \infty$; S_c is the contour contact area; and b, ν, and r_1 are constants that characterize the nature and geometry of the contacting surfaces; and $\dot{\Omega}$ is the rate of elastic response.

Kosterin (63) has advanced a theoretical and experimental analysis of the relation of the real contact area and the friction force to the stationary contact time. He has shown that in some cases experimental results are well described by the relation

$$S = (S_c b)^{m/(m+b)} \left(p^{(\nu+m)/B} \right)^{\nu/(\nu+m)} [1 - \exp(-\delta_1 p_1 t)]^\nu \tag{2.13}$$

where B, δ_1, and p_1 are constants that characterize rheological properties of the material, and m is the strain hardening exponent. This formula is based on the relation of approach to contact time:

$$\varepsilon_t = \varepsilon_\infty [1 - \exp(-\alpha t)], \tag{2.14}$$

where α is the constant of the material.

2.3 THEORIES OF SOLID FRICTION

By the law of friction here and below is understood the dependence of the friction force upon normal load, $F_f = F(P)$. The law of friction can be derived from Eq. (2.3), taking c = const. Here it is assumed that the relation F(p) is determined by the relation $S_r(p)$. In fact, experimental results (30, 37, 40, 45) provide a basis for confirming that the law of friction is determined by the analytical form of the function S = S(P).

In solid-friction theory, the most common laws are monomial and binomial relations. Amonton's law is a widely known empirical monomial relation

$$F_f = \mu P \tag{2.15}$$

Bowden, Tabor (15), and other foreign scientists believe that the friction force under plastic contact conditions can be described by the relation

$$F_f = YS_r = [Y/(HB)]P. \tag{2.16}$$

Therefore, the friction coefficient is equal to the ratio of specific shear strength Y to hardness HB. Thus, the friction coefficient does not depend upon experimental conditions, but is determined by only the properties of the material. Equation (2.16) is the simplest basis of Amonton's law.

As will be shown below, in general the friction coefficient is not of constant magnitude, and Eq. (2.16) only approximately characterizes the friction force. We shall not examine in detail all experimental and theoretical studies on the law of friction. A complete survey is provided elsewhere (14-17). Here we focus on only those theoretical works whose results can be applied to polymer external friction.

Taking as a model a crystalline lattice whose atoms are immobile and incompressible, and examining the relative motion of the two crystalline surfaces, Deryagin (17) states that the friction force can be expressed as

$$F_f = \mu_o P, \tag{2.17}$$

where μ_o is the friction coefficient, determined by the tangent of the angle of inclination for atomic planes. Equation (2.17) is the theoretical basis for Amontons's law of friction. The inclusion of the force of intermolecular interaction N produces a binomial relation of the friction force to the normal load:

$$F = \mu_o (P + N). \tag{2.18}$$

In deriving Eq. (2.18) it was assumed that the normal load P is additive to the intermolecular attractive force N. Consequently, the estimated friction coefficient is $\mu = \mu_o (P/N + 1)$. In view of the fact that $N = p_m S$, where p_m is the specific molecular attraction force we obtain

$$F = \mu_o (P + p_m S). \tag{2.19}$$

Equation (2.19) differs from Coulomb's law,

$$F_f = A + BP, \tag{2.20}$$

in that in Eq. (2.19) the dependence of A upon the real contact area is provided for, and consequently so is its dependence upon the normal load. As Tolstoy has pointed out (23, 24), the load dependence of the friction force depends linearly upon load when the contact area does.

According to Eqs. (2.18) and (2.19), the linear relation of the friction force to the normal load also exists under the conditions of stable real contact area (or for the infinitely small $\partial S/\partial p$). Therefore, Eq. (2.18) has been submitted to experimental verification under conditions of S = const and $\partial S/\partial p \approx 0$ (32). This has been achieved either by full preliminary adjustment of the real contact area to its nominal value or by application of boundary lubricants and plastic bodies, for which S_r = const.

In studies by Lazarev, Deryagin, and Denisov (32, 33), the applicability of Eq. (2.18) with and without lubricants is shown by the crossed-filaments method. As an example, Fig. 2.5 shows the relation of the friction force to normal load for quartz filaments. As is evident, the linear relation observed can be described by Eq. (2.18).

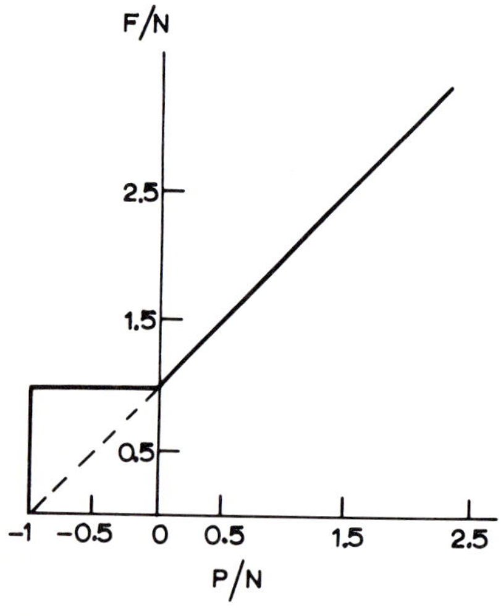

Fig. 2.5 Relation of the relative friction force F/P to the relative load P/N for quartz thread surfaces in air (17).

Kragelsky's theory (25, 26) is the most advanced development in solid friction theory. He has proposed that the friction force can be represented as the sum of molecular and mechanical components:

$$F = f_{mol}S_{mol} + f_{mech}S_{mech} \quad (2.21)$$

where f_{mol} and f_{mech} are the molecular and mechanical components respectively of the specific friction force, S_{mol} is the molecular contact area, and S_{mech} is the actual mechanical contact area.

According to Deryagin's law of friction (2.18), the molecular component of the specific friction force can be written

$$f_{mol} = P_{ma} + \gamma P_r,$$

where $p_r = P/S_r$, p_{ma} is the normal molecular interaction stress, and γ is the coefficient that characterizes molecular roughness. The mechanical component of the friction force can also be written as a binomial relation:

$$f_{mech} = s + \beta_2 p_r,$$

where s is the shear stress in the absence of a normal load, and β_2 is a coefficient that characterizes the material's strengthening under a normal load.

For the condition $S_{mol} = S_{mech} = S_r$, Eq. (2.21) takes the form

$$F = \alpha S_r + \beta_3 P = (\alpha + \beta_3 p_r)S_r, \quad (2.22)$$

where α and β_3 are constants that depend upon the surface geometry and nature of the material. Equation (2.22) reflects the dual molecular-mechanical nature of solid friction.

The following equation has been obtained for the friction force (27) derived from the superposition of the deformation and adhesion components of the friction force:

$$F = (h/d)(Y/\sigma_N) + (Y/P)S_r, \quad (2.23)$$

where h is the depth of the deformed contact area, d is the mean diameter of a single friction contact, σ_N is the mean compressive stress in a materail at the contact point, Y is the shear strength, and S_r is the real contact area.

Equation (2.23) gives a clear physical representation of the coefficients α and β_3 in the binomial friction law (2.22). These coefficients are associated with contact geometry and the stressed state of the material (Y/σ_N). The second term in Eq. (2.23), as experiment demonstrates, is always less than the first for metal solids. Its value depends upon the strength of the adhesion bonds, that is, upon the first term. This shows the dual molecular-mechanical nature of friction contact.

The equation for the friction coefficient that is derived from (2.22) is the generalized (according to Kragelsky) law of friction:

$$\mu = (aS_r)/P + \beta_3. \tag{2.24}$$

For plastic contact, taking $S_r = P/\Sigma_h$,

$$\mu = \alpha/\Sigma_h + \beta_3 = \text{cost}, \tag{2.25}$$

where Σ_h is yield pressure for indentation. Equation (2.25) can be taken as the theoretical basis for Amontons's law.

As has already been indicated above (see Fig. 2.1), there are several types of breaking contact in friction; therefore, the friction force can be represented in the form of the following sum:

$$F_f = \sum_{i=1}^{4} F_i,$$

where F_1 is the resistance due to the material's elastic deformation, F_2 is the resistance due to the material's plastic displacement, F_3 is the resistance due to the material's shear, F_4 is the resistance due to film shear. For each of these forces, Kragelsky (14) obtained the following equations theoretically.

(1) The friction coefficient for elastic contact is

$$\mu_1 = k\delta' (h_{max}/r)^{\nu/(2\nu + 1)} (p_c/E)^{1/(2\nu + 1)} \tag{2.26}$$

where ν, h_{max}, r are friction surface characteristics; k is a constant; E is the modulus of elasticity; p_c is the specific contour pressure; and δ' is the hysteresis loss coefficient. Equation (2.26) for $\gamma = 0$ is analogous to Greenwood and Tabor's formula (6) obtained by rolling a steel ball on a vulcanized rubber and assuming no adhesion force:

$$\mu_1 = \delta' p_c \left[\frac{9\pi}{64} \frac{1 - X}{E}\right], \qquad (2.27)$$

where X is Poisson's ratio.

(2) The friction coefficient for plastic displacement is

$$\mu_2 = (k \tan \varepsilon)(C_x/C_z),$$

where C_x and C_z are strengthening coefficients, and k tan ε is the geometric characteristic of friction contact.

(3) The tangential resistance coefficient during cutting, to a first approximation, is determined by the previous equation for μ_2. However, it is necessary to take into account that the coefficient C_z in this case is larger than for plastic displacement.

(4) The film fracture coefficient is determined according to Eq. (2.24).

Kragelsky and Mikhin (34) have shown that for friction of rough surfaces, in which micro-asperities are simulated by spheres, along a plastically deformed surface the friction coefficient increases parabolically with increasing indentation, but linearly with increasing adhesion:

$$\mu = \Sigma_b/p_r + 0.91\ \bar{m} + b(h_d/R)^{1/2}, \qquad (2.28)$$

where Σ_b is the shear strength of an adhesion bond for N = 0, b is a constant, h_d is the depth of roughness indentation with steady sliding, R is the radius of a sphere, and $\bar{m}$ is the strengthening coefficient.

Equation (2.28) gives a quantitative correlation between the deformation volume ($b\sqrt{h_d/R}$) and adhesion ($\Sigma_b/p + 0.91\ \bar{m}$) components of the friction coefficient. With equality of the adhesion and deformation components, when p_r is practically equal to Y (with a small correction coefficient c), we have

$$h_d/R = [(\Sigma_b/cY + 0.91\ \bar{M})/b]^2 \qquad (2.29)$$

This equation allows us to evaluate each component's share in the friction coefficient. Equation (2.28) has been confirmed by experimental studies (14, 34, 64). On the basis of this equation, keeping in mind that term h_d depends upon the normal load, it is generally possible to obtain (for elastic and plastic contact) the relation of the friction coefficient μ to the real pressure p_r. From Fig. 2.6, it follows that for elastic contact, (that is, for

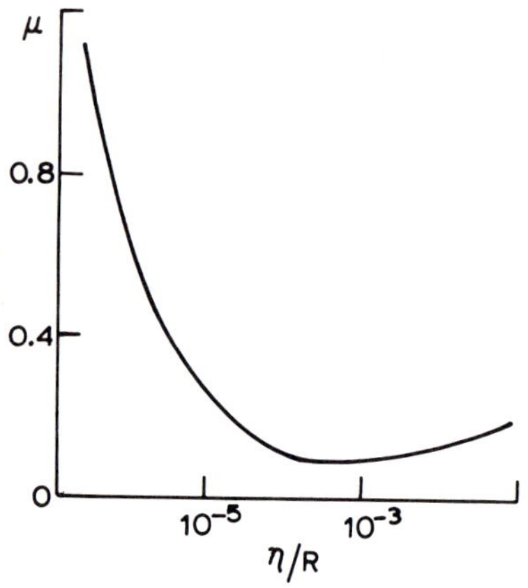

Fig. 2.6 Relation of the friction coefficient μ to ratio h_d/R (64).

small values of p_r or h_d/R), the friction coefficient decreases, while for large values of p_r or h_d/R it increases with load. The left side of the curve can be interpreted in terms of Coulomb's law, and the right side in terms of Amonton's law.

On the basis of carefully conducted research, Epifanov (28-31) has demonstrated that external friction of clean metallic surfaces results in plastic shear of a thin film of the weaker member in the friction pair. In this case, the friction force is proportional to the shear area:

$$F = YS_r.$$

Here $Y = Y_o + \beta\sigma_p$, where Y_o is the shear strength for $P = 0$, and σ_p is normal stress. Then the friction force can be expressed as

$$F = Y_oS_r + \beta_4P. \tag{2.30}$$

In this case, the dependence of the friction force upon the normal load is determined by the dependence of the shear area upon the load $S_c(P)$. For a constant shear area, the friction force depends upon the load only through an increase in the shear strength of the metal under pressure.

Figure 2.7 shows Epifanov's results (30) on the free cutting of copper by blades with working edges of different widths. With large normal loads, the contact area does not depend upon load, because the friction force is proportional to the cutting area. The data in Fig. 2.8 give further corroboration, from which it follows that the

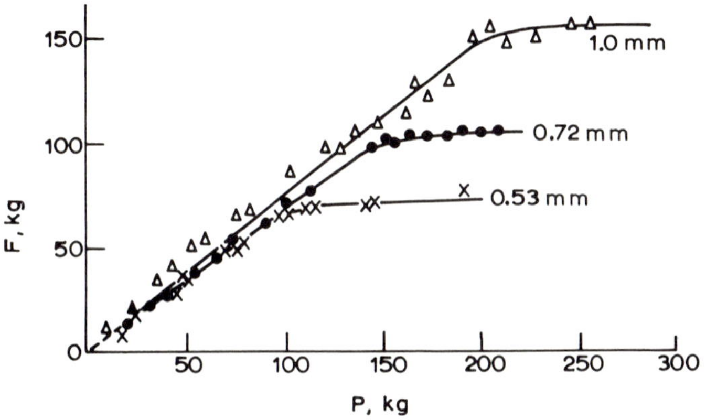

Fig. 2.7 Relation of the friction force to normal load during free cutting of copper by blades (30). Width of the working edge of the blade is shown on the curves.

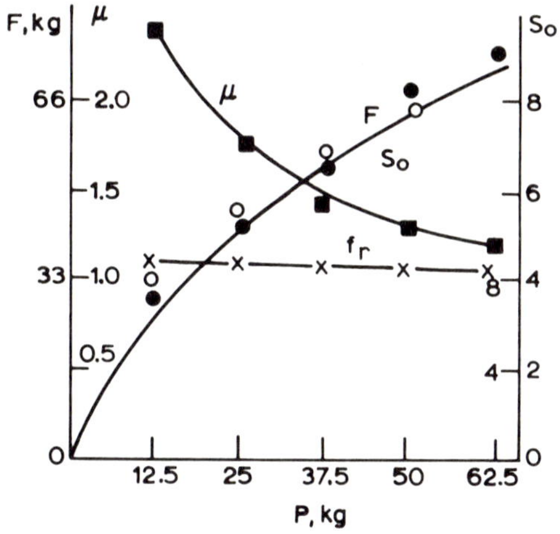

Fig. 2.8 Relation of the friction area S (●), the friction force F (O), the friction coefficient μ (■), and the specific friction force f_r (×) to normal load p (30).

contact area completely determines relation of the load to the friction force. The decrease in friction coefficient with increasing load is explained by the lack of proportionality of the real contact area to the normal load.

From the above, the significance of the real contact area for determining the friction force is clear. Therefore, it is preferable to determine neither the friction force nor the nominal friction force (that is, the friction force that refers to the nominal contact area), but rather the specific real friction force, that is, the friction force based on the real contact area. At present, studies on the measurement of the specific real friction force are not numerous at all, and this is no surprise, since experimental results obtained by various researchers in the field of solid friction often are not comparable.

Let us develop a brief analysis of the abovementioned relation of the friction force to normal load. First of all, it is necessary to point out that these relations do not allow for temperature and sliding velocity effects. Some equations take these parameters into account only indirectly within the constants in the friction laws. Moreover, the range of applicability of equations obtained for kinetic friction is not clear. Even in Deryagin's theory, it is postulated that the friction force is independent of temperature and sliding velocity.

The friction laws examined above, with the exception of (2.16), have a binomial form, but this does not mean that they reflect a single external friction mechanism for solids. Equations (2.22) and (2.30) have much in common, because they display in the real contact area the load dependence of the friction force. Consequently, it can be said that the analytic expression of the real contact area as a function of load determines the law of friction. Amonton's law and Bowden's formula are particular cases of the abovementioned binomial laws of friction. By contrast, the friction force in Eq. (2.18) is determined mainly not by the real contact area, but by molecular roughness, that is, by term $\mu_0 p$. A great advantage of the friction laws in Eqs. (2.16), (2.22), (2.28), (2.30) is the possibility of calculating the magnitude of the friction force from the mechanical properties of the material. Under the assumption that the friction force is determined primarily by the real contact area, it becomes important to consider the shear area, the plastic displacement area, the adhesion bonding area, etc.

The above allows us to conclude that if during application of the normal load the real contact area is changed, the friction force is determined mainly by the load dependence. When the contact area for one reason or another does not increase under load, the normal load can affect the friction force not through the real contact area, but only directly, according to Deryagin's law.

2.4 FRICTION FORCE AND FRICTION COEFFICIENT

Effect of stationary contact time

Considering that the friction force is proportional to the real contact area [Eq. (2.3)], it is possible to obtain the relation of the static friction force to the duration of stationary contact (62):

$$F = F_\infty - (F_\infty - F_0)\exp(-kt), \tag{2.31}$$

where F_∞ is the friction force at $t\to\infty$, F_0 is the friction force at $t = 0$, and k is a constant that characterizes the material's rheological properties and the surface geometry.

Figure 2.9 gives experimental results for different friction pairs. As is evident, the curves have a form that is predicted by the theory--see Eq. (2.31). Analogous results have been obtained for other materials (65).

The relation of the static friction force to the time of stationary contact is especially important to take into account for investigation of relaxation oscillation (63). As Kosterin has pointed out, in place of Eq. (2.31) the following empirical formula can be used advantageously:

$$F = ap^c[1 - \exp(-kt)], \tag{2.32}$$

where a, c, k are constants.

According to Kudinov (66), Eq. (2.31) is a particular case of the more general relation

$$F = (\mu P)/[(T_p + 1)(T_N p + 1)]$$

where p is the differential operator, and T_N and T_p are coefficients that have the dimensions of time. Kudinov (66) also showed that Eq. (2.31) can be obtained from an examination of dynamic contact

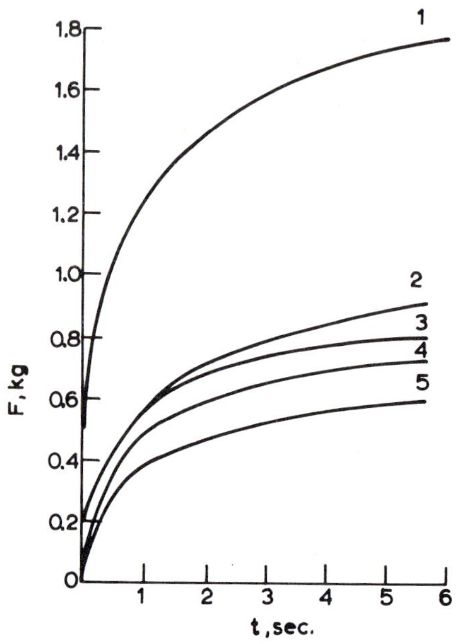

Fig. 2.9 Relation of the friction force to time of stationary contact for friction pairs (N = 3 kg) (14): (1) copper-lead; (2) magnesium-steel, type 45, heat-treated; (3) Plexiglas-copper; (4), (5) hard metals (magnesium, Armcoiron, and others).

properties. In this case,

$$F = P\mu[1 - \exp(-t/T)], \tag{2.33}$$

where T is a constant, equal to the surface-friction translation time at a constant sliding velocity.

Equation (2.33) formally agrees with Eq. (2.32). The physical difference between these equations consists in the fact that the first equation takes into account the formation of the contact area as a result of creep and elastic responses that occur under static conditions, and the second takes into account translation in the contact zone at a constant sliding velocity. It is interesting to note that Eqs. (2.32) and (2.33), which are obtained by examination of static and dynamic contacts, lead to equivalent relations of the friction force to the contact time. This is one way to get from the time dependence of the friction force to its speed dependence (14, 16).

Effect of sliding velocity

Kragelsky and Shchedrov (67) have examined sliding friction when deformation of a hard surface's asperities depends upon time and the deformation process of a material can be represented by the equation

$$\dot{f}_e + rf_e = E\dot{u} + E\dot{\Omega}u,$$

where f_e is the elementary tangential force, r is the relaxation rate, E is the modulus of elasticity, $\dot{\Omega}$ is the rate of secondary action on the surface film, and u is the deformation. The final result of the theory is expressed by the equation

$$F = f - (\alpha + \beta_5 v)\exp(-\eta' V), \qquad (2.34)$$

where α and β_5 are constants.

With v = 0, the friction force is equal to F_0. Equation (2.34) has maximum value at the sliding velocity

$$v_{max} = (\beta_5 - \alpha\eta')/(\beta_5\eta'), \qquad (2.35)$$

equal to

$$F_{max} = F_f' + (\beta_5/\eta)\exp[(\alpha\eta')/\beta - 1]. \qquad (2.36)$$

The term F_f' corresponds to the friction force for an unlimited increase in the sliding velocity, the parameter α depends weakly upon pressure and essentially upon the material, the parameter β_5 depends upon the material's viscosity, and the parameter η' represents the product of the clamping decrement and the viscosity coefficient. To determine the parameters that enter into Eq. (2.34), it is necessary to determine the following: the modulus of elasticity, the relaxation rate, the rate of response; and the elastic oscillations that develop during sliding (67).

The circumstances of the maximum condition (2.35 are physically clear for $\alpha = 0$. Then $\eta' v = 1$. If η' is treated as the relaxation rate, then the condition $\eta' v = 1$ specifies the usual reasonance for dynamic tests.

For high sliding velocities (in relation to relaxation rates), deformations of a solid are practically elastic. Roughness also leads to the formation of elastic waves in the contact zone during sliding. Adirovich and Blokhintsev's theory (68) considers this case and shows that the dissipative forces decrease as the sliding velocity increases. Analyzing this theory, Akhmatov (16) notes that his experimental results indicate that elastic-wave generation cannot be the sole cause of friction; it is sometimes necessary to take into account other forms of interaction.

Adirovich and Blokhintsev's friction theory can explain polymer friction on a hard surface at high velocities, because the relaxation rate of micro-asperities is significantly less than their interaction rate. Actually, as will be shown below a maximum in the dependence f the friction force's on sliding velocity is observed experimentally. In this case, the right side of the curve F(ν) can be interpreted by Adirovich and Blokhintsev's theory, but theirs is not the only possible interpretation.

Experimental studies give different relations of the friction to sliding velocity. This is caused by the high sensitivity of function F(v) to the physico-chemical state of sliding surfaces (16).

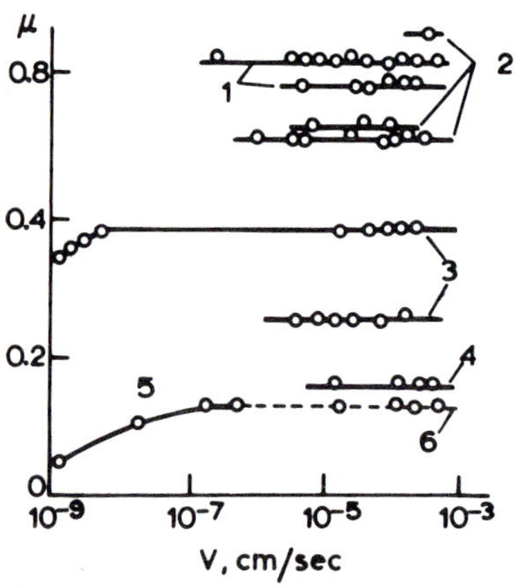

Fig. 2.10 Relation of the friction coefficient to velocity with damping (1-4) and without damping (5, 6) (69):
(1) hemisphere on a surface without lubricant;
(2) surface on surface without lubricant;
(3) hemisphere on a surface with lubricant;
(4) surface on surface with lubricant;
(5) the same, without a damper, smooth sliding;
(6) the same, with auto-oscillations (plotted at the mean ordinate).

At the present time, the friction force at low sliding velocities has been well studied. Figure 2.10 shows the results of measuring the friction coefficient of metallic surfaces with and without lubricant. These data were obtained with specially developed equipment that allowed damping of the friction surface's normal oscillations (69, 70). The results shown in Fig. 2.10 persuasively demonstrate that the great effect of normal oscillations on the kinetic character of the friction coefficient does not depend upon the sliding velocity, whereas without damping of normal oscillations, the friction coefficient increases with increasing velocity. Thus, the velocity dependence of solid friction is determined by normal oscillation. Some scientists (11, 71) conclude that freedom of normal oscillations of the slider is a necessary condition for both a falling friction-force rate and frictional auto-oscillations. The theory of frictional auto-oscillations is based on normal transfer freedom, as described in studies by Soviet scientists (11, 12).

Kunin and Lomakin (72) have shown that the dependence of the friction force on the sliding velocity can be both falling and rising; in this light, the oscillatory processes that accompany external friction have a larger significance. For ductile metals (47), the relation F(v) has a maximum (Fig. 2.11).

Akhmatov (16), Kostetsky(65), Granovsky (73), and others have also obtained a relation of the friction force to the sliding velocity which has a maximum. Levitsky (74) has provided a theoretical basis

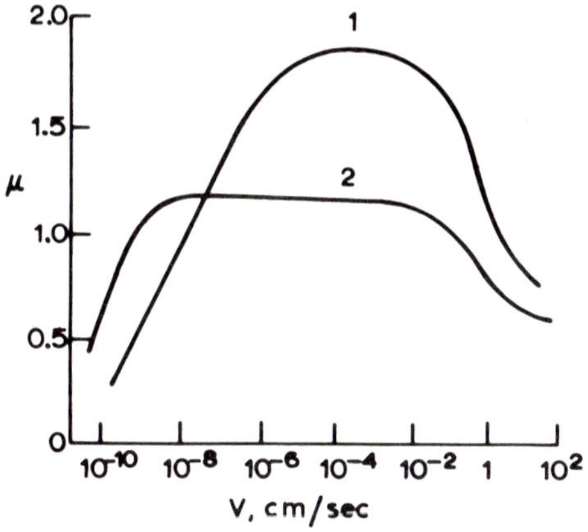

Fig. 2.11 Relation of the friction coefficient to sliding velocity for a steel hemisphere on indium (1) and on lead (2) (47).

for this relation, by considering that the formation of adhering bridges conforms to the law of monomolecular reaction and Eyring's theory of absolute rates.

Kuznetsov (75) proposed that the relation of friction force to sliding velocity should be explained by the effect of asperities that are formed during sliding. It appears that the rigidity of asperities is important. If the rigidity of the asperities increases, then the friction force increases with increasing sliding velocity, and decreases with decreasing rigidity of the asperities.

Flerov's dissertation work (76) was devoted to the formation of asperities during sliding. He showed that the normal load affects the character of the friction force's relation to the sliding velocity. For small and medium normal loads the friction coefficient is constant, or changes according to the curve with a maximum. The presence of a maximum in the relation F(v) is explained by the formation of asperities, whose development depends primarily upon surface temperature.

In our opinion, it is impossible to explain the relation of friction force to sliding velocity solely in terms of the formation of asperities, although sometimes that can be the deciding factor. According to data provided by Kragelsky (14), the speed dependence of the friction force depends upon normal load. From Fig. 2.12, it follows that for large pressures a falling relation is observed and for small pressures a rising one.

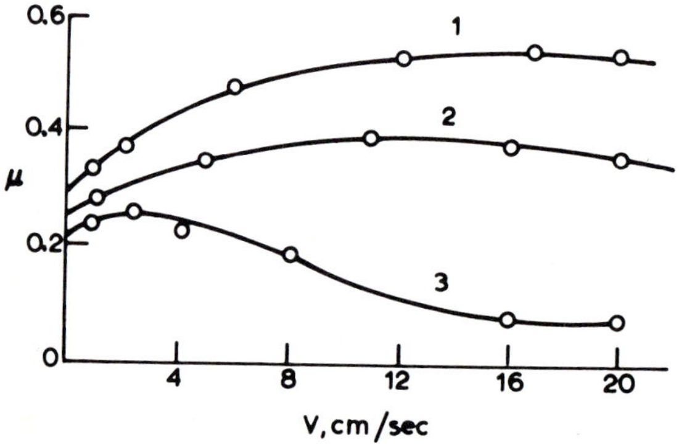

Fig. 2.12 Relation of the friction force to sliding velocity under different normal loads (14): (1) small loads; (2) medium loads; (3) large loads.

Akhmatov's (16) complete summary of dependences of friction force on sliding velocity for solids is given in Fig. 2.13.

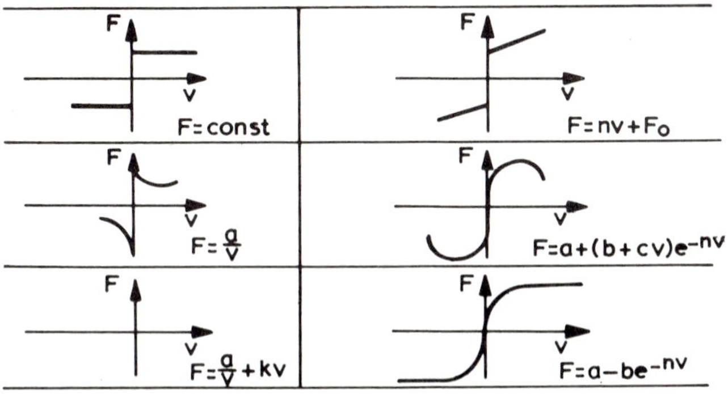

Fig. 2.13 Relations of friction force to sliding velocity for solids (16).

Effect of free, normally oriented contact vibrations

Since 1963 Tolstoy and co-workers have shown that no comprehensive treatment of the mechanism of sliding friction is possible without considering the mutual relation between friction and the free contact vibrations produced by sliding, which affects the sliding friction force. As a matter of fact, no experimental proof is needed to substantiate the phenomenon of normally oriented free contact vibrations that accompany sliding friction. The nonlinearity and asymmetry of these vibrations are also beyond question; they are due to the lifting of an unrestrained slider above the supporting surface as the amplitude increases.

The following reasoning leads to these two conclusions: Let us consider the contact layer between the slider and the support. This layer differs from any parallel bulk layer in that it is only partly filled with the solid material, and hence the total tangential cross-sectional area of the projections is but a very small fraction of that of the bulk layer. Therefore, if we compare the contact layer with a bulk layer of the same nominal area and thickness, the mechanical compliance of the former will be enormous by comparison. In other words, the specific compliance (1/K')/h of the contact layer (where K' is the stiffness and h is the layer thickness) is much greater than that of the bulk layers. As for the total compliance 1/K', it is usually also much greater for the contact layer than for the whole thickness of the slider.

For instance, the stiffness of the contact layer under a steel slider (30 # cm^{-2} tangential nominal area, ≈ 1 micron mean deviation of contact profile from the median plane, load 0.25 kg) was 2.5 x 10^9 dynes/cm (77), whereas the bulk stiffness of the same slider, 1.5 cm in height, was E S/h ≈ 5 x 10^{13} dynes/cm, so that $K'_{bulk}/K'_{cont} \approx 10^4$. The ratio K'_{bulk}/K'_{cont} decreases as the load and the height of the slider are increased, but even for a slider 10 times as thick, the ratio will be of the order of 100.

Under such conditions, the slider may be approximated by a perfectly rigid solid supported on a system of microsprings that simulates the behavior of the microasperities. Any impressed impulse possessing a normal component is bound to produce free vibrations of such a system in the normal direction. Inequality of the contact and bulk stiffnesses, however, is not essential for the effect of these vibrations on friction. It only simplifies the evaluation of the frequency. The bulk vibrations of the slider, if any, affect the amplitude of the contact vibrations just as any coupled vibrational systems affect each other. But any considerable difference in natural frequencies makes the effect of bulk vibrations on the contact vibrations negligible.

The essential difference between contact and bulk vibrations is that nonlinearity and asymmetry characterize the former, linearity and symmetry the latter. The nonlinearity and asymmetry of contact vibrations occur because the contact stiffness steadily increases as the contact layer is compressed, gradually bringing smaller and smaller projections into contact and flattening the bigger ones. The linearity and symmetry of the bulk vibrations, on the other hand, is due to the near-constancy of stiffness because the strain in a thick slider is small. Symmetrical vibrations cannot directly affect the mean height of a slider over a support. The asymmetry of nonlinear vibrations is absolutely essential for the lifting effect, as well as the thermal expansion.

It follows that a direct vibrational effect on friction may be produced only by contact vibrations. An increase in their amplitude lifts the slider, and the frictional force must decrease (77-82). The relation between these vibrations and friction, however, is not a one-sided one. Free, normally oriented contact vibrations are inevitably produced by sliding (77-82), viz., by the normal components of the impulses imparted to the slider asperities by the substrate asperities. To summarize: the contact vibrations must be produced by sliding, and in their turn must affect friction.

Is the amplitude of these vibrations in the course of sliding sufficient to produce a life resulting in a measurable decrease in the frictional force? A number of experimental observations give an unambiguously positive answer to this question. Perhaps one of the most convincing facts, apart from direct oscillograms, is the falling frictional-force-velocity relation (77-82). As the sliding velocity increases, the normal components of the impulses imparted to the asperities become stronger, the slider is lifted higher, and the frictional resistance (which unambiguously depends on the slider height (83-85)) decreases.

If the driving system is not stiff enough, the instability due to the falling force-velocity relation results in stick-slip.* The latter may of course be suppressed by imposing sufficiently heavy viscous damping that acts tangentially (88). Using such a system, we can obtain a force-velocity curve of negative slope, at least for $10^{-7} < v < 10^{-3}$ cm/sec.

The most convincing observation, however, is as that whenever damping is imposed in the normal direction (Fig. 7.28) (79-81), not only is the stick-slip suppressed, but the negative slope of the force-velocity relation also disappears. The force of friction becomes independent of the sliding velocity. This means that the decreasing force-velocity relation is due to the normally directed free contact vibrations produced by sliding, which affect the friction force.

Another convincing fact is that the friction force under normal viscous damping is not only higher than the mean dynamic friction force without normal damping, but even appreciably higher than the static friction force without normal damping (79-81). This shows that the static friction force is also somewhat lowered by the free

---

*R. Courtel and his co-workers (86) have published a series of important papers showing the role of the front ridge (bourrelet frontal) in producing tangential and normal vibrations with eventual lowering of friction. But the effects involved are associated with the transverse vibrations of the body of rodlike sliders rather than contact vibrations.

contact vibrations excited in a state of apparent rest by microvibrations whose amplitude fluctuates between 0.01 and 2.00 microns.

The friction-force-velocity relation at very low velocities, however, may be a rising one. This may be accounted for as follows (79-81): The overall effect of sliding velocity on friction is a superposition of two opposite effects: (1) an increase in sliding velocity increases the amplitude of free contact vibrations, lifts the slider, and thus tends to reduce friction; (2) an increase in sliding velocity increases the rate of shear wherever the material is sheared at the rubbing surfaces, and thus tends to increase the total frictional resistance. The total derivative of the tangential resistance with respect to sliding velocity is therefore given by

$$dF/dv = \partial F/\partial y(dy/dv) + \partial F/\partial v, \tag{2.37}$$

or

$$dF/dv = \tau[(\partial S/\partial y)(dy/dv)] + S(\partial s/\partial v), \tag{2.38}$$

where y is the relative height of the slider over the substrate, S_r is the real contact area, and s is the shearing stress. The first term is essentially negative because $\partial Sr/\partial y$ is negative, dy/dv being positive. The second term is necessarily positive. The sign of dF/dv therefore depends on the relative magnitude of the two terms on the right-hand side of Eqs. (2.37) and (2.38). At very low sliding velocities, the microimpulses on the colliding asperities are too small to produce intense free vibrations that result in an appreciable lifting effect represented by dy'/dv; moreover, the lower the velocity v, the smaller the stress s. For these two reasons, the first (negative) term may be smaller than the second (positive) one. Besides, S_r is larger for smaller velocities.

At higher velocities S_r is diminished, whereas s is steadily increased, so that at sufficiently large velocities the first (negative) term exceeds the second, and a decreasing force-velocity relation results.

If external viscous damping is imposed on the slider, however, dy/dv is accordingly reduced, and the absolute magnitude of the first negative term may be reduced to that of the second so as to make dF/dv zero, or even positive.

If frictionally excited free contact vibrations capable of affecting friction did not exist, there could be no sharp resonance effect in the lowering of frictional force produced by force vibrations of normal direction as observed by Grigorova and Tolstoy (71) (Fig. 2.14).

The resonance frequency exhibited by the friction force coincides within about 10 to 15 per cent with the rough estimate given by $(1/2\pi)(\sqrt{k/m})$ (where k is the contact stiffness measured directly and m is the mass of the slider). No better agreement could be expected in view of the nonlinearity of the contact vibrations.

The depth of the resonance minima of the friction force (Fig. 2.14) under the action of an oscillator consuming only ≈ 1 watt is not comparable with the slight resonance effect observed by Volossov and Boguslavski (88). The natural frequency of the contact vibrations for steel on steel of medium roughness is of the order of 10^3 Hz, whereas the slight friction minima in their case occurred at 80 Hz,

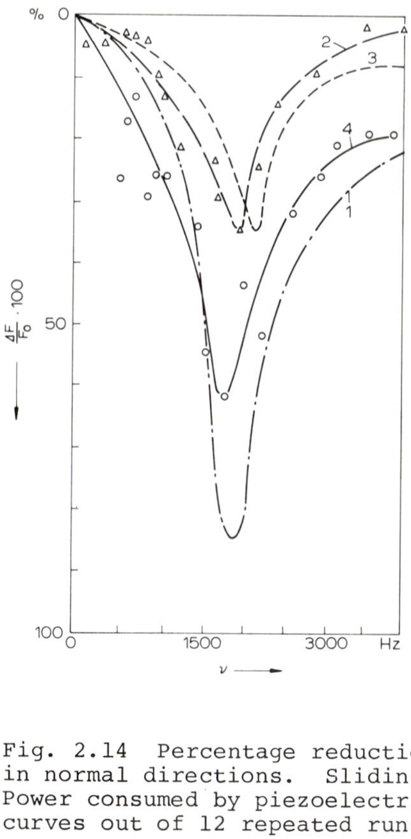

Fig. 2.14 Percentage reduction of friction force by forced vibrations in normal directions. Sliding fraction of clean flat steel surfaces. Power consumed by piezoelectric oscillator ≈ 2 watts. Four typical curves out of 12 repeated runs are shown. Slider mass ≈ 1.5 kg.

which was the natural frequency of the supporting construction. As mentioned above, any linear symmetrical vibrations may affect friction only through their effect on asymmetrical contact vibrations. The knowledge of the existence of linear symmetrical vibrations and their approximate natural frequency enables the true optimum frequency of forced vibrations to be used, with the result that an incomparably greater effect is obtained at a much lower power consumed by the oscillator.

Another resonance effect of force vibrations was observed in suppressing stick-slip vibrations (Fig. 2.15)(89). Here also a

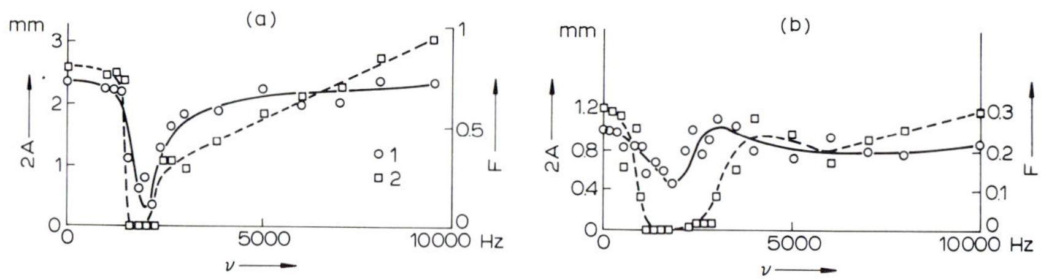

Fig. 2.15 Sliding friction force (solid lines) and peak-to-peak amplitude of tangential stick-slip oscillations (dashed lines) vs normal oscillator frequency in Hz. Cleaned surfaces on the left, stearic acid boundary lubrication on the right.

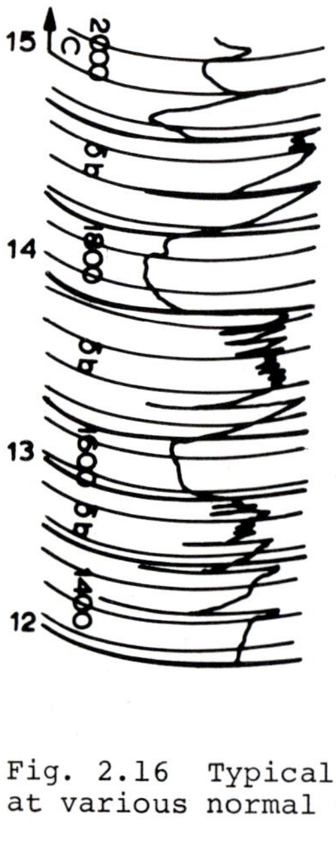

Fig. 2.16 Typical oscillogram of tangential stick-slip oscillations at various normal vibrator frequencies.

sharp resonance optimum was observed at frequencies on the order of $(1/2\pi)(\sqrt{k/m})$, estimated from the contact stiffness and mass of slider.

The above effects on friction cannot be interpreted without acknowledging the existence of free, normally directed contact vibrations, which have been observed directly in oscillograms (Fig. 2.16)(89).

The oscillographic records were obtained by two methods. One of them consisted in recording the oscillations of the electrical resistance of the contact layer, the oscillograms being photographed on a cathode-ray oscillograph screen. The ordinate axis was calibrated to read the normal displacements of the slider by imparting normal impulses of known magnitude to the latter and recording the oscillograph deflections. This conductometric method made it possible to record the normal slider amplitudes to tenths of a micron. Such amplitudes of free contact vibrations occur only at the rather high sliding velocities obtainable during the slip stages of tangential stick-slip vibrations. A typical loop oscillograph record of the latter is given in Fig. 2.17. Each x slip, represented by an almost

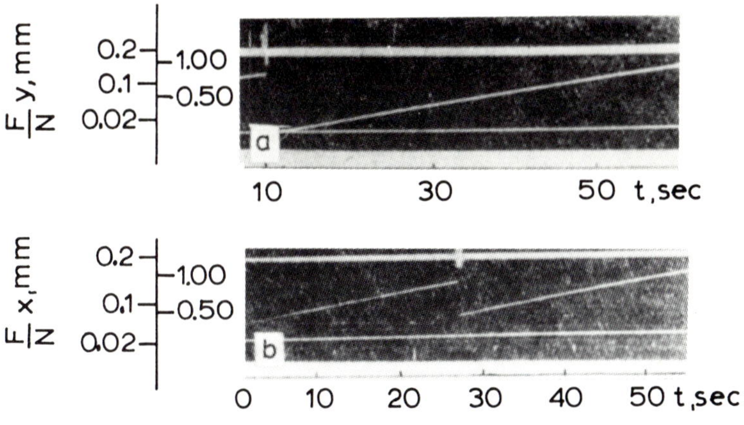

Fig. 2.17 Typical oscillograms obtained during stick-slip sliding on apparatus shown elsewhere without damping. The slanting straight lines are the stick stages of tangential stick-slip oscillations. The nearly vertical slip stages are indistinguishable, their duration being of the order of a millisecond. The short asymmetric pulsed crossing the upper horizontal line represent the sequence of free contact oscillations in the normal direction excited during the rapid tangential slips. They are expanded along the time axis in Fig. 2.18.

vertical line directed along the x axis, is accompanied by a normal bump, represented by a short Z segment asymmetrically crossing the horizontal line that was traced during the stick stage. When stretched out along the time axis on the cathode-ray oscillograph screen, each Z segment resolves into a Z(t) oscillogram of the type shown in Fig. 2.18 # (82).

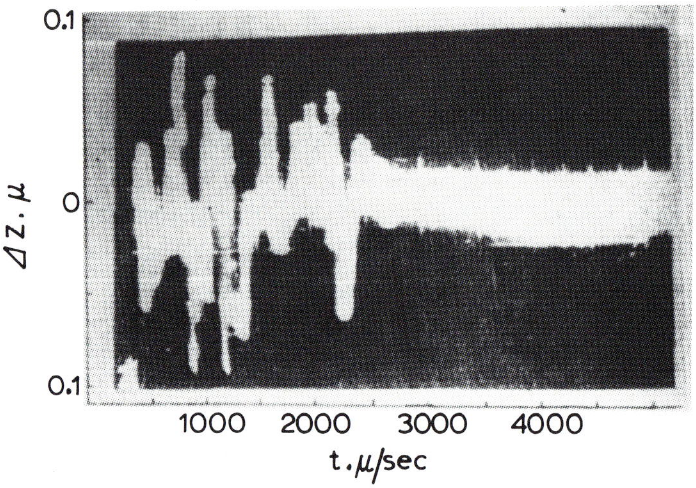

Fig. 2.18 Cathode-ray oscillogram showing a burst of free, normally directed vibrations which were excited by a slip during the course of tangential stick-slip oscillations of a slider.

Another method of oscillographing the free contact vibrations was based on using a minute inductance pickup embedded in the bearing surface of the slider at its center. The sensitivity of this system was about 10 times as high as that of the above system, so that even the slider vibrations that were excited by the microvibrations in a state of apparent rest could be recorded, including those excited by steady sliding at the low sliding velocities attainable by the apparatus shown in Fig. 7.28.

At low sliding velocities the mean amplitude of these vibrations is a steadily increasing function of the velocity, as might be expected from the proportionality of the normal impulse components to the sliding velocity. Evaluation of the frequencies of the free vibrations involves appropriate analytical methods, the vibrations being, as mentioned above, essentially nonlinear.

Summarizing the results obtained, it may be considered an established fact that the free frictional contact vibrations in the normal direction do exist and substantially affect friction. Moreover, their existence offers the possibility of using forced vibrations of a slider to reduce friction and to suppress stick-slip sliding with an incomparably higher efficiency than that attainable at frequencies other than those of the free contact vibrations. We believe that the above discussion can help to explain the dependence of friction force on sliding velocity.

If we consider the rate relation F(v) as the result of surface mechanical interaction, then for elastic-plastic contact (for which the static friction force is time-dependent), F(v) can be explained in the following manner. First, with increase in velocity (for small v values) viscous resistance increases on contact, but with large specific loads, anomalous viscous flow leads to a more abrupt increase in the viscosity coefficient with velocity. Secondly, with low sliding velocities an insignificant increase in the real contact area is observed because of the effect of "smearing" and machining of asperities; this should appear especially at high pressures. Consequently, for low sliding velocities, the friction force should increase with increasing velocity. This has been observed experimentally, particularly for friction of plastic materials (14, 16, 47).

For high sliding velocities,* an elastic-plastic body behaves as an elastic body, but for elastic contact, the friction force depends weakly upon sliding velocity. Increase in elasticity leads to substantial diminution in the real contact area and consequently to a reduction in the friction force. Moreover, for high sliding velocities, the contact time is short. This leads to an additional diminution of the friction force. However, it is necessary to add that at high velocities it is impossible to disregard the heatup which can lead to either rise or a fall of the friction force, depending on the nature of the contact. In the interstitial velocity zone, competition among the above factors leads to the appearance of a maximum in the curve.

---

*The relation of the friction-bond formation rate to the sliding velocity can serve as a criterion of whether a rate is high or low.

Effect of temperature

The relation of the friction force to temperature is principally significant for the study of the solid-friction mechanism. This relation has been little studied, mainly because under static conditions the temperature determines the kinetics of the formation of the real contact area and the static friction force, whereas under dynamic conditions the temperature at the contact depends in addition upon the sliding velocity (90, 91). In contrast to studies of the relation F(T), work on measuring the temperature rise due to friction is well advanced (90, 92-98). Such studies have great practical significance and are listed elsewhere (91).

Troyanovskaya (90) investigated the temperature gradient in friction and its effect on the friction coefficient. According to her data, the friction coefficient depends upon sliding velocity:

$$\mu = [(\kappa I\alpha_0)/(pv)](dT/dz), \qquad (2.39)$$

where κ is the thermal conductivity coefficient, dT/dz is the temperature gradient, α_0 is the coefficient of heat-flux distribution, p is the pressure, I is the friction thermal power, and v is the sliding velocity.

Besides the above, solid external friction is characterized by dependence upon roughness and upon the external environment (e.g. films, and boundary layers on friction surfaces), and by the effect of surface-active environments and the like (5, 7, 8, 25, 31, 34, 99-106). We have considered only the most important relations and characteristics, paying attention to the basic goal of this chapter, i.e., an examination of fundamental principles of metal friction in order to analyze the general friction mechanism, and the possible application of these principles to interpret the mechanism of polymer friction. From this chapter, it is evident that the change in the real contact area during friction is especially important. Unfortunately, the overwhelming majority of friction studies have not considered change in the real contact area, because of experimental difficulties.

The external-friction mechanism is intimately related to the nature of internal friction and, more specifically, to inelasticity and viscous resistance of the contact material. From this point of view, external friction should be characterized by a dynamic modulus, the friction activation energy, the relaxation time, and other factors. Viscous resistance of the contact material, which is determined by plastic deformation of single-crystal and polycrystalline solids, is associated with dislocation processes (107). Thus, structural defects that arise during deformation at contact points inhibit dislocation (108). The magnitude of static friction for single crystals and polycrystals is associated with the creep limit (18, 19). Although relaxation processes and other elastic imperfections in solids have been adequately studied, most studies of external friction do not include such data. Nevertheless, all notions of qualitative differences between external and internal friction are unsound. Essentially, external friction differs from internal friction only by the presence of a surface, and its effect may be taken into account through the real contact area.

REFERENCES

1 N. Gralen and B. Olofsson, Proc. Swed. Text. Res. Inst., 17(1947) 488; 20(1950)476; 23(1953)623.
2 R.F. King and D. Tabor, Proc. Phys. Soc., B 66(No. 9)(1953)728.
3 K.V. Shooter and D. Tabor, Proc. Phys. Soc., B 65(1952)661.
4 B.V. Deryagin, S.B. Ratner and M.F. Futran, Dokl. Akad. Nauk SSSR, 92(1953)1137.
5 S.B. Ratner, Dokl. Akad. Nauk SSSR, 93(No. 1)(1953)47; 83(No. 3) (1952)443.
6 J.A. Greenwood and D. Tabor, Proc. Phys. Soc., 71(1958)989.
7 N. Gralen, Proc. Roy. Soc., A 212(1952)491.
8 H.G. Howell, Proc. Swed. Text. Res. Inst., 23(No. 8)(1953)589.
9 V.V. Lavrentev, Vysokomol Soedin, 4(No. 8)(1962)1151.
10 A.Yu. Ishlinsky and I.V. Kragelsky, Zh. Tekh. Fiz., 14(No. 4-5) (1944).
11 B.V. Deryagin, V.E. Push and D.M. Tolstoy, Zh. Tekh. Fiz., 26(No. 6) (1956).
12 V.A. Kudinov, Issledovaniye Kolebanii Metallorezhushchikh Stankov pri Trenii (Study of Metal-cutting Machine Tools' Oscillation During Friction), AN SSSR, Moscow, 1958.
13 Yu.I. Kosterin, Mekhanicheskiye Avtokolebaniya pri Sukhom Trenii (Mechanical Auto-Oscillations During Dry Friction), AN SSSR, Moscow, 1960.
14 I.V. Kragelsky, Treniye i Iznos (Friction and Wear), Izd. 2-ye, Mashinostroyeniye, 1968.
15 F.P. Bowden and D. Tabor, The Friction and Lubrication of Solids, Clarendon Press, Oxford, Part I, 1950; Part II, 1964.
16 A.S. Akhmatov, Molekulyarnaya Fizika Granichnogo Treniya (Molecular Physics of Boundary Friction), Fizmatgiz, 1963.
17 B.V. Deryagin, Shto Takoye Treniye? (What is Friction?), AN SSSR, Moscow, 1963.
18 I.V. Obreimov and L.V. Shubnikov, TRFKhO, 58(1926)817.
19 M.V. Klassen-Neklyudova, Plasticheskiye Svoistva Metallov (Plastic Properties of Metals), Gostekhizdat, 1933.
20 J.T. Burwell and E. Rabinowicz, J. Appl. Phys., 24(No. 2)(1953) 136.
21 G.M. Bartenev and G.I. Epifanov, Inzh. Fiz. Zh., 1(No. 2)(1934).
22 B.V. Deryagin, Dokl. Akad. Nauk SSSR, 3(No. 2)(1934)93.

23 D.M. Tolstoy, Dokl. Akad. Nauk SSSR, 90(No. 5)(1953)819.
24 D.M. Tolstoy, Sbornik Trudov Posvyashchennykh Akademiku P.P. Lazarev (Studies Dedicated to Scholar P.P. Lazarev), AN SSSR, Moscow, 1956, p. 183.
25 I.V. Kragelsky, Trudy 2-ei Vsyesoyznoi Knoferentsii po Treniyu i Iznosu Machinakh (Transactions of the Second All-Union Conference on Friction and Wear in Machines), AN SSSR, Moscow, Vol. 3, 1949, pp. 178-183.
26 I.V. Kragelsky and B.V. Deryagin, ibid., p. 159.
27 I.V. Kragelsky, Dokl. Akad. Nauk SSSR, 140(No. 5)(1961)1048.
28 G.I. Epifanov, Dokl. Akad. Nauk SSSR, 114(1957)4; 118(1958)6.
29 G.I. Epifanov, Sbornik Issledovanii po Fizike Tverdogo Tela (Studies in Solid Physics), AN SSSR, Moscow, 1957, pp. 60-70.
30 G.I. Epifanov, in, Sukhbye treniye (Dry Friction), Izd. AN LATV SSR, 1961, pp. 29-40.
31 G.I. Epifanov and N.I. Minayev, IZV. Vyssh. Uchebn. Zaved., Fiz., (No. 1)(1959).
32 V.P. Lazarev and B.V. Deryagin, Trudy 2-ei Vsyesoyznoi konferentsii po Treniyu i Iznosu v Mashinakh (Transactions of the Second All-Union Conference on Friction and Wear in Machines), AN SSSR, Moscow, Vol. 3, 1949, p. 106.
33 P.V. Denisov, Dissertation, Stankin, 1955.
34 I.V. Kragelsky and N.M. Mikhin, in Teoriya Treniya i Iznosa (The Theory of Friction and Wear), Nauka, Moscow, 1965, p. 30.
35 I.V. Kragelsky, Zh. Tekh. Fiz., 12(No. 11)(1942).
36 S.B. Einbinder, Izv. Akad. Nauk SSSR, Otd. Tekh. Nauk, Mekh. (No. 6)(1962).
37 N.V. Demkin, Fakticheskaya Ploshcad Kontakta (The Actual Contact Area), AN SSSR, Moscow, 1965.
38 I.V. Kragelsky and G.Ye. Lazarev, in Peredovoi Nauchnotekhnicheskii i Proizvodstvennyi Opyt (Advanced Scientific-Technical and Industri Experience), No. 3, tema 30, TSITEI, 1960.
39 G.Ye. Lazarev and N.V. Demkin, Trudy 1-ei Konferentsii Molodykh Nauchnyk Sotrudnikov (Transactions of the first Young Scientific Workers' Conference), TsNIITMASh—ONTIM, 1961.
40 P.Ye. D'yachenko et al., Ploshchad Fakticheskogo Kontakta Sopryazhennykh Poverkhhostei (The Actual Contact Area of Conjugated Surfaces), AN SSSR, Moscow, 1963.
41 N.B. Demkin, Kontaktizovanie Sherochovatikh Poverkhnostze, Nauka, Moscow, 1970.
42 N.A. Konstantinova, V.V. Lavrentev, G.M. Bartenev and N.S. Korevenskaya, Mekh. Polim., 2(1966)263.
43 N.S. Korenevskaya, V.V. Lavrentev, S.M. Yagnatinskaya, V.G. Payevsk and S.S. Voyutsky, Vysokomol Soedin, 8(No. 7)(1966)1247.
44 F.P. Bowden and D. Tabor, Proc. Roy. Soc., A 169(1939)391.
45 J.F. Archard, Proc. Roy. Soc. A 243(1957)190.
46 R.B. Finoh, J. Text. Res., 21(1951)383.
47 J.T. Burwell and E. Rabinowicz, J. Appl. Phys., 24(No. 2)(1953) 136.
48 J. Huffington, J. Res., 12(1959)10.
49 Ye.M. Shvetsova, in Treniye i Iznos v Machinakh (Friction and Wear in Machines), No. 7, Izd. AN SSSR, Moscow, 1951, pp. 12-33.
50 V.A. Zhuravlev, Zh. Tekh. Fiz., (No. 7)(1940).
51 S.B. Einbinder, Pranch, in Teoriya Treniya i Iznosa (Theory of Friction and Wear), Nauka, Moscow, 1965.
52 N.B. Demkin, in Teoriya Treniya i Iznosa (Theory of Friction and Wear), Nauka, Moscow, 1965.
53 D. Tabor, Proc. Roy. Soc., 251(1952)378.
54 D. Tabor, Proc. Roy. Soc., A 251(No. 1266)(1959); Wear, 4(1961) 391.
55 J.A. Greenwood and D. Tabor, Proc. Phys. Soc., 71(No. 989)(1958).

56 I.V. Kragelsky, Zh. Tekh. Fiz., 13(1943)145.
57 I.V. Kragelsky, in Treniye i Iznos v Mashinakh (Friction and Wear in Machines), No. 5, AN SSSR, Moscow, 1950, p. 103.
58 O.T. Ilchenko, Vestn. Machinostr., (No. 10)(1958).
59 A. Rubinstein, Proc. Roy. Soc., B 69(No. 441)(1956)9.
60 R. Holm, Electrical Contacts, H. Gerbers, Stockholm, 1946, p. 398; J. Appl. Phys., 19(No. 4)(1948)361-366.
61 F.P. Bowden and D. Tabor, Proc. Roy. Soc., A 169(No. 938)(1939) 391-413.
62 I.V. Kragelsky, Zh. Tekh. Fiz., 11(1942)12; 14(No. 4)(1944).
63 Yu.I. Kosterin, Trudy 3-ei Vsyesoyuznoi Konferentsii po Treniyu i Iznosu v Machinakh (Transactions of the third All-Union Conference on Friction and Wear in Machines), No. 2, AN SSSR, Moscow, 1960, p. 65.
64 I.V. Kragelsky, in Nauka i Chelovechestvo (Science and Mankind), Znaniye, 1966, p. 383.
65 B.I. Kostetsky, in Treniye i Iznos pri Rezanii Metallov (Friction and Wear during Metal-Cutting), 1955, pp. 81-101.
66 V.A. Kudinov, in Sukhoye Treniye (Dry Friction), Izd. AN LATV SSR, 1961, pp. 55-64.
67 I.V. Kragelsky and V.S. Shchedrov, Zh. Tekh. Fiz., 18(No. 6) (1948).
68 E. Adirovich and D. Blokhintsev, J. Phys. (USSR), 7(No. 1)(1943) 29.
69 D.M. Tolstoy and Pan'-bin Yao, Dokl, Akad. Nauk SSSR, 114(1957) 1231.
70 D.M. Tolstoy, Stanki Instrum. (in Russian)(Machine Tools and Equipment), (No. 1)(1961).
71 S.P. Grigorova and D.M. Tolstoy, Dokl. Akad. Nauk SSSR, 167(No. 3) (1966); Stanki Instrum. (in Russian), (No. 9)(1969)17.
72 I.A. Kunin and G.D. Lomakin, Zh. Tekh. Fiz., 23(1955)1996.
73 M.B. Granovsky, in Treniye i Iznos pri Rezanii Metallov (Friction and Wear during Metal-Cutting), 1955, pp. 14-31.
74 M.P. Levitsky, in Povysheniye Iznosostoikosti i Sroka Sluzhby Mashin (Increase of Durability and Period of Service in Machines), Kiev, 1953.
75 V.D. Kuznetsov, in Razvitiye Teorii Treniya i Iznashivaniya (Development of the Theory of Friction and Wear), AN SSSR, Moscow, 1957, p. 74.
76 V.I. Flerov, Dissertation, Tomsk, 1957.
77 D.M. Tolstoy, (Tolstoi), Dokl. Akad. Nauk. SSSR, 153(No. 4)(1963).
78 D.M. Tolstoy and R.L. Kaplan, Novoe v Teorii Trena (Collected papers in Russian), Nauka, Moscow, 1966, p. 42.
79 D.M. Tolstoy, Wear, 10(1967)199-213.
80 D.M. Tolstoy, Schmierst. Schmierungstech., (No. 32)(1969)57-67.
81 D.M. Tolstoy, G.A. Borisova and S.R. Grigorova, O Prirode Trenie Tverdyh Tel (Collected papers in Russian), Izdatelstve Nauka i Tehnika, Minsk, 1971, pp. 116-126.
82 D.M. Tolstoy, G.A. Borisova and S.R. Grigorova, Dokl. Akad. Nauk, SSSR, 206(No. 1)(1972).
83 I.V. Kragelsky, Trenie i Iznos (in Russian), Izdatelstve Mashinevedenie, Moscow, 1968, p. 282.
84 N.M. Mihin, Trenie Tverdyh Tel (Collected papers in Russian), Nauka, Moscow, 1964, pp. 52-61.
85 G.Y. Ramishvili, Seebshchenia Akad, Nauk. Gruzinskei SSR. (C. R. Acad. Sci. of Georgian Soviet Socialist Republic, in Russian), Tbilizi, 34(1954)53-60.
86 R. Courtel, C. R. Acad. Sci., Paris, 253(1961)1758-1760; ibid., 1906-1908; M. Barquins and R. Courtel, ibid., 260, pp. 1805-1088; M. Barquins and R. Courtel, ibid, 267(1968)H. 664-667; M. Barquins and R. Courtel, ibid., seance du 24 fevrier (1969); ibid., 266 A

(1968)1068-1085; R. Courtel, Bull. (Fr.), (No. 40)(1965); Met. Corros. Ind., (Nos. 273-274)(1965); R. Baglin, P. Rougier and R. Courtel, C. R. Acad. Sci., Paris, 268(1969)666-669.
87 Leeng Fu Sheng and Pang Beeng Yaw, Issledovania po Poverhnostnym Silam (Investigation of Surface Forces, Collected papers in Russian), AN SSSR., Moscow, 1961, p. 131.
88 S.S. Volosov, Tehnologicheskie i Metrologicheskie Osnevy Tcchnesti Regulirevania Razmerov v Mashinestroenii (in Russian), Moscow, 1964 Osnevy Technesti Aktivnege Kentrolia Razmerov (in Russian), Moscow, 1969.
89 S.R. Grigorova, D.M. Tolstoy and A.V. Chichinadze, Dokl. Akad. Nauk SSSR, 202(No. 1)(1972).
90 G.I. Troyanovskaya, Dissertation, IMASH AN SSSR, Moscow, 1956.
91 A.V. Chichinadze, Raschet i Issledovaniye Vneshnego Treniya pri Tromozhenii (Calculation of and Research into External Friction During Frictional Action), Nauka, Moscow, 1966.
92 A.V. Chichinadze, Dissertation, IMASh, AN SSSR, Moscow, 1954.
93 V.S. Shchedrov, in Treniye i Iznos v Machinakh (Friction and Wear in Machines), No. 10, AN SSSR, Moscow, 1955.
94 J.F. Archard, Wear, 2(No. 6)(1959)438-456.
95 H. Blok, Inst. Mech. Eng. (London), 2(1937)14-20.
96 A.I. Kuyun, Dissertation, Kiev, 1954.
97 B.I. Kostetsky, Iznososto'kost'detalei Mashin (Durability of Component Machines), Mashgiz, 1950.
98 M.P. Levitsky, Zh. Tekh. Fiz., (No. 9)(1949).
99 G.I. Epifanov, Dissertation, IFKh, AN SSSR, Moscow, 1954.
100 D.N. Garkunov and I.V. Kragelsky, Dokl. Akad. Nauk SSSR, 113 (No. 2)(1957)326-327.
101 P.A. Rebinder, Fisiko-Khimicheskaya Mekhanika, Publ. H., Moscow, 1958.
102 A.P. Semenov, in Teoriya Treniya i Iznosa (Theory of Friction and Wear), Nauka, Moscow, 1965.
103 I.M. Lyubarsky, in Teoriya Treniya i Iznosa (Theory of Friction and Wear), Nauka, Moscow, 1965.
104 G.V. Vinogradov et al., Wear, 4(1961)274; 6(1963)202.
105 G.V. Vinogradov et al., in Treniye i Iznos v Mashinakh (Friction and Wear in Machines), No. 15, AN SSSR, Moscow, 1962, p. 420.
106 G.V. Vinogradov et al., Teoriya Smazochnogo Deistviya i Novyye Materialy (The Theory of Lubrication Action and New Materials), Nauka, Moscow, 1965.
107 A.H. Cottrell, Dislocation and Plastic Flow in Crystals, Oxford University Press, London, 1963.
108 Ya.I. Frenkel, Vvedeniye v Teoriyu Metallov (Introduction to Metal Theory), Fizmatgiz, 1958.

Chapter 3

POLYMER FRICTION IN THE GLASSY AND CRYSTALLINE STATES

At ordinary temperatures rigid polymers (plastics) exist in the glassy and crystalline states, and have a modulus of elasticity from two to three factors of ten higher than polymers in the rubbery state. On the other hand, in comparison with metals the modulus of elasticity for rigid polymers is small and extremely sensitive to temperature change. For example, for polymethyl methacrylate in the temperature range from 0° to 100°C, the modulus of elasticity for stretching changes from 28 x 10^3 to 5 x 10^3 kg/cm^2.

All basic mechanical characteristics of rigid polymers have a pronounced relaxation character (see Chap. 1) that dominates frictional effects in these materials. Under ordinary conditions, molecules of glass materials possess little mobility; however, the deformation properties depend in large measure upon stress. This special characteristic predetermines several specific external-friction properties of rigid polymers. It is significant, too, that glassy polymers are deformed in the contact zone and are practically elastic (1). Thus, rigid polymers represent a special group of materials that differ from metals in friction properties but share with them, as will be shown below, a number of regularities.

3.1 NATURE OF RIGID-POLYMER FRICTION

As was shown in Chap. 2, generally the friction force is proportional to the real contact area:

$$F = C_f S_r \quad . \tag{3.1}$$

This equation can also be used in the study of rigid-polymer friction. However, there is no agreement with regard to the proportionality coefficient that characterizes this friction.

Tabor and others (2,3) think that the proportionality factor C_f should be Y and that it is essential to understand the strength of adhesion "setting junctions" by analog to such junctions in metals (see Chap. 2). In connection with this understanding, the friction coefficient can be expressed by the relation $\mu = Y/HB$. The hardness HB can be replaced by the cold-flow limit Σ_f or the yield strength Σ_y. Usually we use all three to estimate μ (2-13).

Tabor and co-workers (2,10) investigated the relation between the friction coefficient and Y/Σ_y, for polyethylene, Plexiglas, polytetrafluoroethylene (PTFE), and polyester resins. The results of this study are given in Fig. 3.1. As shown in the figure, all polymers studied, except for PTFE, produce larger friction-coefficient values than the theory predicts. The authors believe that high values of μ are connected with an increase in adhesion bond strength as a result of high local stresses in the contact zone. Therefore, resistance to bond shear exceeds resistance to bulk yield shear, as the theory demands. The divergence for PTFE is explained by its exceedingly low adhesion to metal and by a shear transfer process on the plastic-metal interface (14).

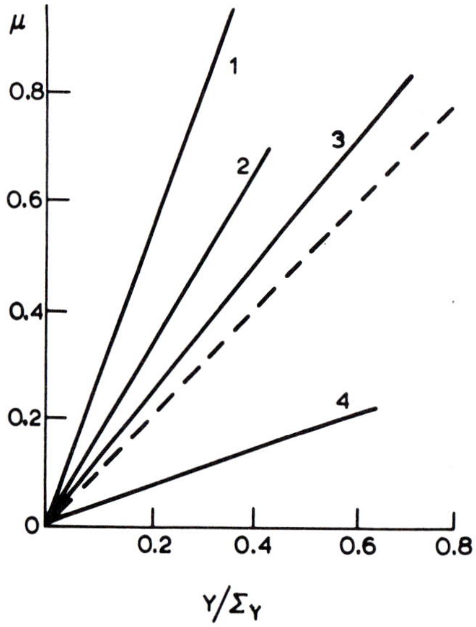

Fig. 3.1 Relation of the friction coefficient μ to the ratio Y/Σ_y on steel: (1) Plexiglas (Hirax); (2) polytrifluorochloroethylene; (3) linear polyethylene; (4) PTFE. Broken line was plotted with Bowden's formula $\mu = Y/\Sigma_y$ (2).

Einbinder and co-workers (15-17) have shown that polymers in the contact zone behave as strengthened, elastically-deformed materials. Young's modulus E of polymeric materials depends upon hydrostatic pressure (Fig. 3.2). For example, with rise in pressure to

2000 kg/cm^2, the modulus of elasticity for glassy polymers is increased two to three times (17). The limits of strength and of cold flow increase considerably. Poisson's ratio changes comparatively little (1.1 times). According to Einbinder, the increase in polymer mechanical characteristics with rise in pressure is determined by a decrease in specific volume and by an increase in energy interactions among polymer molecules.

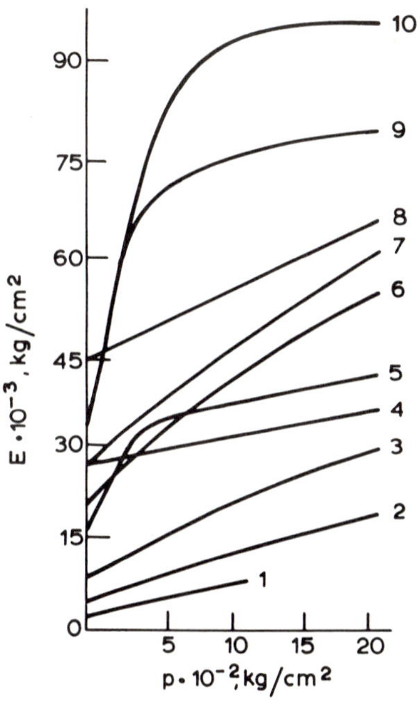

Fig. 3.2 Relation of Young's modulus E to hydrostatic pressure (14): (1) polytehylene; (2) polytetrafluoroethylene (PTFE); (3) polycapolactam; polymer K-78-51; (4) vinyl plastic; (5) Plexiglas (Hirax); (6) polymer K-17-2; (7) amino-plastic ΦKp-1; (8),(9) metals.

During sliding between two materials of different hardness, the friction coefficient is determined by the shear strength of the softer material (2). A direct relation between the shear stress and the friction force was also obtained by other researchers (2,3,6,8,9). King and Tabor (10) showed that a correlation between μ and Y/Σ_y is observed for some polymers over a rather wide range of temperatures (Fig. 3.3).

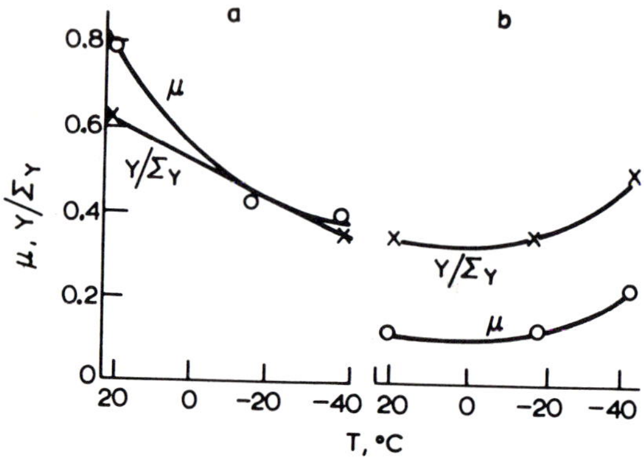

Fig. 3.3 Relation of the friction coefficient μ and the ratio Y/Σ_Y to temperature (10): (a) polyethylene; (b) polytetrafluoroethylene (PTFE).

From what has been said, it follows that Bowden's adhesion theory requires qualification to be applicable to plastics. Adhesion bonds serve only for the localization of shear formation in the bulk of the material. Consequently, the deformation component plays a dominant role in the total friction force. The question of the relationship between deformation and adhesion components in the friction force has been examined by Kuritsyna and Istomin (4), and we shall give their study more detailed attention. They investigated the friction of a spherical slider against a smooth plane. The sphere they used was a little chromium-hardened steel ball 6 mm in diameter, with surface treatment (10). By the use of the track width w, the real contact area was defined as $X = w^3/12R$, where R is the radius of the ball. The adhesion component of the friction force was determined from the reduction of the overall friction force due to surface lubrication. Under identical loading, the track width is the same for friction with and without lubrication. The deformation component F_d is proportional to w^3, because the volume of the material displaced by the sample is proportional to w^3 (Fig. 3.4). The slope of a straight line characterizes the specific deformation component. As seen in the figure, rising temperatures lead to a reduction in the friction force's deformation component. Its adhesion component is determined from the difference $F-F_d$. Its relation to the square of the friction track width, w^2, is shown in Fig. 3.5. It is

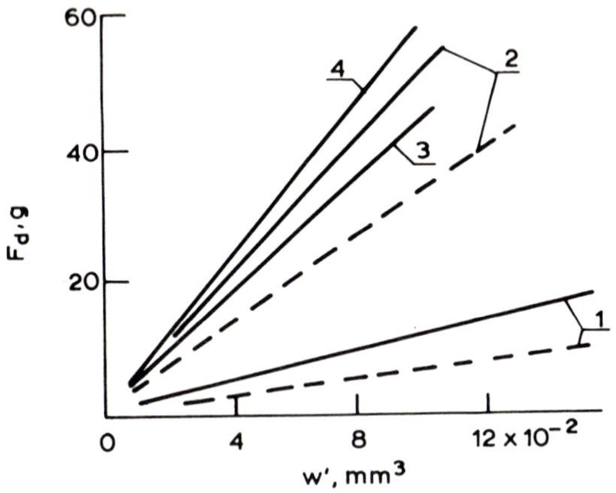

Fig. 3.4 Relation of deformation component of the friction force to width of friction path [—— at 20°C; --- at 70°C]. (1) PTFE; (2) polycaprolactam; (3) polyethylene (linear); (4) Polyamid.

possible to determine the friction force's specific adhesion component from the slope.

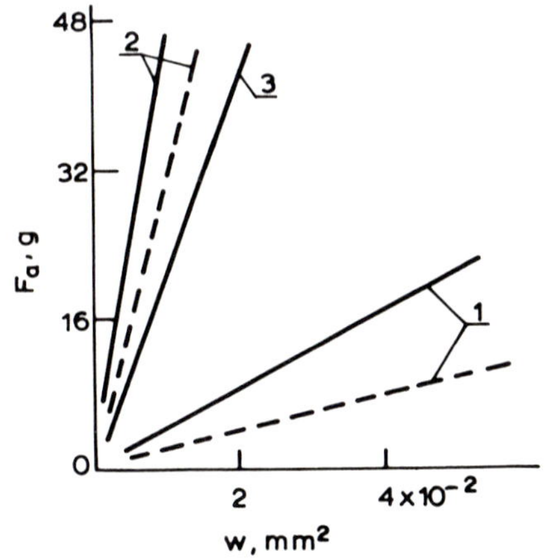

Fig. 3.5 Relation of adhesion component of friction force to square of path width [—— at 20°C; --- at 70°C]. (1) PTFE; (2) polycaprolactam; (3) polyethylene (linear).

It has been shown (4,18) that there is a linear relation between the friction deformation component and hardness, and an inversely proportional relation between the adhesion component and the contact angle θ_α (Table 3.1). The ratios of the friction force's adhesion to deformation components for certain polymers are: PTFE, 0.025;

TABLE 3.1

Adhesion and deformation components of friction force for polymers

Polymer	T (°C)	$F_{def.}$	Hardness[a] (kg/mm^2)	$F_{adh.}$	Contact angle θ_{α} (°)
PTFE	20	4.0	3.1	0.1	105
	70	2.2	1.5	0.05	---
Polyethylene	20	15.6	3.5	0.51	86
Nylon	20	17.0	7.7	1.02	65
	70	12.0	4.2	0.79	--
Polyamid-68	20	20.2	8.5	----	--

[a]Under load for 5 min.

polycaprolactam resin and fiber, 0.06; and polyethylene, 0.032. Thus, the adhesion component is equal to a fairly constant percentage of the deformation component. However, it is necessary to keep in mind that for relatively smooth surfaces with asperities of small radius of curvature the adhesion component's role increases significantly. The relation between adhesion and deformation components has also been studied by others (2,5-7).

The small-slider-on-plane research method used by Istomin and Kuritsyna is very common and simple, but it serves as a friction model only for surfaces that have large asperities. According to Kragelsky's theory, the deformation component of the friction force is significantly less than the adhesion component for elastic contact (see Chap. 2).

The experiments carried out by Istomin and Kuritsyna pertain to conditions in which the friction track forms as a result of slow, cold flow deformation. With such deformation, there are rather large mechanical losses. In some cases there is hysteresis during elastic deformation, and these losses may be seen in the friction characteristic of glassy polymers. For metals such losses are small. For friction on smooth surfaces, deformation losses are insignificant in comparison with losses that occur in breaking adhesion bonds.

Tanaka (14), by using a hemispherical polymer indenter, showed that rigid-polymer friction is related to deformation properties in the contact zone, but that the friction force is proportional to the contact area. Under a static normal load a wave is formed beside the indenter. Applied tangential force produces an increase in the wave ahead of the indenter (Fig. 3.6). Using a simple model (the Kelvin model) for describing stress-deformation relations, Tanaka obtained the following relationship between the sliding velocity v, the relaxation time τ, and the contact geometry:

$$\tau v = (r_1^2 - r_2^2)/(2r_2) \qquad (3.2)$$

where r_1 is the size of the contact in front of the indenter, and r_2 is the wave surface radius behind the indenter.

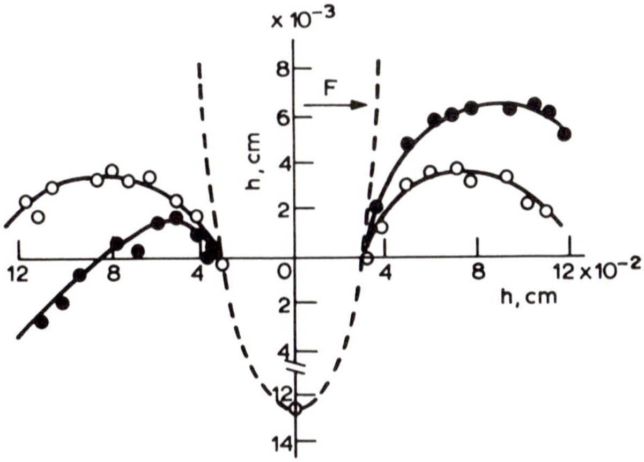

Fig. 3.6 Effect of the tangential force F on deformation of a Plexiglas (Hirax) surface by an indenter under a load of 350 g (14). Tangential force F: 0, 0 g; ●, 500 g.

Values for τ from Eq. (3.2), for different polymers, are as follows: for polystyrene, 9.8×10^{-4} sec; for PTFE 1.5×10^{-4} sec; for phenolic resin, 3.7×10^{-4} sec. These values seem to be within reasonable limits. According to these results, the deformation component is larger than the adhesion component, which agrees with the experiments by Istomin and Kuritsyna (4). Thus, polymer friction properties are determined by deformation properties in the contact zone. Thus friction losses are conditional on mechanical losses in the same polymers (19). But this conclusion does not hold in general, because it cannot be extended to smooth-surface friction.

Bowers, Clinton, and Zisman (20) have studied the role of purely adhesive interactions. They studied friction for polyethylene and its halogen derivatives. The replacement of hydrogen in polyethylene atoms by fluorine brings about a reduction in the friction coefficient as a result of a decrease in surface energy. In his work on polycaprolactam resin and fiber friction, Mitrovich (21) noted the relation of the friction force to intermolecular interactions.

Consequently, the relationship between F_a and F_d is determined by the type and nature of the polymer and the base layer, the surface roughness, and the testing conditions. Furthermore, as will be shown below, F_a and F_d are interdependent, in agreement with Kragelsky's friction theory (see Chap. 2).

3.2 REAL AREA OF CONTACT

The relation of the friction force in plastics to the normal load (the law of friction) is determined primarily, as in metals, by change in the real contact area upon loading. Consequently, we can write $F(P) = c_f S_r(P)$, where c_f is a constant that does not depend upon loading. However, for large loads, according to Deryagin and Toporov (22), the load directly affects the value of c_f, as follows from Deryagin's law (see Chap. 2).

First let us examine comparatively small loads. Archard (1), using Mekhau's optical method, demonstrated that for loads to 100 kg the friction force is proportional to the real contact area (Fig. 3.7). In addition, the relationship $F = c_f S_r$ is often

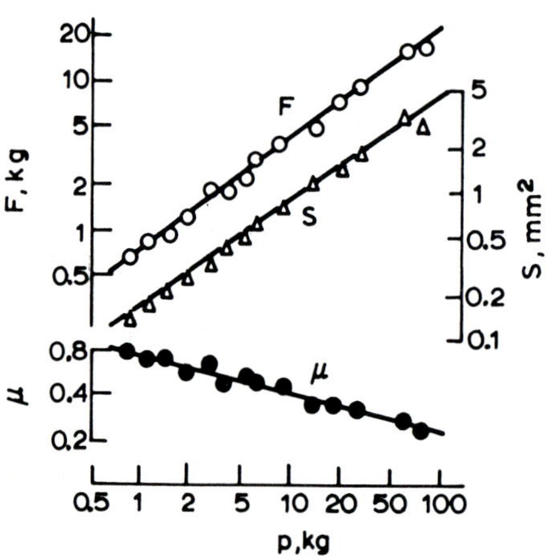

Fig. 3.7 Relation of the friction force, real contact area, and friction coefficient to normal load N for Plexiglas (Hirax) on Plexiglas (1).

studied by measuring the impression made on the polymer by a metal hemispherical indenter after application of a fixed normal load (2-4). However, such a measurement of the contact area is only qualitative, because (owing to relaxation processes) the size of the impression depends upon the load time and the impression measurement time.

Figure 3.8 shows the change in the real contact area S_r as a function of load time, according to Bilik's data (23). As seen from the figure, the contact area for PTFE depends significantly

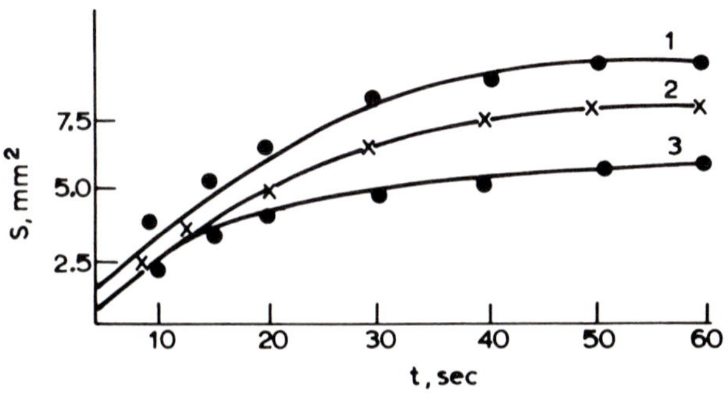

Fig. 3.8 Relation of the actual contact area to time of stationary contact for p = 36 kg/cm^2 (23): (1) polytetrafluoroethylene (PTFE); (2) Polyamid-68; (3) vinyl plastic.

more upon stationary-contact time than do Polyamid-68 and vinyl plastic. Such a relation $S_r(t)$ is completely to be expected because it is related primarily to polymer yielding, which is higher in PTFE. From these data, one cannot say whether the formation of the real contact area has been completed, because the saturation of the contact area in the prolonged-contact zone has not been clearly shown. Important in this study is the conclusion about the relation of contact-area formation to the normal pressure and polymer roughness.

Bilik (23) conducted extensive experimentation using Mekhau's method to investigate the relation of the static real contact area in rigid polymers to pressure. Figure 3.9 shows the resulting relations S(p) for four polymers. For PTFE under specific loads to 120 kg/cm^2, contact-area saturation is observed in practice. For values greater than 120 kg/cm^2, the saturation contact area increases but slightly at the expense of nominal contact-area growth.

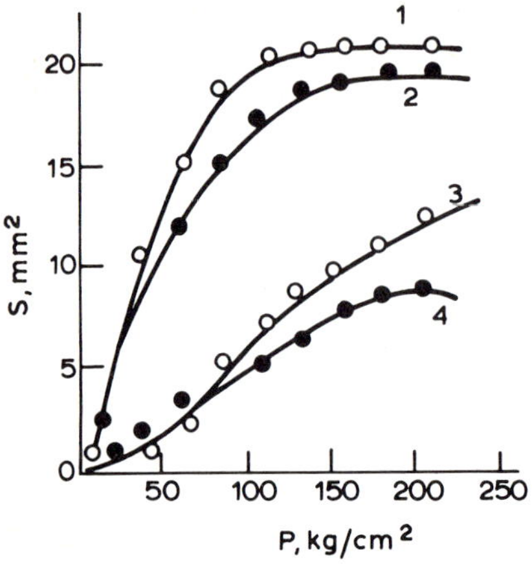

Fig. 3.9 Relation of the real contact area to pressure (23): (1) polytetrafluoroethylene (PTFE); (2) polycarbonate; (3) vinyl plastic; (4) Polyamid-68. Surfaces of the specimens were burnished with a polished disk at p = 1.5 kg/cm^2 and v = 0.3 m/sec.

To explain the results obtained, Bilik proposed the empirical relation

$$S_r = \alpha P_r^{\beta} \exp(\gamma/P_r) \ , \qquad (3.3)$$

where α, β, γ are constants. For polytetrafluoroethylene, $\alpha = 22.4$, $\beta = 0$, and $\gamma = 0.18$; for polamide, $\alpha = 13.8$, $\beta = 0.098$, and $\gamma = 0.07$. These values were chosen so that the curve for Eq. (3.3) would describe the experimental data. These constants lack physical content, and Eq. (3.3) thus has no physical significance.

In deformation, the contact of micro-asperities has an elastic character. This was demonstrated above. Ignoring the form and size of asperities, we shall consider micro-asperities as distributed randomly along the polymer surface. The roughness of a hard surface is significantly less than that of a polymer.

We have examined (24) polymer (normal) contact in terms of the probability of encounter between roughness and a smooth boundary. When a solid's surface is completely saturated, $S_r = S_n$. The probability of bond formation, or the probability of encounter between an asperity and an undisturbed surface, is equal to zero. On the other hand, initially a solid's surface is undisturbed and the probability of encounter is equal to unity. Consequently, the probability of contact formation is proportional to the free surface $(1-\phi)$. Then the rate of real contact area change is

$$d\phi/dp = \alpha(1-\phi) \qquad (3.4)$$

where α is contact constant that depends upon the polymer's roughness and elastic properties. Let us assume in the first instance approach that $\alpha = \gamma/E$, where γ is the roughness coefficient and E is the polymer's modulus of elasticity. Then it follows from Eq. (3.4) that

$$\phi = 1 - (1-\phi_0)\exp(-\gamma p/E) \ , \qquad (3.5)$$

where $\phi_0 = S_0/S_n$; S_0 is the real contact area in the unloaded state for $p = 0$. For elastic contact S_0 comprises very little, (almost none) of the nominal contact. The existence of S_0 is explained as resulting from imperfections in the polymer's elastic properties (due to preliminary surface grinding, etc.) and from adhesion forces between surfaces (reactions to full elastic recovery after removal of the load). Equation (3.5) satisfies distinct physical boundary conditions: for $p \to \infty$, $S_r = S_n$, and for $p \to 0$, $S_r = S_0$.

Let us examine the physical significance of the coefficient γ. Between absolutely smooth and plane parallel surfaces, the contact area for any load is equal to the nominal area, which corresponds to $\gamma = \infty$. For pronounced roughness, contact is formed only on a few surface points; therefore, S, and consequently γ too, will be comparatively small. For such surfaces, the growth of the actual contact area with increase in load (dP/dp) will be extremely small, corresponding to the smallest value $\gamma = \gamma_{min}$. For low pressures, when $p << E/\gamma$, it can be confined to a linear approach in expansion of the exponents. As a result, Eq. (3.5) converts into

$$\phi = \phi_0 + (1-\phi_0)\ (\gamma p/E) \ . \qquad (3.6)$$

For $\phi_0 = 0$, from Eq. (3.6) we have

$$\phi = \gamma p/E \ . \qquad (3.7)$$

The contact process that is characterized by Eq. (3.5) results in an increase in number of contacts with their size of each remaining constant. Actually, for contact of glassy polymers and vulcanized rubbers, an increase in contact area occurs because of an increase in the number of contacts (1,4,9,23-27). Then, too, for large loads some increase of size in the contact points themselves takes place (26).

We transpose Eq. (3.5) to a form convenient for experimental verification:

$$\ln(1-\phi) = \ln(1-\phi_0) - \gamma p/E \ . \qquad (3.8)$$

Figure 3.10 shows the relation of the real contact area to pressure with coordinates that conform to Eq. (3.8). As seen in the figure, the experimental points lie fairly well on a straight

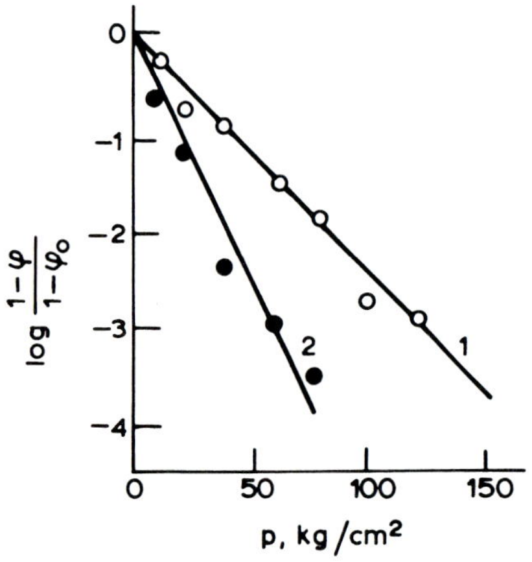

Fig. 3.10 Relation of the real contact area to pressure with coordinates conforming to Eq. (3.8) (24): (1) polyethylene; (2) PTFE.

line, which corroborates the applicability of the equation obtained. The results Bilik obtained for PTFE (Fig. 3.9) also satisfy Eq. (3.6). For rigid polymers (Polyamid-68 and vinyl plastic), Eq. (3.8) describes the experimental results inadequately. This can be explained, according to Einbinder's data (15), by a change in the modulus of elasticity E with increase in pressure. For a constant load, the contact area for these polymers is small (Fig. 3.9), but consequently there are significant local pressures in the contact zone, which bring about a change in the modulus of elasticity.

Other researchers have also noted a change in the modulus of elasticity for rigid polymers in the contact zone (2,10).

If we take into account the increase in the modulus of elasticity in the contact zone, then Eq. (3.8) can describe experimental results for a wider class of polymers. Moreover, for glassy polymers it is necessary to take into account change in the nominal contact area for high pressures, when the pressure is comparable to value E/γ. The relation (3.6) also holds true for a wide class of polymers under dynamic conditions, because the sliding velocity affects the real contact area weakly (24,26). For reciprocating motion of the sample at low frequency (3.2×10^{-3} sec^{-1}), as has been pointed out elsewhere (27), an increase in the contact area occurs over time as a result of abrasion of the polymer surface and the leveling of asperities.

Studies of the real contact area, described below, were made on a polymer-glass contact (Mekhau's method). One expects that during sliding on different surfaces the real contact area can depend upon speed. On this point Fort's work (28) has a bearing through its investigation of the speed dependence of the friction force for rigid polymers.

Temperature has a significant effect on the contact area. According to data obtained by other researchers (24,26), a sudden increase in the contact area is observed at the glass temperature T_g. Figure 3.11 shows relations of the contact area to temperature for linear polyethylene and for polyvinyl chloride. For temperatures

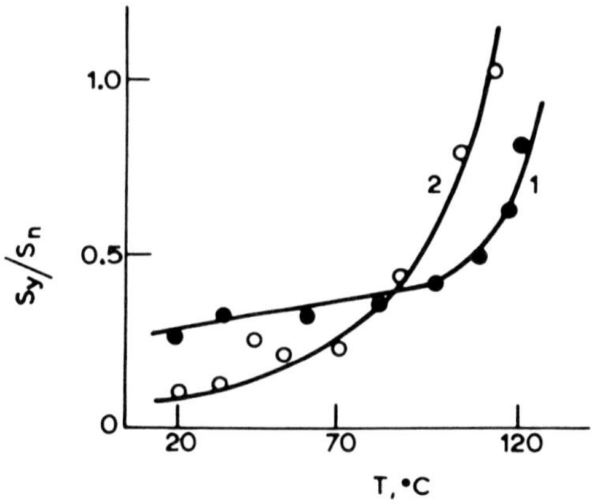

Fig. 3.11 Relation of the relative real contact area S_r/S_n to temperature (25): (1) polyethylene; (2) polyvinylchloride.

below T_g, an increase in the real contact area is observed with rise in temperature. For polyvinyl chloride, the glass temperature T_g = 70°C, but linear polyethylene has a melting point for the crystalline state (90 per cent of the polymer's whole bulk) of 120°C. With rise in temperature the crystalline content in polyethylene is diminished. For temperatures close to the melting point, plastic deformation accompanies cold-flow deformation. In this temperature range, the real contact area grows not only because of the increase in the number and size of contact points, but also because of an increase in the nominal contact area (26).

In conclusion, one must note that little attention has been given to the study of the real contact area, which is the fundamental characteristic of friction and wear. At present, usually the optical method is used to study it, but this method allows only a qualitative judgment about the molecular contact values. Studies on the physics of small deformations, applicable to polymer materials, are but slowly developing. All this significantly hampers the development of our understanding of the friction properties of rigid polymers.

3.3 FRICTION FORCE AND COEFFICIENT OF FRICTION

Effect of normal load (law of friction)

It is necessary to state, once and for all, that in the study of the law of friction for rigid, glassy polymers there is not consensus on the relation to load of either the friction force or the friction coefficient, nor are there precise values for the friction coefficient. In our opinion, this is due to two circumstances: first, to the use of different investigation methods (loading cycles, sliding velocity, external conditions, etc.), and secondly, to sharp differences among original physico-mechanical characteristics in the polymers that have been studied. Let us take as an example the well-studied PTFE. The degree of crystallinity of this material varies with the preparation technique from 0.45 to 0.80 (29). Noting that this material's melting point is 327°C, but that the glass temperature of its amorphous portion is about 120°C, one can see clearly that in such a wide range the physical properties of the original state can change. PTFE takes different forms in the crystalline state (30). Cold flow under constant stress is an extremely important property of PTFE. Other polymer materials also have a wide range of physico-mechanical properties (see Chap. 1).

It is fully understandable that without a calculation of special structural features and physico-mechanical properties, it is difficult to establish regularities in friction.

Now let us examine the main experimental results that have been obtained from investigating the law of friction for rigid polymers. The range of loads studied can be divided tentatively into three sections: small specific loads, from 10 to 20 kg/cm^2; medium loads, from 20 to 200 kg/cm^2; and large loads, from 200 kg/cm^2 up. The most widespread research method, especially outside the Soviet Union, is the indenter-plane method (1-14).

Several researchers (1-3,10,11,13,31,32) have established that the friction coefficient μ for plastics does not depend upon load. Semyonov and Pozdnyakov (31), through much experimentation, showed that μ = const; that is, Amonton's law is true. Figure 3.12 presents some of their results. Polyethylene (μ = 0.137) and PTFE (μ = 0.049) have the lowest friction coefficients on steel. For the friction

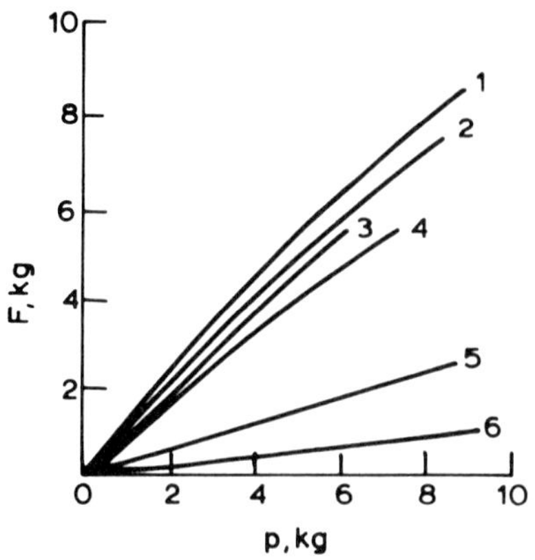

Fig. 3.12 Relation of the friction force to normal load for polymers on steel (31): (1) laminated fabric; (2) polycaprolactam; (3,4) nylon; (5) polyethylene; (6) polytetrafluoroethylene (PTFE).

pair PTFE on PTFE, μ = 0.037, but for the pair polycaprolactam on polycaprolactam, μ = 0.62. The authors do not give the reason for such a change. With elastic materials, too, the friction coefficient of a polymer-on-polymer pair is always higher than for friction on metal surfaces.

However, it has also been shown (8,9,12,33-36) that deviations from Amonton's law exist. In this connection, exponential ($F \propto p^n$), binomial, Coulomb's-law-type, and other empirical relations have been proposed. These do not possess theoretical significance, and we shall not discuss them.

For rigid polymers, the friction coefficient depends upon the nature of the rigid surface. Figure 3.13 presents the results of calculating polymer static friction on four solids (37). The

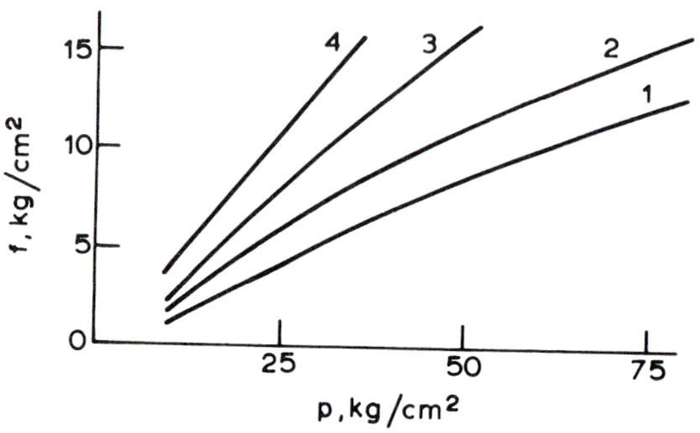

Fig. 3.13 Effect of a hard surface (∇9) on the relation of the friction force to pressure for polyvinylchloride on four solids: (1) bronze; (2) steel; (3) aluminum; (4) polyvinylchloride.

highest friction coefficient is observed for polymer-on-polymer friction ($\mu = 0.41$). This is dependent on a large real contact area developing as a result of rheological processes on the boundary. The change in the friction coefficient indicates that the adhesion component of the friction force generally may be disregarded.

Kragelsky (38) conducted a simultaneous study of the relations of the real contact area and the friction force to load. Static friction on Plexiglas was measured and had a modulus of elasticity of 2×10^4 kg/cm^2. Figure 3.14 presents his experimental results in terms of the following variables: the specific real friction force $f_r = F/S_r$, and the actual pressure $p_r = P/S_r$. As seen in the figure, a linear relation is observed, corroborating the so-called Kragelsky's elementary law (2.22). It is interesting to note that with increase in load, the real pressure rises (Table 3.2), which in the author's opinion attests to Plexiglas' work-hardening capacity.

Deryagin and Toporov (22) investigated the friction force of some polymers at high pressures. Samples of polymer films in the thickness range 40-70 μm were tested by placing them between steel slabs (see Chap. 7). Figure 3.15 presents the results obtained.

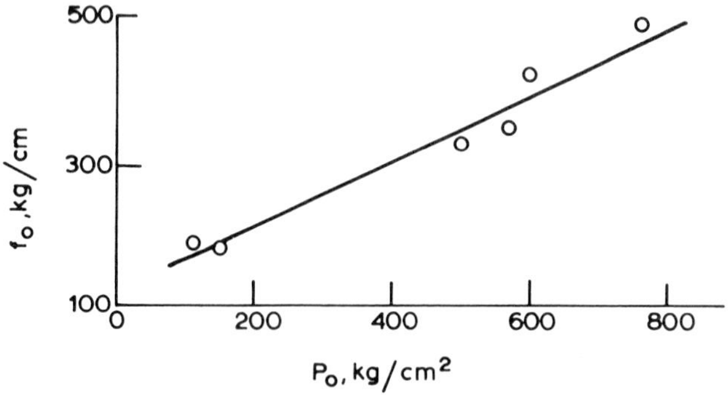

Fig. 3.14 Relation of the specific real friction force to pressure for a polymer on steel (∇10) (38).

TABLE 3.2

Change in the real specific friction force and the contact area in relation to pressure (38)

P(kg)	F(kg)	$S_r(10^{-3}\ cm^2)$	$f_r(kg/cm^2)$	$p_r(kg/cm^2)$
0.2	0.16	9.1	176	220
0.4	0.26	16	175	250
0.95	0.45	14	320	600
1.95	1.0	29	345	670
2.95	1.65	34	486	870
4.20	2.50	59	425	710

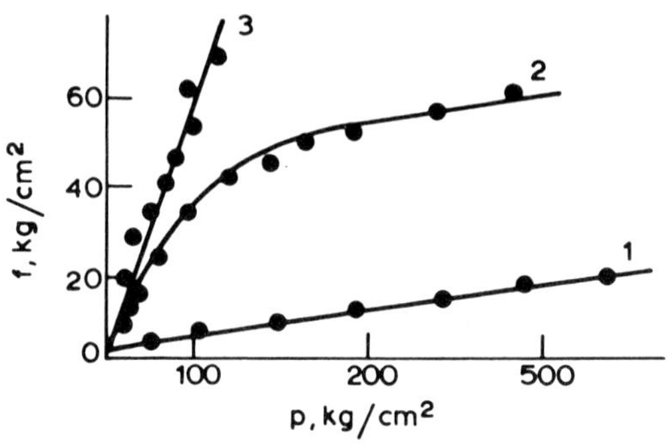

Fig. 3.15 Relation of the specific friction force to pressure for polymers on steel (22): (1) PTFE; (2) polyethylene; (3) polyamide.

For PTFE, the relation of the specific friction force to pressure can be expressed, according to Deryagin's law (2.19), in the following form

$$f = \mu_0(p - p_r\phi) .$$

It has been noted (22) that for polyethylene up to the specific normal load 140 kg/cm^2 there is a change in the relative real contact area ϕ, and the friction force increases nonlinearly with rise in pressure. Above 140 kg/cm^2 (the cold-flow limit for PTFE), a linear relation f(p) is observed, determined (where S = const) by the first term in Deryagin's equation. An analogous relation is observed, too, for friction of polyamide. Polyethylene behaves differently. Its relation f(p) is nonlinear. This, apparently, is related to the fact that contact-area formation in polyethylene occurs at high loads. Probably under these conditions polyethylene's crystalline structure changes. On the other hand, it is possible that change in the material's elastic properties, related to its work-hardening (an increase in the modulus of elasticity), is significant. For polyethylene, this effect is quite strong (see Fig. 3.15).

Unfortunately, Deryagin and Toporov (22) do not note whether change in the modulus of elasticity or in film thickness occurs with an increase in load. According to the data in Fig. 3.15, for PTFE μ = 0.02, and for polyamide μ = 0.6. There is no doubt that with the chosen method of investigation the contact area essentially does not change.

We studied external friction of PTFE on steel with hydrostatic compression applied (see Chap. 7). First, we showed that methods of investigation using a polymer in the form of a thin film pasted on a rubber cylinder sample, and in the form of a bulky cylinder, are equivalent. This means that linear thermoplastic polymers like PTFE, which undergo cold flow at room temperature, under hydrostatic compression behave as high-viscosity liquids. In other words, for a sufficient loading time, pressure is distributed evenly throughout the bulk.

Figure 3.16 shows the results of investigation into the relation of the specific friction force to pressure. As seen in the figure, up to approximately 400 kg/cm^2 the friction force depends non-linearly upon pressure. Comparison with the relation of real contact area to pressure for PTFE (Fig. 3.9) allows us to assume

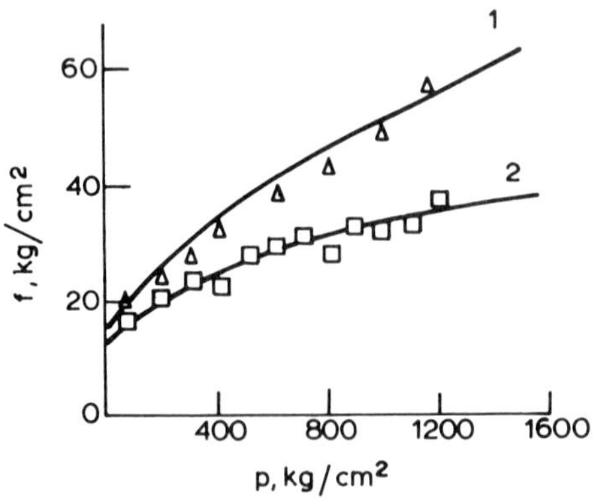

Fig. 3.16 Relation to the specific friction force to pressure for polymers on steel under compression: (1) polyethylene; (2) PTFE.

that increase in the friction force with rise of pressure to approximately 250 kg/cm^2 is explained by an increase in the real contact area. The discrepancy in pressure values can be explained both by different stress modes and by an overestimation of the saturation zone for the contact area according to the data obtained by the optical method.

Thus, for all pressure ranges, the experimental data can be interpreted by the equation

$$F = cS + c'p \,, \tag{3.9}$$

where c' is a constant. When S = const, the friction force depends directly upon pressure. Further, the linear relation F(p), or the consistency of the friction coefficient for high pressures, can be explained either as due to molecular roughness (Deryagin) or as the direct effect of pressure on the intermolecular interaction force (Bartenev).

Between Deryagin's and Bartenev's formulations there is, however, a difference of principle. According to Deryagin, the first term, related to adhesion forces, is the smaller, whereas in Bartenev's view, the second term, related to the direct effect of pressure on the friction force, is the smaller. When for significant adhesion magnitudes it is impossible to ignore change in the real contact area with rise in pressure, it is preferable to use Eq. (3.9).

In general, the law of friction for polymer can be written on the basis of Eq. (3.9), supplemented by Eq. (3.5):

$$f = c_1[1 - (1-\phi_0)\exp(-\alpha p)] + c'p \ . \qquad (3.10)$$

Unfortunately, at the present time neither Deryagin's equation nor Eq. (3.10) allows a calculation of the friction force, because the constants that enter into these equations are unknown. Further development of the theory of friction must contribute to the solution of this problem.

Effects of contact time, sliding velocity, and temperature

Polymer relaxation properties clearly appear in the relation of the friction force to the stationary contact time. This relation is determined by the time dependence of the contact-area formation. Kragelsky (38,39) already in 1939 has noted the relation of the friction force for metals to contact time (see Chap. 2). This relation can also be used to interpret polymer data. Figure 3.17 shows the results of investigating the relation of the friction force for Plexiglas on steel to the stationary contact time for three normal loads. The limit of the friction force F_∞ is determined by the strain limit ε_ℓ of micro-asperities and depends upon load.

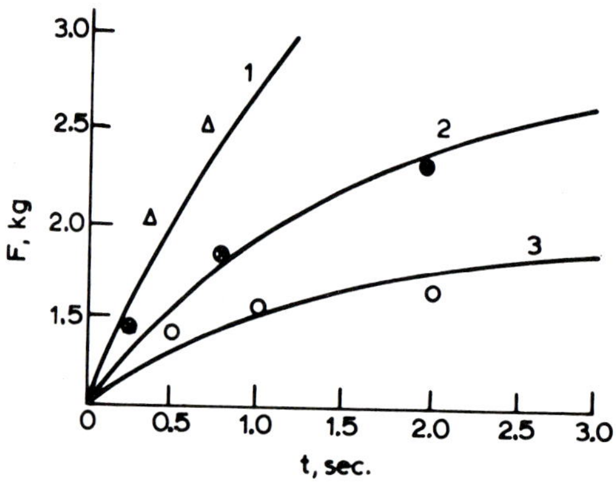

Fig. 3.17 Relation of the friction force to time of stationary contact for Plexiglas on steel (38). Normal loads: (1) 10 kg; (2) 5 kg; (3) 3 kg.

This relation is explicitly represented by Eq. (2.31). It is important to note that the exponent of the factor does not depend upon load and is determined by the material's properties only. Analogous results have been obtained for other polymers too (23,26).

The time dependence of the friction force shows up in change of this quantity upon application of tangential force (39). According to polymer physics, the rate dependence of the friction force (in essence the time dependence) is the most important characteristic of polymer friction properties. Some friction theories, especially Bowden's (25), do not take into account the relation of the friction force to sliding velocity. However, it is clear even in his theory (see Chap. 2) that plastic deformation generally proceeds through time (40,41), and therefore in the friction of plastic materials a velocity dependence ought to appear. If a solid changes its contact area with time, then the friction force definitely should depend upon the sliding velocity. Thus, the time dependence of friction force also causes a speed dependence (38).

According to other studies (9,42), the friction force for plastics does not depend upon sliding velocity. However, still other studies, which were carried out for a wide range of velocities, indicate that the friction force for plastics depends upon sliding velocity (13,18,43).

In accordance with the dual molecular-mechanical nature of friction, the relation of friction force to sliding velocity will be determined by the rate of establishing molecular adhesion contact and by the bulk deformation rate for the material's bulk. The temperature of the process and the nature of the material have large influences. As we have already seen, polymer external friction depends (for rigid polymers) mainly on the deformation properties of the polymeric material itself.

There is no doubt that discrete parts of a polymer's surface are deformed periodically. Consequently, the velocity dependence of the friction force is determined by the polymer's dynamic properties. To formulate a polymer friction theory, an evaluation of the frequency or rate of a material's deformation is essential.

Now let us examine some experimental facts. Tanaka (14), during the sliding of a glass hemispherical indenter, photographed Newton's rings, which characterize a polymer's deformation in the contact zone with great precision. With the passage of time (a succession of exposures), an increase in the friction force was observed until the moment of the sample's distuption or shear (Fig. 3.18). Negative numbers correspond to the rest after application of the load; positive, to the time after the start of sliding. From Fig. 3.18, it is evident that the tangential force grows proportional to time until the first motion of the indenter, then falls sharply at the

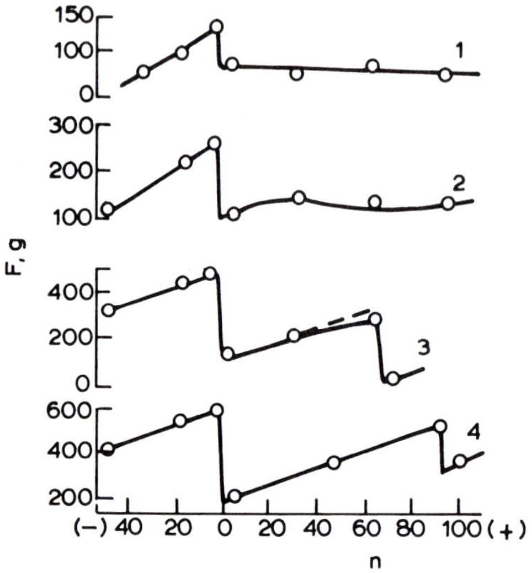

Fig. 3.18 Relation between the friction force for polymers and the frame number n (14): (1) PTFE (N = 350 g); (2) polyethylene (N = 350 g); (3) polyvinylchloride (N = 56 g); (4) Plexiglas (Hirax) (N = 350 g).

first instant of slip, then grows until the next slip, and so on. For low-adhesion polymers, smooth sliding is observed after initial slip. For elastic polymers like polyvinylchloride and Plexiglas (Hirax), intermittent slip is observed all the time during sliding. As these experiments show, the friction force is almost proportional to the depth of the indenter's indentation. Change occurs in the depth h of the indenter's indentation (Fig. 3.19).

Observation of the friction track showed that it was most clearly expressed (deepest) in polymers that slid smoothly, that is, in PTFE and polyethylene. The friction track depth decreases with time, but in PTFE this reduction is very small. It is striking that during sliding the modulus of elasticity, E, increases. It is also interesting that sliding can be observed when the tangential force is less than the maximum static friction.

Bueche and Flom (19) studied the deformation-loss part of polymer friction and its relation to mechanical losses in the bulk. The relation of the friction force to the sliding velocity (without lubrication) has a maximum (Fig. 3.20) whose position depends upon temperature. Under these conditions, the friction force generally is determined by adhesion and deformation interactions. With lubrication, the friction curves forms as shown in Fig. 3.21. It is possible to determine the periodic deformation frequency with an

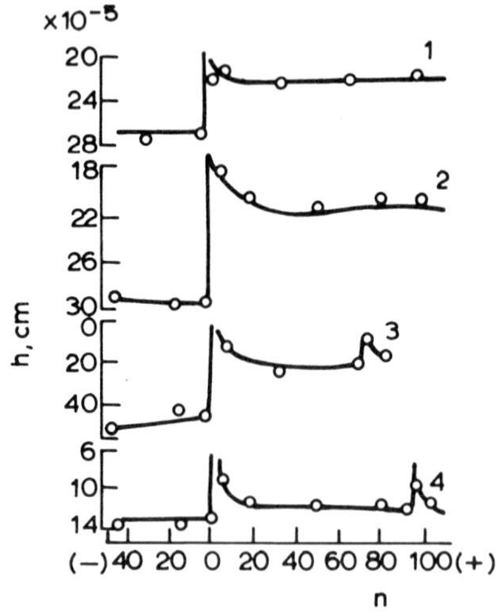

Fig. 3.19 Relation of the contact depth h of the indenter's indentation to the ordinal frame number n (14): (1-4) see Fig. 3.18.

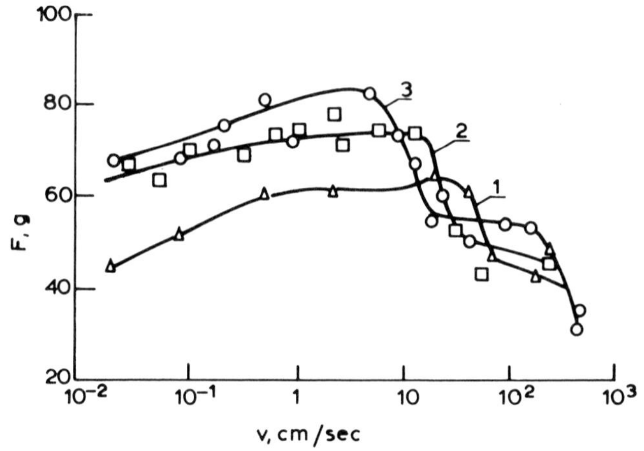

Fig. 3.20 Relation of the friction force for Plexiglas on steel to the sliding velocity (without lubrication) (19): (1) 25°C; (2) 60-70°C; (3) 100-110°C.

indenter, by noting the time necessary for the indenter to traverse the diameter of the contact area. A deformation frequency of 25 Hz corresponds to a 0.4-mm-diameter indenter sliding at a velocity of 1 cm/sec. Comparison of measurements shown in Fig. 3.21 with the

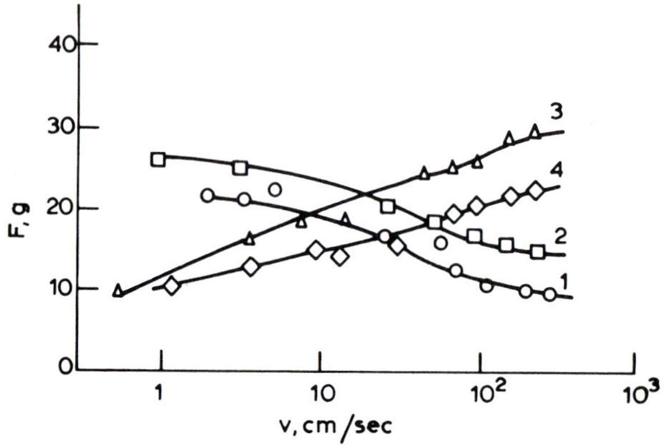

Fig. 3.21 Relation of the friction force for Plexiglas (Hirax) on steel to the sliding velocity (with lubrication) (19): (1) 25°C; (2) 46°C; (3) 78°C; (4) 105°C.

change in tan δ (Fig. 3.22) in this frequency range allowed us to explain the curves F(v) and discover the relation between the loss tangent and polymer friction properties. As seen in Fig. 3.22, good correspondence is observed between frequency relations, tan δ, and the friction force.

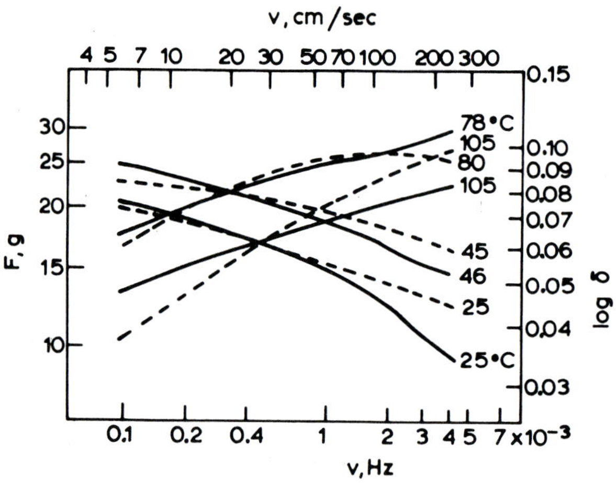

Fig. 3.22 Frequency dependence (19) of the friction force for steel on Plexiglas with lubrication (——) and the loss tangent for Plexiglas (----).

Figure 3.23 shows the results of friction for Plexiglas on steel (37). The friction-force maximum is located in the sliding-velocity

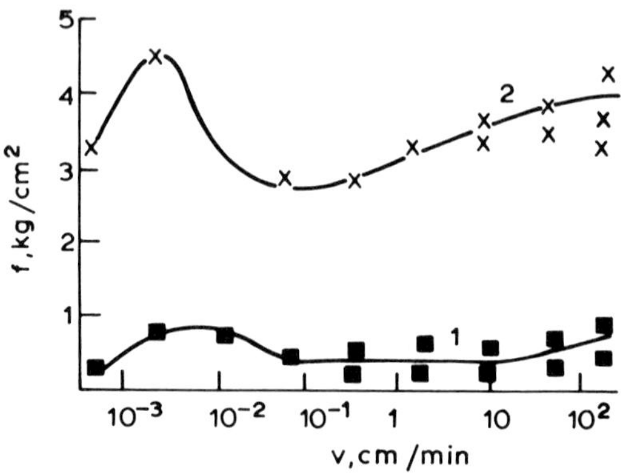

Fig. 3.23 Relation of specific force to sliding velocity for Plexiglas on steel (37). Pressure: (1) 3.42 kg/cm^2; (2) 26.32 kg/cm^2.

range 5×10^{-3} cm/min. Its position is practically independent of pressure, but the maximum is more clearly evident under larger pressures. A steel surface was treated to the eighth class of surface finish with micro-asperities of about 3 μm. The deformation frequency 0.33×10^4 Hz corresponds to a sliding velocity of 1 cm/sec. Because $V_{max} = 5 \times 10^{-3}$ cm/min, the position of the maximum agrees with the deformation frequency of micro-asperities, 1.5×10^2 Hz. According to the data obtained by Bueche and Flom (19), for friction of Plexiglas on steel, the maximum conforms to a frequency of 1.25 x 10 Hz. The correspondence is not precise; however, such a calculation is only a semiquantitative estimate.

The results for friction of polyvinylchloride on steel permit us also to discuss the relaxation nature of friction. With change in temperature, a merging of maxima is observed (Fig. 3.24). These results permit us to predict (37) the applicability to polymer friction of the Williams-Landel-Ferry principle of time-temperature equivalence (see Chap. 1). For PTFE, the friction force is practically independent of the sliding velocity.

McFarlane and Tabor (13) obtained analogous results for polyethylene, polypropylene, and polyamide. Thus, crystalline polymers also have a maximum in the curve of F(v). McFarlane and Tabor note that the maximum becomes less apparent with increase in the polymer's rigidity. Thus, for example, in thermosetting polymers there is practically no maximum. Mitrovich (21) studied the relation P(v) for polycaprolactam in the velocity range from

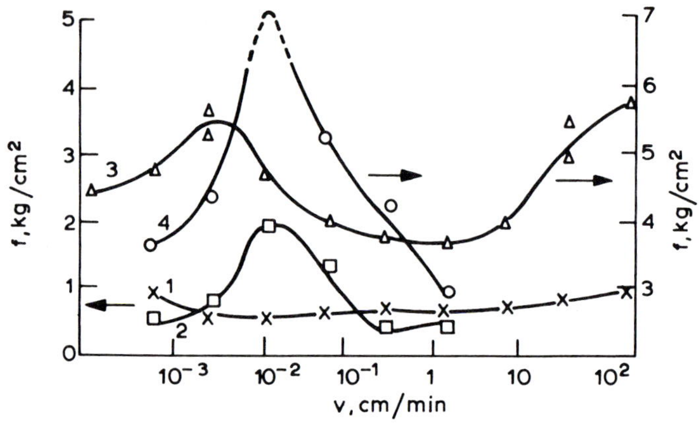

Fig. 3.24 Relation of specific friction force to sliding velocity for polyvinylchloride on steel (37). (1) p = 3.4 kg/cm^2, T = 57°C; (2) p = 3.4 kg/cm^2, T = 50°C; (3) p = 26.3 kg/cm^2, T = 17°C; (4) p = 26 kg/cm^2, T = 50°C.

8×10^{-4} to 80 m/min. In this range the curve P(v) has a minimum. Mitrovich believes that increase in the friction force is associated with surface heating and increase in the contact area.

We (24) studied the relation of the specific real friction force to sliding velocity for polyethylene (Fig. 3.25). The weak dependence of the specific real friction force upon the sliding

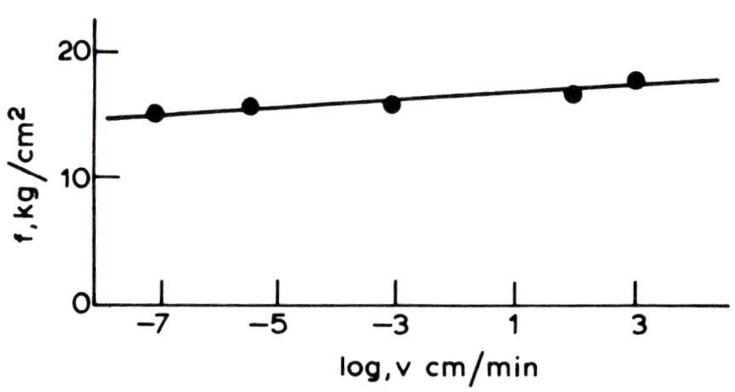

Fig. 3.25 Relation of the specific real friction force to sliding velocity for polyethylene on glass (24).

velocity indicates that the adhesion component of the friction force for plastics also depends little upon the sliding velocity. Some increase in the adhesion component is explained by the effect of adhesion of the bond-breaking rate (44). As will be shown in Chap. 4, the molecular-kinetic theory of friction leads to a similar kind of relation.

In some cases (45) the friction force for plastics decreases with an increase in the sliding velocity. This decrease is explained by surface plowing during sliding. The relation F(v) for plastics, as is evident, has not been studied adequately. With the exception of two or three works, a systematic study of the velocity dependence of the friction force has not been made. Generally, published results have been obtained either for manufactured articles (bearings, etc.) or for composite materials (usually with fillers) that are usable in industry.

Experimental results we have examined indicate that rigid-polymer relaxation properties also determine the form of the relation F(v) and the location of the maximum. In this connection, the higher the glass temperature and the temperature of the test, the more mobile are the polymer's molecules, and higher is the velocity of the friction-force maximum. The location of the friction-force maximum for rigid polymers lies in the range of 10^{-2} cm/min, which is three times lower than for crosslinked rubbers and three to four times higher than for metals. This agrees qualitatively with the mechanical properties of the materials studied.

The effect of temperature is equivalent to that of the sliding velocity. According to the Williams-Landell-Ferry (WLF) principle, a decrease in temperature is equivalent to an increase in sliding velocity (see Chap. 1). Between the mechanical transition temperature T_M and the cyclic-force frequency ν there is the relation

$$1/T_M = B - C \ln \nu$$

where B and C are constants.

The glass temperature T_g for rigid polymers lie above room temperature (Table 3.3). This quite thoroughly hampers investigation into the velocity dependence of the friction force and makes research on the temperature dependence preferable for an analysis of polymer friction properties. Temperature also determines the polymer's physical state (see Chap. 1). In view of these two points, let us examine the effect of temperature on the friction properties of rigid, glassy polymers.

TABLE 3.3

Some polymer mechanical characteristics

Polymer	Strength (10^{-2} kg/cm^2) Under compression	Strength (10^{-2} kg/cm^2) Under tension	Relative elongation at break $\Delta \ell/\ell$ (%)	Elastic modulus E (kg/cm^2)	Glass temperature T_g (°C)	Brinell hardness HB (kg/cm^2)
Polyethylene						
Low pressure	2.25	2.2-3.9	15-100	40-105	120-130*	1.8
High pressure	----	0.7-1.6	90-150	12-25	105-110*	---
PTFE	1.2	1.4-3.2	200-400	47-85	327**	3-4
Polyvinylchloride	5.6-9.1	3.5-6.3	2-40	246-422	65-75	---
Polystyrene	8-11.2	3.5-6.3	1.0-2.5	280-350	80-100	---
Polymethyl Methacrylate (not plasticized)	8.4-12.5	4.9-7.7	2-10	315	115-120	14.0
Polyamid-68	4.7-8.8	7.2-8.4	25-320	105-250	210-215	---
Epoxy Resin	10.5-15	2.8-9.15	3-60	336	110-140*	17.0
Ebonite	----	7.0	3	250	80	---

*The softening temperature.

**The melting temperature.

King and Tabor (10) studied external friction for polyethylene, polymethylmethacrylate, PTFE, and polytrifluorochlorethylene on glass in the same temperature range from -100 to +80°C. Figure 3.26 shows the results of their research. As shown, with rise in temperature the friction coefficient increases for all the polymers with the exception of PTFE. The friction coefficient for polyethylene around -60°C follows a curve with a maximum.

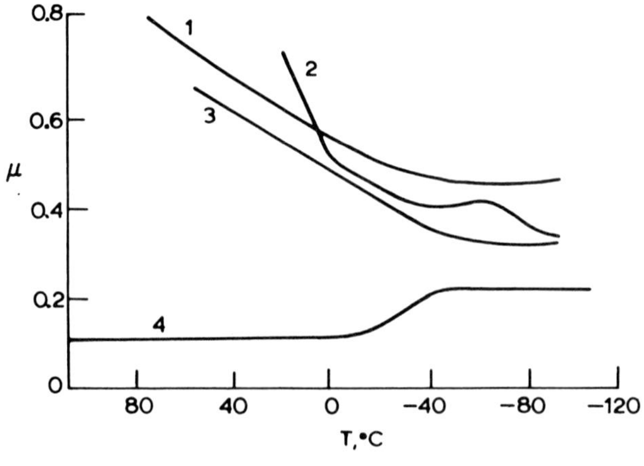

Fig. 3.26 The temperature dependence of the friction coefficient for polymers on glass (10): (1) polymethylmethacrylate; (2) polyethylene; (3) polytrifluorochlorethylene; (4) PTFE.

To evaluate the applicability of Bowden's theory, King and Tabor conducted a parallel study of the ratio Y/Σ_y and of the friction coefficient μ. The best agreement between them is observed for polyethylene. These experiments qualitatively confirm Bowden's adhesion theory if we assume that

$$\mu = \mu_0 - k\ Y/\Sigma_y\ ,$$

where μ_0 is the friction coefficient that is not related to adhesion, and k is a constant (usually $k \geq 1$), because in the contact zone the material is under high hydrostatic pressure, which, as is known, raises the shear strength.

For PTFE, k = 0.3, which can be explained by this material's extremely low adhesion. At the softening temperature of PTFE, the friction coefficient rises to high values. For PTFE, the relation of Σ_y to temperature can be expressed by the equation $\Sigma_y = 0.53 - 483.64/T(°K)$. Down to 200°C, the friction coefficient for PTFE on steel and on PTFE is practically constant. But for sliding of steel on PTFE, with rise in temperature an increase in μ is observed (Table 3.4).

TABLE 3.4

The effect of surface temperature on the friction coefficient of PTFE

T (°C)	PTFE on PTFE	PTFE on Steel	Steel on PTFE
20	0.04	0.04	0.09
50	0.04	0.04	0.09
80	0.04	0.04	0.10
100	0.04	0.04	0.10
150	0.04-0.05	0.04	0.11
200	0.05	0.04	0.14

The relation between μ and σ_c/σ_T as the temperature varies is also noted in the other studies (3,21). In studies that Mitrovich (21) conducted on the temperature dependence of the friction coefficient for polymers, it was shown that μ increased with temperature (Fig. 3.27). It was noted that the character of the

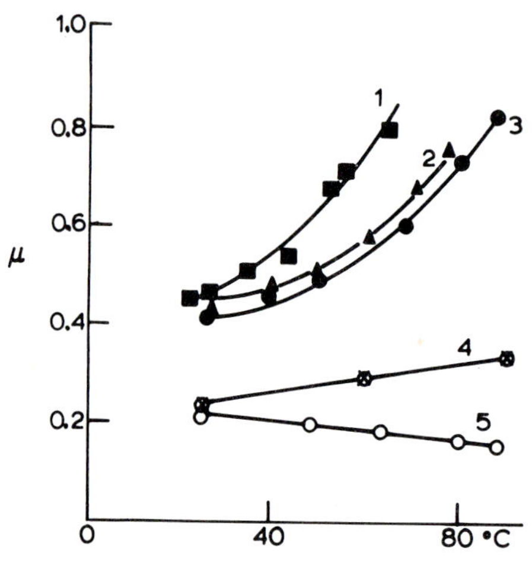

Fig. 3.27 The temperature dependence of the friction force for polymers on steel for v = 13.2 m/min and N = 1 kg (21): (1) polycaprolactam; (2) Polyamid-68; (3) Polyamid AK-7; (4) polyethylene; (5) PTFE.

change in μ depends on the form of molecular interaction between polymer and solid, and not on the change in the real contact area. But it is evident that a rise in temperature should lead to a decrease in intermolecular forces (adhesion), and not to an increase. Moreover, the real contact area increases with temperature. All this throws doubt on the present explanation of the increase in μ with temperature.

According to our studies (25), the relation f(T) essentially determines the change in the real contact area (Fig. 3.10), which depends upon the change in the polymer's mechanical properties. Figure 3.28 shows f(T) for some polymers. For polystyrene and polyvinylchloride, a sharp increase in f is noted at a temperature of 70°C. For polyamide at T = 115-120°C, a smooth change in f is observed with rise in temperature.

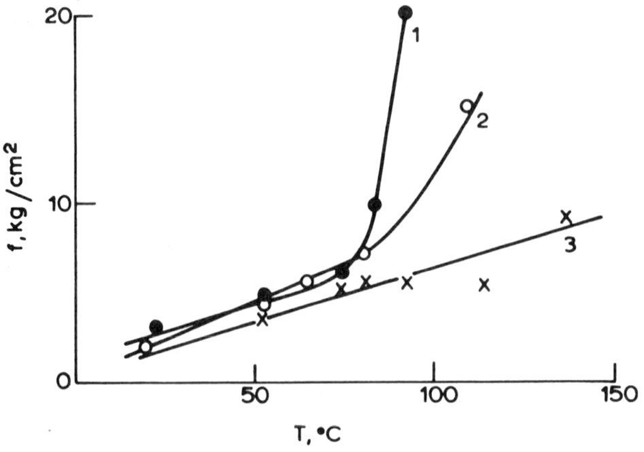

Fig. 3.28 Temperature dependence of the specific friction force for polymers on steel at p = 10 kg/cm^2 (37): (1) polystyrene; (2) polyvinylchloride; (3) polyamide.

The temperature dependence of the friction force for such rigid polymers appears different than, (for example) for ebonite, for which the specific friction force decreases with increasing temperature (Fig. 3.29). This decrease can be explained by the increase in mobility of a network polymer's molecules.

Thermosetting resins are an important class for rigid polymers. The most widely used are epoxy resins, used as adhesives in glass-reinforced plastics (46). These resins have a high modulus of elasticity below the softening point and a weak friction-force temperature dependence down to the softening temperature (Fig. 3.30).

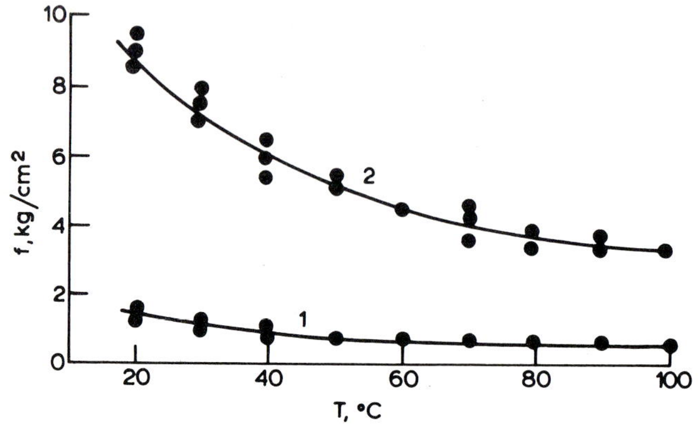

Fig. 3.29 Relation of the specific friction force to temperature for ebonite on steel. Pressure: (1) 1.34 kg/cm^2; (2) 20.6 kg/cm^2.

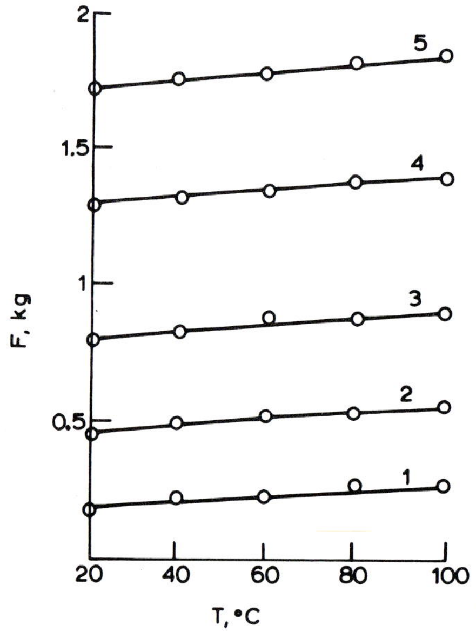

Fig. 3.30 Relation of the friction force to temperature for epoxy resin EDT-10 on glass. Load: (1) 1 kg; (2) 2 kg; (3) 3 kg; (4) 4 kg; (5) 5 kg.

In connection with the low thermal conductivity of plastics, a temperature rise due to friction which affects polymer physico-mechanical properties and structure at the surface film has great significance. As a result of the low thermal conductivity, high temperatures can develop even with low sliding velocities and pressures. Fundamental work in this field has been done by the Soviet scientists Kragelsky, Shchedrov, Kudinov, Chinchinadze, and Troyanovskaya (47-53). Of non-Soviet scientists above all Archard (54) and Blok (55) should be mentioned.

The calculation of the rising friction temperature rise due to sliding is determined by the following basic quantities:

1. The specific friction power, expressed as the product of the specific friction force times the sliding velocity ($W = fv$);
2. The thermal flow dispersion coefficient, which indicates the proportion of a thermal current that arrives at one of the contacting bodies;
3. The reciprocal period of contact, which is equal to the ratio of the friction pair's contacting areas (51);
4. The contact thermal conductivity, which is determined by the material's thermo-physical contacts.

Systematic study of the contact boundary temperature, the temperature field, and the friction coefficient was carried out by the Chinchinadze and Troyanovskaya (56), and they demonstrated that a high friction coefficient corresponds to a large temperature gradient $\partial T/\partial z$ (see Chap. 2). Consequently, the surface temperature does not unambiguously determine the friction coefficient, because the bulk temperature field leads to change in the material's mechanical properties in the bulk [Eq. (2.41)].

With a constant friction force, the temperature field depends upon the reciprocal period of contact and heat-transfer conditions. Increase in the reciprocal period of contact leads to a significant rise in surface temperature. Change in heat-transfer conditions affects surface temperature only a little, but shows up in heat dissipation in the bulk (56).

The fatigue resistance of a polymer-metal pair under intensive heat generation is affected not only by the friction temperature, but also by the duration of sliding, because the polymer is degraded with the passage of time. In connection with this, Bilik (57) recommends the following maximum allowable temperatures for polymer-metal friction pairs:

	T (°C)
Polyvinylchloride	60-95
Polystyrene	60-95
Polyolefins	70-105
Polyamides	80-110
Epoxy resins	80-135
Polycarbonates	100-135
Fluorine-containing elastomers	135-175
Fluorine-containing rigid polymers	155-200

The thickness of the polymer or of the polymer film shows up in the contact thermal conductivity. It is known (20,57-60) that, for example, the antifriction material PTFE works better against hard materials as a thin film than in bulk.

Effects of other factors

External friction in large measure depends upon testing conditions and the surfaces of the rubbing pair. In this section, we shall examine the effect on the friction force of the following basic factors: the method of calculating the friction force, the states and treatments of the surfaces, moisture content, lubrication (qualitatively), environment, structures and properties of the friction pair, polymer fillers, and electrostatic charges.

For given testing conditions, the friction coefficient depends upon choice of method, i.e., the ways of measuring force and temperature, the manner of tangential-force application, etc. For example, the static friction coefficient for PTFE, obtained by different methods, varies from 0.01 to 0.3. It is noteworthy that the friction coefficient often depends upon which of the friction pair is the mobile element (Table 3.5). As shown, the friction coefficient of polyethylene does not depend upon whether it is the stationary or mobile element of the friction pair, for polyvinylchlori the difference is small, and for polymethylmethacrylate the differenc is significant. In all probability, this is associated with the transfer of a softer material to the surface of the harder friction-pair element. An analogous phenomenon is also observed for plastic bearings (61). It has been discovered that the fatigue resistance for polycaprolactam bearings depends upon whether the polymer is the stationary friction pair element or not. If the bearing is statio ary, its wear is minimal.

The data in Table 3.6 confirm the conclusion that the softer elemen of the friction pair controls its friction coefficient when sliding on hard surfaces (62).

TABLE 3.5

The effect of surface treatment for the friction coefficient of polyethylene on other polymers (cylinder-plane method)

Treatment of the surface	Sliding element		
	Polyethylene	Polymethylmethacrylate	Polyvinylchloride
Buffing	0.45	0.34	0.39
Mechanical treatment	0.46	0.41	0.36

Treatment of the surface	Stationary element		
	Polyethylene	Polymethylmethacrylate	Polyvinylchloride
Buffing	0.45	0.64	0.38
Mechanical treatment	0.46	0.57	0.41

TABLE 3.6

Effect of polymer type on the friction coefficient

Polymer	Polymer on polymer	Polymer on steel
Polyethylene	0.25	0.25
Polyvinylchloride	0.4-0.45	0.35-0.4
Polymethylmethacrylate	0.4-0.6	0.5
PTFE	0.04	0.04

For polymer friction, the state of the surface does not exert such a large effect on the friction coefficient (20,62) as for metal friction. This circumstance is associated primarily with the low adhesion capacity of polymer surfaces and with the significant deformability of polymer. However, according to data (31), the friction coefficient is sensitive to traces of contamination.

Usually, polymer products are obtained by molding or by mechanical processing. As experiments show (21), the surface quality thus obtained affects the friction coefficient (Table 3.7). This has been confirmed elsewhere (20,32).

TABLE 3.7

Effect of surface treatment on the friction coefficient of nylon (66)

Treatment of the surface	Nylon on nylon	nylon on steel	steel on nylon
Molding at 30°C	0.70	0.42	0.31
Molding at 110°C	0.63	0.41	0.31
Mechanical treatment	0.46	0.41	0.31

Matveyevsky (63) conducted special studies on friction of steel on Polyamid AK-7 through five polymer surface-cleaning methods: (1) washing by alcohol and by ethyl ether; (2) the same, but with subsequent drying in an oven at 60°C over a period of 5 minutes; (3) washing in sequence by benzene, acetone, alcohol, ethyl ether, and carbon tetrachloride; (4) the same, but with subsequent drying in an oven at 60°C over a period of 5 minutes; (5) continuous flow. The highest friction coefficient was observed during washing by method 5 ($\mu = 0.68$), and the lowest by method 1 ($\mu = 0.43$). Method 4 was the best ($\mu = 0.64$).

Moisture affects friction in two ways: by forming a film on the surface and by being absorbed by the polymer itself.* Polyamides are classified according to moisture absorption in the following order: P-68, 0.48 per cent; P-54, 0.97 per cent; polycaprolactam, 2 per cent; AK-7, 30 per cent (64). With increase in moisture content from 0 to 10 per cent, the friction coefficient for polyamides increases from 0.9 to 1.19 (65). Polymers vary in their sensitivity to moisture. According to some data (12), the friction coefficient for polyamide fibers also increases with moisture. For other polymers (polypropylene, polyethylene, and especially PTFE), the friction coefficient depends very weakly upon moisture (65,66).

The effects of lubrication of polymer friction will be examined only qualitatively. The hydrodynamic theory of lubrication cannot satisfactorily explain the experimental data for polymer friction (67). A special characteristic of polymers is their self-lubrication ability (62). All this greatly hampers an analysis of experimental data in this field.

The effectiveness of lubrication on polymers is significantly lower than on metals, as a result of the small number of active centers on a polymer's surface which are capable of retaining a monomolecular lubrication film. Moreover, cases are observed (20) in which the introduction of lubrication leads to an increase in the static friction coefficient. This is shown in Table 3.8.

Polar and nonpolar lubricants behave in total conformity with the polymer's adhesion capacity (63). So, for example, for friction of polyethylene and PTFE on steel, the introduction of polar additive into the lubricant does not affect the magnitude of the friction force. For polyamides, polar lubricants are more effective than nonpolar (21). With polar lubricants, smooth sliding and good constancy of the friction coefficient are assured. Polar lubricants affect friction in thermosetting resins analogously.

In several cases, a polymer is swollen by lubrication; such swelling deteriorates its mechanical properties and can lead to an increase in wear (62). This is explained by the fact that lubricants diffuse into the polymer and reduce the surface-film strength, and this, in turn, leads to an increase in the contact area. Table 3.8 shows that lubrication of nylon by water leads to an increase in the static friction coefficient. It is interesting

*Moisture absorption for plastics is determined according to All-Union State Standard 4650-65.

that the relation of the friction force to load, sliding velocity, and temperature is not affected by lubrication (68).

TABLE 3.8
Effect of lubrication on the friction coefficient for nylon (66)

Lubrication	Nylon on nylon		Steel on nylon	
	Static friction	Kinetic friction	Static friction	Kinetic friction
Without lubrication	0.42	0.35	0.37	0.34
Water	0.52	0.30	0.23	0.19
Glycerol	0.35	0.19	0.23	0.18
Ethylene glycol	0.58	0.18	0.20	0.16

Graphite and molybdenum disulfide are applied to polymers as solid lubricating materials, and they are often introduced into polymers as fillers. Fillers play a dual role: they improve a polymer's mechanical properties and act as lubricants (69). The friction coefficient on steel for PTFE, filled with graphite (30 per cent) is stable and is equal to 0.03 over a wide range of pressures and sliding velocities.

At the present time, there is wide industrial use of filled polymers for sliding bearings and similar products. Graphite and molybdenum disulfide are also used as polymer base films (antifriction greases) to coat hard surfaces.

For friction and wear of metallic materials, physico-chemical processes that take place on friction surfaces have great significance (70-73). Here, the environment plays a fundamental role. For rigid polymers, the environment's effect still has not been adequately studied. According to data obtained by Vinogradov and co-workers (62,65), the most active environmental agent is atmospheric oxygen. Its effect is associated with an interaction at the polymer-metal boundary. If metal wear occurs (as for rigid plastic, paired with soft metal), then a decrease in the oxygen content promotes adhesion between the metal and the plastic surface to which the metal is transferred. Where there is no transfer of the metal, oxygen does not significantly affect polymer friction.

It is known that for polymer friction, static electrification is observed, which in manufacturing (for example, in the paint and varnish industry) leads to dangerous consequences. In spite of the importance of research into frictional electrification, there have been few studies published in this field (57,64,66,74-76).

Deryagin ascertained the nature of electrical charge formation (77). He showed that on the boundary between the solids an electrical double layer arises under static conditions.

Yelkin and Nikolaev reported (75) that for friction on polymethyl methacrylate an additional friction force arose, conditional on an electrical charge field on the friction surfaces. In a vacuum, the size of the charge that arises on friction surfaces is significantly lower; therefore, significant additional increase in the friction force is not observed as a result of surface electrification.

We should note that the friction force depends not only upon the polymer's nature and type, but also upon the special structural characteristics of the friction-pair elements. The structure of the friction junction naturally also determines the friction mode and conditions. Studies of the effect of structure on the fatigue resistance of the friction pair were conducted by Bely and co-workers (30,78), Bilik (23), Matveyevsky (63), Chichinadze (49), and others.

3.4 Selection of Friction Pairs

The needs of mechanical engineering are simulating the development of two groups of polymer materials and polymer-base compositions: those with very low friction coefficients and those with very high ones. These materials are called, respectively, antifriction and friction materials. The use of one or the other kind of material is determined not only by the value of the friction coefficient, but also by its physico-chemical properties.

Let us examine fundamental requirements of antifriction and friction polymer materials, according to Kragelsky (38).

Antifriction materials must have:

1. a low friction coefficient (the general condition);
2. weak adhesion to a counter body and small mechanical losses;
3. good thermal conductivity;
4. requisite mechanical properties;
5. positive gradients in mechanical properties gradients throughout their depth;
6. the capacity to firmly retain boundary lubrication films.

Some of the abovementioned conditions are in conflict. For example, low adhesion properties, which cause low boundary-film adhesive strength (as in PTFE and polyethylene), promote a low friction coefficient. Structurally, high strength is essential, but polymers generally have low strength compared with metals. It is also important to note the poor thermal conductivity of polymer materials.

The above considerations point to the necessity of specially developing and producing antifriction polymer compositions. In this area, serious studies are being conducted in groups under the direction of Khrushchov (74) and Kragelsky (38). Improvements in antifriction polymer compositions are characterized by the development of new polymers and polymer compositions that have low adhesion in conjunction with high mechanical strength and high thermal conductivity.

Friction materials must meet the following fundamental requirements:

1. a high friction coefficient (the general requirement);
2. high thermal conductivity and good heat resistance;
3. adequate strength (the structural requirement).

Important qualities of friction materials are their long lives, consistency of friction coefficient, and superior resistance to wear.

Friction materials can also be created by synthesis. Under Kragelsky's direction a new friction material was created--Retinax. The friction coefficient of this material (of different qualities) lies between 0.33 and 0.4, and the wear between 0.03 and 0.07 mm/hour. It is produced with an asbestos base and is one of the best friction materials. Other promising materials are being developed on asbestos-rubber and ceramic bases. Metal-polymer compositions are also being developed. Fundamental ways of improving friction polymer compositions include the preparation of new polymers and compositions with high thermal conductivity and resistance to wear derived from mechanical strength.

In industry, polymer friction materials generally are applied in braking systems, and antifriction polymer materials in sliding bearings and packings.

PTFE is widely used because of its low friction coefficient. It is applied as a thin film to be a filler and as an independent construction material. For large specified loads, a porous metal (steel, bronze) is used, whose pores are filled with PTFE. Such materials are used both dry and with lubrication. The useful range for products in severe environments is guaranteed to temperatures of 200-250°C and pressures to 100-150 kg/cm^2.

Promising compositions include PTFE with graphite and molybdenum disulfide. Introduced as fillers, they lower the friction coefficient. Such materials have better mechanical properties than pure PTFE.

Sliding bearings made of polyamides, which work with or without lubrication, have a wide application. By contrast with PTFE, they have high mechanical strength. For friction with lubrication, polyamide compacts operate to a pressure of 1500 kg/cm^2 with a friction coefficient of 0.001.

REFERENCES

1 J.F. Archard, Proc. Roy. Soc., A, 243(1957)190.
2 K.V. Shooter and D. Tabor, Proc. Phys. Soc., B, 65(1952)661.
3 J. Chapman, M. Pascoe and D. Tabor, J. Text. Inst. Proc., 46(No. 13)(1955).
4 N.P. Istomin and A.D. Kuritsyna, Mashinovedenie, (No. 1)(1965)104.
5 D. Atack and D. Tabor, Proc. Roy. Soc., 246(1958)539.
6 B. Olofsson, Proc. Swed. Inst. Text. Res., 20(1950)476.
7 E. Lord, J. Text. Res., 46(1955)41.
8 M.W. Pascoe and D. Tabor, Proc. Roy. Soc., A, 235(No. 1291)(1956)210.
9 N. Gralén, B. Olofsson and J. Lindberg, Proc. Swed. Inst. Text. Res., 23(1953)629.
10 R.F. King and D. Tabor, Proc. Phys. Soc., B, 66(No. 9)(1953).
11 K.V. Shooter, Proc. Roy. Soc., 212(1952)488.
12 B. Lincoln, Brit. J. Appl. Phys., 3(1952)260.
13 J.F. McFarlane and D. Tabor, Nature, 197(1963)856.
14 K. Tanaka, J. Phys. Soc. Jpn., 16(No. 60)(1961)2003.
15 S.B. Einbinder, Issledovaniya Treniya i Stsepleniya Tvordykh Tel (Studies on Solid Friction and Adhesion), Riga, 1966.
16 S.B. Einbinder and E.F. Pastrigina, Vysolcomol Soedin, 5(No. 9)(1966).
17 S.B. Einbinder, N.G. Laka and I.Yu. Maiors, Mekh. Polim., (No. 1)(1965).
18 A.D. Kuritsyna, Vestn. Machinostro., (No. 4)(1967)40.
19 A. Bueche and D.G. Flom, Wear, 2(1958)168.
20 R.C. Bowers, W.C. Clinton and W.A. Zisman, Mod. Plast., 31(1953)131; 41(1963)139.
21 V.P. Mitrovich, in Plastmassy kak Antrifriktsionnyye Materialy (Plastics as Anti-Friction Materials), AN SSSR, Moscow, 1961, p. 46.
22 B.V. Deryagin and Yu.P. Toporov, Dokl. Akad. Nauk. SSSR, 146 (No. 6)(1962)1356.
23 M.M. Bilik, Pary Treniya Metall-Olastmassa v Machinakh i Mekhanizmakh (Metal-Plastic Friction Pairs in Machines and Mechanisms), Mashinostroyeniye, 1966.
24 G.M. Bartenev, V.V. Lavrentev and N.A. Konstantinova, Mekh. Polim., (No. 6)(1967)1078.
25 G.M. Bartenev and V.V. Lavrentev, Dokl. Akad. Nauk. SSSR, 141 (No. 2)(1961)334.
26 V.V. Lavrentev, Vysolcomol Soedin, 4(No. 8)(1962)1151.
27 V.V. Lavrentev, G.Ye. Lazarev and A.I. Serebrennikov, Proizvodstvo i Pererabotka Plastmass, No. 8, Izd. NIIPM, 1967.
28 T. Fort, Jr., J. Phys. Chem., 66(1962)1136.
29 B. Golding, KhimiyaPi Tekhnologiya Polimernykh Materialov (The Chemistry and Processing of Polymer Materials), IL, 1963.
30 V.A. Bely et al., in Teoriya Treniya i Iznosa (The Theory of Friction and Wear), Nauka, Moscow, 1965, p. 298.
31 A.P. Semyonov and V.V. Pozdnyakov, in Plastmassy kak Antifriktsionyye Materialy (Plastics as Anti-Friction Materials), AN SSSR, Moscow, 1961, p. 60.
32 D. Summers-Smith, Res. Correspond., 8(1955)515.
33 L. Lodge and H.G. Howell, Proc. Phys. Soc., B, 67(1954)89.
34 N. Gralén, Proc. Roy. Soc., A, 212(1952)491.

35 H.G. Howell, Proc. Swed. Inst. Text. Res., 23(1953)589.
36 J. Huffington, J. Res., 12(1959)10.
37 V. Lavrentev, Plaste Kautsch., (No. 6)(1962)282; (No. 10)(1963)72.
38 I.V. Kragelsky, Treniye i Iznos (Friction and Wear), Mashgiz, 1968.
39 I.V. Kragelsky, Trudy i Vsyesoyuznoi Konferentsii po Treniyu i Iznosu (First Proceedings of All-Union Conference on Friction and Wear), Vol. 1, 1939.
40 L.M. Kachanov, Plastichnost' (Plasticity), AN SSSR, Moscow, 1948.
41 A.A. Ilyushin, Ibid.
42 N. Cook, ASME Paper No. 58 (1958).
43 J.V. Vinogradov, Wear, 5(No. 6)(1962).
44 H. Vickers, J. Appl. Polym. Sci., 6(1962)316.
45 T. Fort, Jr., J. Phys. Chem., 66(1962)1136.
46 G.D. Anreyevskaya, Vysokogrpchnyye Orientirovannyye Stekloplastiki (Visco-Stable Oriented Glassy Plastics), Nauka, Moscow, 1966.
47 I.V. Kragelsky, G.Ye. Chupiloki and A.V. Chichinadze, in Protsessy Treniya v Aviatromozakh (Aviation Brake Friction Processes), AN SSSR, Moscow, 1955.
48 I.V. Kragelsky, A.V. Chichinadze and G.I. Troyanovskaya, in Voprosy Stoitel'stva Ridroelektrostantsii (Hydroelectric Station Construction Problems), AN SSSR, Moscow, 1954.
49 A.V. Chichinadze, Dissertation, IMASh, AN SSSR, Moscow, 1954.
50 V.S. Shchedrov, Treniye i Iznos v Mashinakh (Friction and Wear in Machines), 10, AN SSSR, Moscow, 1955, pp. 155-296.
51 V.S. Shchedrov and A.V. Chichinadze, in Povysheniye Effektivnosti Tormoznykh Ustroistv (Increasing Brake Apparatus Fatigue Resistance), AN SSSR, Moscow, 1959, pp. 180-183.
52 G.I. Troyanovskaya, Dissertation, IMASh, AN SSSR, Moscow, 1956.
53 V.A. Kudinov, in Sukhoye Treniye (Dry Friction), AN Latv. SSR, 1961.
54 J.F. Archard, Wear, 2(No. 6)(1959)438-56.
55 H. Blok, Proc. Gen. Disc. on Lubrication, Inst. Mech. Eng., 11(1937)4-30.
56 A.V. Chichinadze and G.I. Troyanovskaya, in Peredovoi Nauchno-Tekhnicheskii i Proizvodstvennyi Opyt (Advanced Scientific-Technical and Industrial Experiments), Topic 20, No. M-57-127/6, VINITI, 1951.
57 M.M. Bilik, in Iznos i Treniye Metallov i Plastmass (Wear and Friction for Metals and Plastics), Nauka, Moscow, 1964, p. 87.
58 D.G. Flom, J. Appl. Phys., 32(1961)1426.
59 D.G. Flom and N.T. Porile, Nature, 1975(1955)682.
60 L.Yu. Pruzhansky, in Plastmassy kak Antifriktsionnyye Materialy (Plastics as Anti-Friction Materials), AN SSSR, Moscow, 1961.
61 M.F. Stramous, Transportnoye Stroitel'stvo, (No. 12)(1963).
62 V.A. Mustafayev, G.V. Vinogradon and Yu.Ya. Podolsky, in Iznos i Treniye Metallov i Plastmass (Wear and Friction for Metals and Plastics), Nauka, Moscow, 1964.
63 R.M. Matveyevsky, in Plastmassy kak Antifriktsionnyye Materialy (Plastics as Anti-Friction Materials), AN SSSR, Moscow, 1961, p. 22.
64 A.D. Dubinin, Treniye i Iznos v Detalyakh Mashin (Friction and Wear in Component Machines), Mashgiz, 1960.
65 G.V. Vinogradov, Vestn. Akad. Nauk. SSSR, (No. 1)(1961)48.
66 F.P. Bowden and D. Tabor, The Friction and Lubrication of Solids, Clarendon Press, Oxford, Part I, 1950; PART II, 1964.
67 S.B. Ratner, in Treniye Smaxochnogo Deistviya i Novye Materialy (Friction with Lubrication and New Materials), Nauka, Moscow, 1965, p. 26.
68 Konstruktsionnyye Svoistva Plastmass (Structural Properties of Plastics), Khimiya, 1967.
69 G. Dugas, Schweiz. Tech. Z. (No. 8-9)(1961)144.

70 B.V. Deryagin, What is Friction?, AN SSSR, Moscow, 1963, p. 230.
71 A.S. Akhmatov, Molekulyarnaya Fizika Granichnogo Treniya (Boundary Friction Molecular Physics), Fizmatgiz, 1963.
72 G.V. Vinogradov, Trudy III Vsyesoyuznoi Konferentsii po Treniyu i Iznosu v Machinakh (Third Proceedings of the All-Soviet Conference on Friction and Wear in Machines), Issue 3, AN SSSR, Moscow, 1957.
73 G.V. Vinogradov et al., Neftekhimiya, (No. 2)(1961)274.
74 M.M. Khrushchov, in Treniye i Iznos v Machinakh, Issue 6, AN SSSR, Moscow, 1965; in Plastmassy kak Antifriktsionnyye Materialy (Plastics as Anti-Friction on Materials), AN SSSR, Moscow, 1961.
75 A.I. Yelkin and N.N. Nikolayev, Vysokomol. Soedin., 9B(No. 7)(1967)511.
76 G.M. Bartenev, V.V. Lavrentev and N.A. Konstantinova, Mekh. Polim., (No. 2)(1967)309; (No. 6)(1967)1078.
77 B.V. Deryagin, Vestn. Akad. Nauk. SSSR, (No. 7)(1954)10.
78 V.A. Bely, S.V. Shcherbakov and Yu.D. Tereshko, in Primenneniye Plastmass v Machinostroenii i Priborostroenni (The Use of Plastics in Mechanical and Instrument Engineering), Minsk, 1963.

Chapter 4

POLYMER FRICTION IN THE RUBBERY STATE

In the rubbery state the mobility of polymer chains is high, and this determines the general rules for friction in rubbery polymers. The effects of sliding velocity, temperature, and pressure on the friction force are very obvious in this state. Molecular-kinetic concepts of the mobility of polymer molecules on the boundary of contact with a solid constitute the theoretical basis for rules of friction. The general theory of rubbery-polymer friction that has been developed on this basis permits not only the explanation of a number of important experimental relations, but also the prediction of new ones.

Below we shall examine fundamental theoretical works and the most important results in this field. The foundation is the establishment of a relationship between polymer friction properties and structures. In one of the first studies on rubbery-polymer friction properties, Ariano (1) demonstrated that the friction force increases with increase in the sliding velocity. Later Roth, Driscoll, and Holt (2) studied this relation in more detail. Figure 4.1 shows their data, from which it follows that the friction coefficient depends upon the sliding distance and sliding velocity. Stabilized, vulcanized rubbers were studied at a temperature of 21°C for friction on glass and steel. Interesting effects of the sliding distance were found in the transition from static to dynamic friction. With increase in sliding velocity, a corresponding increase in the friction coefficient is observed. With slow sliding velocities (displacement velocities), the friction coefficient does not depend significantly upon the path length, but with a sliding velocity of 0.0001 cm/sec a decrease in the friction coefficient is observed. The initial parts of these curves show a change in the friction coefficient for unsteady sliding. Furthermore, the curves do not show the sudden transition, commonly found in the friction of metals and glassy polymers, which is generally used to define the static friction coefficient. Driscoll and Holt (2) explained this as a special characteristic of elastic materials related to the absence of a static friction force. Later Harry and Prock confirmed this explanation (3).

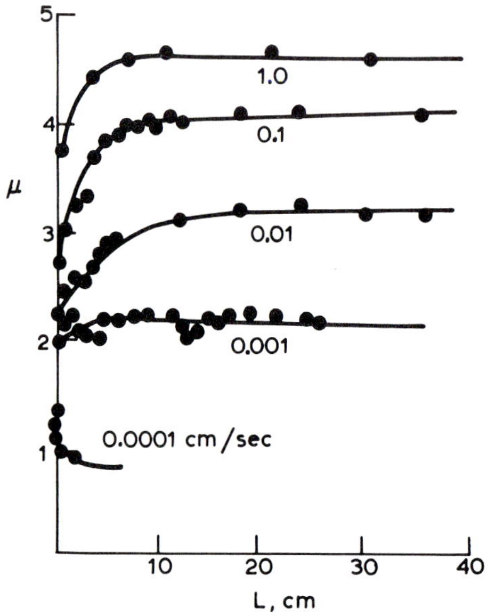

Fig. 4.1 Relation of friction coefficient to sliding distance for a vulcanized rubber at different speeds (2).

It is essential to note that until 1953 studies on rubbery-polymer friction were purely empirical. The mechanism of rubbery-polymer friction was considered analogous to that for metals. In 1953 Schallamach pointed out for the first time that the mechanism of rubbery-polymer friction differs from that for metal friction. He studied the temperature and rate relations of the friction force for filled crosslinked rubbers on glass under small normal loads (≈ 10 kg/cm^2). He chose a range of low speeds (from 10^{-4} to 10^{-2} cm/sec) in order to avoid frictional surface heating.

Figure 4.2 shows the relation of the friction force to the sliding velocity for two temperatures (4). On the basis of this relation, Schallamach proposed that the sliding velocity is an exponential function of the tangential force:

$$v = B_T \exp\{-(U - \Gamma F_f)/kT\} , \qquad (4.1)$$

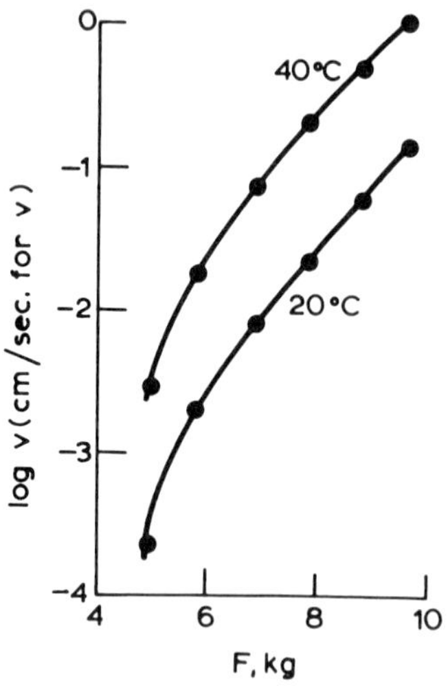

Fig. 4.2 Relation of friction force to sliding velocity for a vulcanized rubber (4).

where B_T is a constant defined later, k is the Boltzmann constant; U is the activation energy of the friction process, ΓF_f is the friction work, and F_f is the tangential force, or friction force.

Measuring the sliding velocity at a constant friction force and different temperatures, Schallamach obtained a linear relation of the velocity logarithm to reciprocal temperature (Fig. 4.3) according to Eq. (4.1). By his data, the activation energy for friction on glass was equal to 19.8 kcal/mole. These results permitted Schallamach to hypothesize that the friction of rubbery materials is molecularly activated, by analogy with viscous flow (Frenkel-Eyring theory). This hypothesis rests on the molecular-kinetic theory of external friction in rubbery polymers (5).

4.1 Nature and Mechanism of Elastomer Friction

Molecular-kinetic theory of friction

Molecular-kinetic concepts of external friction of rubbery materials (vulcanized rubbers, natural rubbers, etc.) were first examined theoretically by Bartenev. He obtained (5) a theoretical relation of the friction force to sliding velocity and temperature. Let us examine the basic points of this theory.

On a polymer surface, there are chain segments (Fig. 4.4) that jump in a disorderly manner on a smooth rigid surface from one

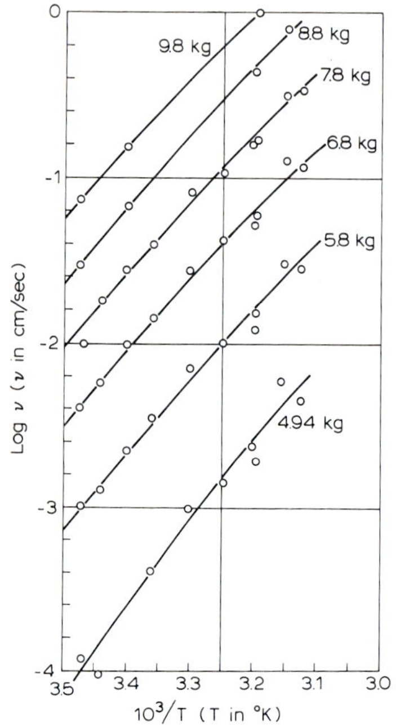

Fig. 4.3 Relation of sliding velocity to temperature at a constant friction force F for vulcanized natural rubber (4).

junction (adhesive point) to another. The time between two successive jumps is defined as the "relaxation time" of a chain and, according to Frenkel (6), can be expressed in the following form:

$$\tau = \tau_o \exp(U/kT) \ , \tag{4.2}$$

where U is the energy barrier, and τ_o is a constant equal to $\sim 10^{-12}$ sec.

The measure of the energy barrier U for external friction is determined by adhesive forces in the polymer-solid pair. Chemical interaction forces are not taken into account in the theory. For a solid's surface to be considered smooth, the mean asperity must not exceed the mean quadratic length of a polymer chain, that is, 100 Å.

In the absence of an external force, the polymer segments on a friction surface jump with equal probability in directions from 0° to 360° with frequency $\nu = 1/\tau$. The jump frequency in the directions between the angles ϕ and ϕ + d is equal to

$$d\nu = (1/2\pi\tau)d\phi = (1/2\pi\tau_o)\exp(-U/kT)d\phi . \quad (4.3)$$

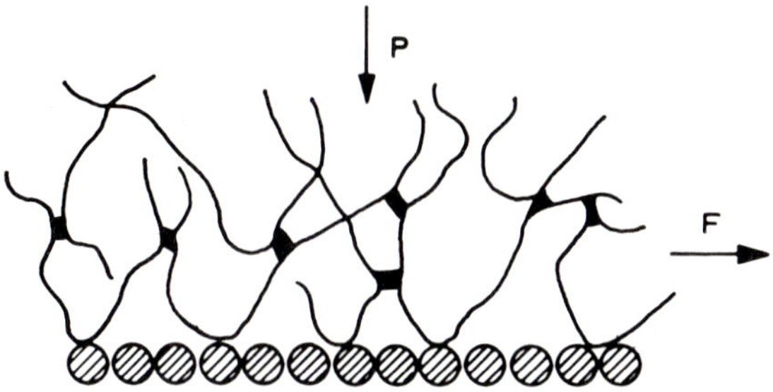

Fig. 4.4 Model of molecular contact: polymer-solid (according to Bartenev) (5).

The work of the external force F changes the energy barrier U in such a way as to reduce its height when the jump is in the direction of the force, and increase its height when the jump is in the opposite direction, by an amount $\delta A = (\lambda/2)\cos\phi\,\delta F$. The quantity δA represents the elementary work done by the external force during the transfer of a polymer segment a distance $\lambda/2$, where λ is the mean jump distance between adjacent bonds in a segment in equilibrium on a hard surface. If δF is the mean tangential force in any one chain, then the composite tangential force or friction force is $F_f = \delta F \cdot n_c$, where n_c is the number of elastomer segments in contact with the rigid surface. Therefore, $\delta A = (\lambda_j/2)(F_f/n_c)\cos\phi$. Setting $\Gamma = \lambda_j/2n_c$, we have $\delta A = \Gamma F_f \cos\phi$. Consequently, the energy barrier is equal to $U - \Gamma\cos\phi$. Then for the jump frequency based on Eq. (4.3) we obtain

$$d\nu = (1/2\pi\tau_o)\exp\{-(U - \Gamma F\cos\phi)/kT\}d\phi =$$

$$(1/2\pi\tau)\exp(z'\cos\phi)d\phi , \quad (4.4)$$

where the parameter z' is defined by the relationship $z' = \Gamma F/kT$.

The sliding velocity for the whole polymer is equal to the mean transfer velocity of a chain in the direction of force. The velocity component in the direction of the tangential force is equal to $dv = \lambda\cos\phi\,d\nu$, but in transverse direction $dv = \lambda\sin\phi\,d\nu$.

Integrating Eq. (4.4), we obtain the polymer specimen's sliding velocity in the direction of the effective force:

$$v = (\lambda/\pi\tau) \int_0^{\pi} \exp(z' \cos\phi)\cos\phi d\phi = -(\lambda/\tau)\, iI_1\,(iz') =$$

$$(\lambda/\tau)\, I_1(z')\ , \qquad (4.5)$$

and in the transverse direction:

$$v' = (\lambda/\pi\tau) \int_{-\pi/2}^{+\pi/2} \exp(z' \cos\phi)\sin\phi d\phi\ , \qquad (4.6)$$

where $I_1\ (iz')$ is the first-order Bessel function of the imaginary independent variable iz; the function $I_2\ (z')$ is the modified first-order Bessel's function. With low friction-force values, when $\Gamma F_f << kT$, the value of $z' << 1$.

Taking into account the first two terms of the Bessel function expansion, we obtain

$$v = (\lambda_j P F_f/2\tau_o kT)\ \exp\ (-\ U/kT)\ . \qquad (4.7)$$

For large force values, when $F_f >> kT$, $z >> 1$, from Eq. (4.5) we have

$$v = (\lambda_j/\tau_o)\ \sqrt{kT/2\tau\Gamma F_f}\ \exp\ \{-(U - \Gamma F_f)/kT\}\ . \qquad (4.8)$$

From Eq. (4.7) it follows that static friction is equal to zero when $v = 0$, $F = 0$.

Equation (4.8) is similar to the empirical equation Schallamach obtained earlier [i.e., (4.1)]. Comparing Eqs. (4.1) and (4.8), we obtain an equation for

$$B = (\lambda_j/\lambda_o)\ \sqrt{kT/2\pi\Gamma F}\ ,$$

which has the dimensions of velocity and is close in value to the velocity of sound v_o in a vitrified vulcanized rubber (≈1000 m/sec). Transforming Eq. (4.8) and setting $B = v_o$, we readily obtain the relation F(v,T) in the form

$$F_f = (U/\Gamma) + (kT/\Gamma) \ln (v/v_o) . \quad (4.9)$$

According to this equation, the friction force linearly depends upon temperature and the logarithm of the sliding velocity. Equation (4.9) can easily be verified experimentally. Since under ordinary conditions always $v << v_o$, Eq. (4.9) can be written

$$\Gamma F_f = U - bT ,$$

where $b = k \ln (v_o/v)$. Extrapolating $F_f(T)$ to the temperature axis ($F_f = 0$ for $T = T_o$) allows us to express the activation energy as

$$U = kT_o \ln (v_o/v) .$$

According to the equation $F_f = c_f S_r$ and Eq. (4.9), the friction constant c_f (or the specific real friction force f_r) is determined by the equation

$$c_f = (2u/\lambda_j S_{mol}) (1 - (T/T_o)) , \quad (4.10)$$

where S is the elementary contact area of a polymer chain with a hard surface, and $T = (U/k) \ln (v_o/v)$.

The friction force, according to a large body of data (4-20), decreases with increasing temperature. Figure 4.5 shows the temperature dependence of the friction force, obtained for different

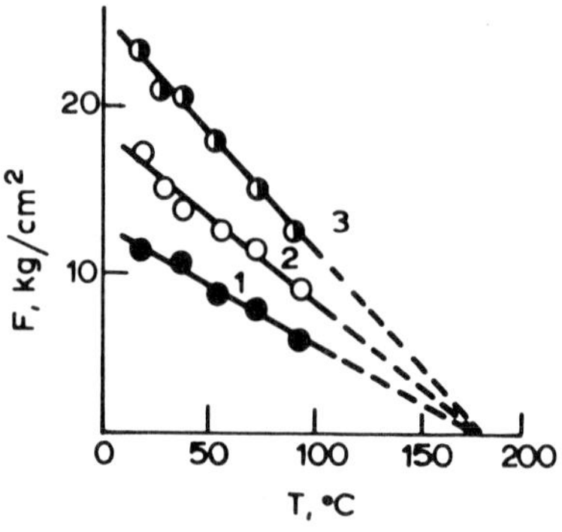

Fig. 4.5 Relation of friction force to temperature for a vulcanized butadiene acrylonitrile rubber (SKN-40) for three specific loads (11): (1) 0.4 kg/cm^2; (2) 1.8 kg/cm^2; (3) 6 kg/cm^2.

pressures (11). We should note first that all the relations are linear and intersect, upon extrapolation to $F_f = 0$, in a narrow temperature range near T_o. This T_o can be used to determine the activation energy. For the given case (friction of the vulcanized SKN-40 butadiene acrylonitrile rubber on steel), U = 12 kcal/mole. A change in the slope of relations F(T) results from a change in the real contact area [the constant γ in Eq. (4.9)]. Therefore, it is easy to plot the results as in Fig. 4.5 on the coordinates according to Eq. (4.10). Then, if the slope actually is determined only by change in the contact area, we obtain a generalized, common relation c(T) for all pressures. Experiments that were conducted on an optical tribometer for a simultaneous determination of F and S (11) permit us to obtain this relation (Fig. 4.6). As shown, Eq. (4.10) conforms entirely satisfactorily to the experimental data. These results not only confirm the theory of friction, but also permit us to suppose that in the equation $F_f = c_f S_r$ the friction constant c_f does not depend upon pressure.

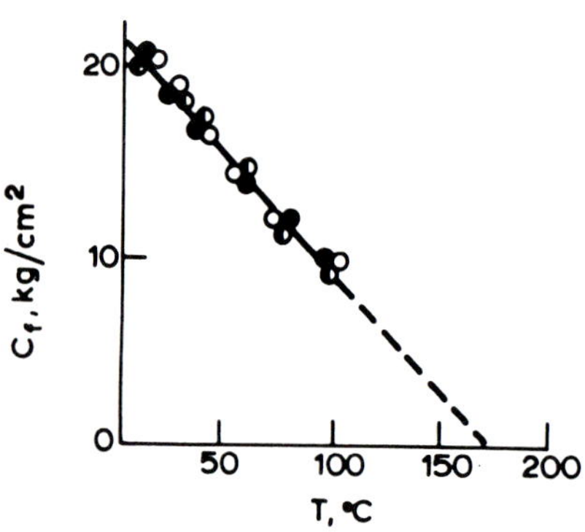

Fig. 4.6 Relation of the friction constant c_f to temperature for vulcanized rubber SKN-40 (butadiene acrylonitrile rubber) for specific loads (11): O, 0.4 kg/cm^2; O, 1.8 kg/cm^2; O, 6 kg/cm^2.

By studying the friction force for elastic material in a broader temperature range, it was shown (7,8) that for high temperatures abnormal behavior of the relation $F_f(T)$ occurred. Here, with rise in temperature the friction force did not decrease as predicted by the theory, but sharply increased (Fig. 4.7). At first, a similar effect was explained as due to a significant increase in the real contact area, but then it was demonstrated (21) that this growth was associated with oxidizing processes on the contact boundary of the polymer-solid pair.

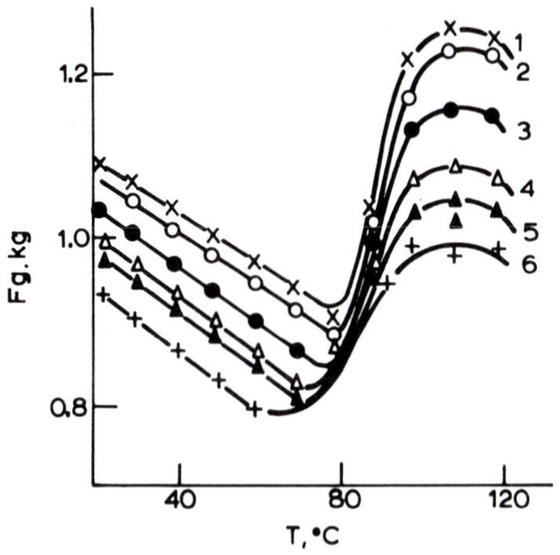

Fig. 4.7 Relation of the friction force to temperature for a vulcanized butadiene acrylonitrile rubber (SKN-40) on steel (8). E = 9.8 kg/cm^2; p = 0.3 kg/cm^2. Sliding velocity: (1) 2.8; (2) 1.0; (3) 0.5; (4) 0.1; (5) 0.05; (6) 0.1 mm/min.

Figure 4.8 shows the relation $F_f(T)$, obtained in a vacuum (10^{-3} mm Hg), where oxidizing processes are significantly reduced under the action of atmospheric oxygen. The nature of the relation $F_f(T)$ also is determined by the type of hard surface. On steel

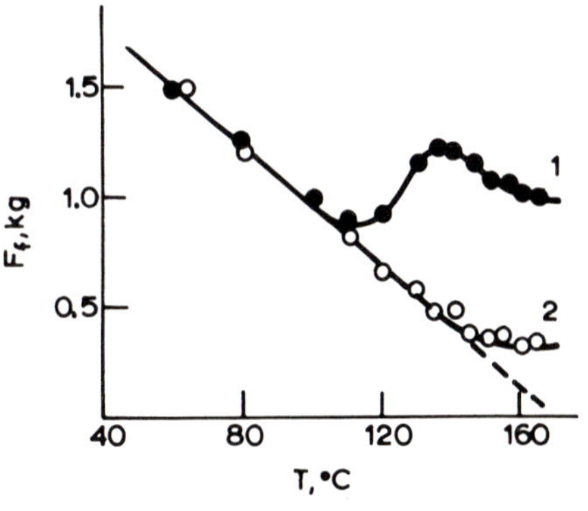

Fig. 4.8 Relation of the friction force to temperature for a vulcanized natural rubber (21): (1) in the atmosphere; (2) in a vacuum of 10^{-3} mm Hg.

surfaces, degradation processes develop more strongly than on glass; therefore, divergence from a straight line was observed earlier.

According to the molecular-kinetic theory of friction, for $F_f >> kT$, the relation between the friction force and the sliding velocity is expressed by Eq. (4.9). The data of Fig. 4.9 corroborate this relation.

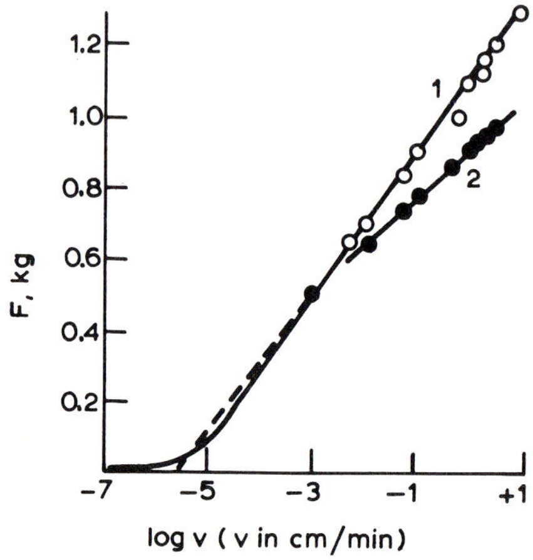

Fig. 4.9 Friction characteristics for a vulcanized natural rubber on polished steel (5). T = 30°C; p = 0.3 kg/cm^2; S_H = 2.34 cm^2. Continuous curves are theoretical, points are experimental data. Modulus of elasticity of the vulcanized rubber: (1) 9.8 kg/cm^2; (2) 31 kg/cm^2.

Styran (18) conducted systematic studies of the velocity dependence of the friction force for rubbery polymers under low pressures. Let us examine his basic results.

Figure 4.10a shows changes in the friction force for vulcanized nitrile rubbers on steel. The relation is linear, and this means that friction in rubbery materials can be explained by the molecular theory. It was also established that the type of hard surface affects the activation energy of the friction process. Thus, for example, for friction of vulcanized natural rubber, v on glass is 2.10, on steel is 18.5, and on ebonite is 24.5 kcal/mole. Similarly, the type of rubber affects the activation energy; with increase in rubber polarity (e.g. nitrile content), the activation energy increases (Fig. 4.10b).

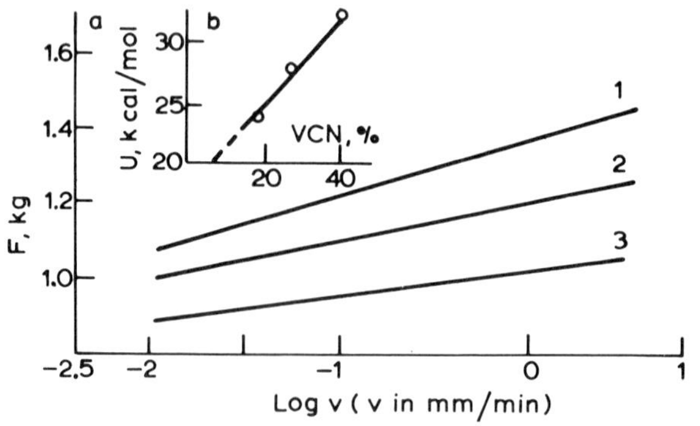

Fig. 4.10 (a) Relation of the friction force to the sliding velocity for three vulcanized butadiene-acrylonitrile rubbers on steel; T = 30°C, p = 0.3 kg/cm^2. (1) Type SKN-18; (2) type SKN-26; (3) type SKN-40. (b) Relation of the activation energy to the polarity of a nitrile rubber (18).

Change in the vulcanized rubber's equilibrium relaxed modulus, (which is an indicator of the dimensions of the vulcanizate's spatial network) from 3.4 to 31 kg/cm^2 brought about a change in the activation energy of only 3 kcal/mole (from 17 to 20 kcal/mole). This is a weak dependence of the activation energy upon the dimensions of the polymer's spatial network. With a change in the modulus, the number of segments that are in contact increases. Thus, with an increase in the modulus from 3.4 to 31 kg/cm^2, the number of chains on a unit of the nominal contact area (in cm^2) changes from 6.1 x 10^{11} to 0.5 x 10^{11} (pressure p = 0.3 kg/cm^2, for a vulcanized natural rubber). Statistical theories of crosslinked polymer friction allow the estimation of the number of segments in 1 cm^3 according to the data on the equilibrium relaxed modulus (see Chap. 1). Consequently, one can estimate the maximum number of segments that exist on the polymer's surface. If the number of segments in 1 cm^3 is equal to n, then the number of segments on the surface is proportional to $n^{2/3}$. Such a maximum number of segments that can be in contact occurs only at very high pressures. According to data obtained by Bartenev and Styran (7,8), the number of segments on 1 cm^2 of the surface changes from 5.5 x 10^{12} to 5 x 10^{13}. The ratio of the number of segments (n_c) to $n^{2/3}$ gives the ratio of the real contact area to the nominal contact area, which is reduced sharply with increase in the modulus--in the given case from 5.5 to 0.3%. On the basis of these studies, the authors concluded that the activation energy is low, and the number of segments strongly depends upon the modulus of elasticity of the vulcanized rubber.

Further, it has been shown (7,8,18) that if one replaces natural rubber with a vulcanized rubber with the same equilibrium modulus (9.8 kg/cm^2), the activation energy is as follows: for natural rubber, 18.5; butadiene rubber (SKBM-20), 20.4; butadiene-styrene rubber (SKS-10), 22.3; butadiene-styrene rubber (SKS-30), 27.2; butadiene-acrylonitrile rubber (SKN-18), 23.8; butadiene-acrylonitrile rubber (SKN-26), 27.9; and butadiene-acrylonitrile rubber (SKN-40), 32.2 kcal/molc. Consequently, with an increase in polarity, the activation energy increases.

The linear relation F_f vs ln v does not continue for sliding velocities of 1 cm/min and higher (Fig. 4.11) (22). The hypothesis

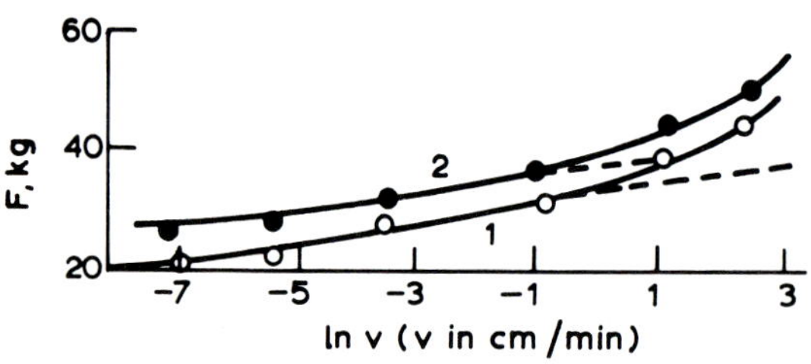

Fig. 4.11 Relation of the friction force to the logarithm of the sliding velocity for vulcanized butadiene-acrylonitrile rubbers (22). (1) Type SKN-18; (2) type SKN-40.

has been advanced that the friction constant c increases nonlinearly with increase in the sliding velocity above 1 cm/min, primarily because the mechanical loss component in the rubber is added to the friction force as determined by the molecular-kinetic theory. Furthermore, for high sliding velocities, according to Bartenev's theory, it is necessary to take into account not only the stationary lifetime τ_1 of a polymer chain on a hard surface, but also the time τ_2 that is expended in the transfer (self-diffusion) of a segment from one contact point to another* (15).

As a result the total cycle time $\tau_t = \tau_1 + \tau_2$ is increased. This increase in time is analogous to the decrease in sliding velocity ($v \approx \lambda_j/\tau_t$); therefore, lower velocity values correspond to the

*In other words, τ_2 is the diffusion time of a polymer segment through the specimen's bulk from the moment of bond breaking to bond formation at a new contact point.

values of c given in Fig. 4.12 for high sliding velocities, as follows from Eq. (4.9). For point A (curve 1), point A′ was obtained experimentally.

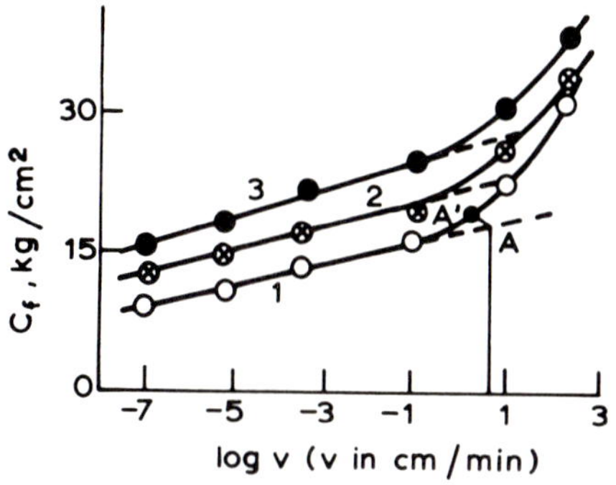

Fig. 4.12 Relation of the friction constant c to the logarithm of the sliding velocity for friction of vulcanized butadiene-acrylonitrile rubbers on glass (22). (1) Type SKN-18; (2) type SKN-26; (3) type SKN-40.

As also would be expected from general theoretical premises, with rise in temperature a divergence from the straight lines is observed when the speed is increased. In other words, at high temperature, Eq. (4.9) is obeyed over a more limited range of velocities.

The effect of pressure on the curve $F_f(v)$ on semilogarithmic coordinates is interesting and unexpected. The divergence shifts to the low-velocity side with rise in pressure, that is, pressure affects the kinetic velocity process to some extent. This effect on the $F_f(v)$ relations can be explained as due to an increase in mechanical losses in the polymer boundary layer with an increase in the modulus of elasticity of the material.

In the friction theory, a calculation of the total time $\tau_t = \tau_2 + \tau_2$ allows a quantitative description of the experimental data and a prediction of the range of divergence in curves F_f vs ln v from a straight line (23). The total time equals

$$\tau_t = \tau_o[\exp(U_1/kT) - \exp(U_2/kT)] ,$$

where U_1 is the activation energy that determines the bond strength of the polymer-solid pair, and U_2 is the activation energy that determines the intermolecular interaction of chains throughout the polymer's bulk and depends upon the polymer's structure.

With this calculation, Eq. (4.3) takes the form

$$d\nu = (d\phi)/(2\pi\tau_t) = (d\phi)/(2\pi\tau_o[\exp(U_1/kT) + \exp(U_2/kT)]) \qquad (4.11)$$

For the sliding velocity we have

$$v = (\lambda_j/\pi\tau_o) \int_0^{\pi} (\cos\phi \, d\phi)/(\exp(U_1 - \Gamma F \cos\phi)/(kT) + \exp(U_2/kT)) ,$$

or

$$v = (\lambda_j/\pi\tau_o a) \int_0^{\pi} (\cos\phi \, d\phi)/(\exp(-z\cos\phi) + c) =$$

$$(\lambda_j/\tau_o a)\ I(z", c) , \qquad (4.12)$$

where

$$a = \exp(U_1/kT);\ c = b/a;\ b = \exp(U_2/kT) .$$

Equation (4.12) differs from Eq. (4.5) by the presence of c in the integrated equation's denominator. It can be shown that the integral in Eq. (4.12) is not expressed by the elementary combinations and specialized functions. This integral I(z", c) can be represented by the Maclaurin series in the vicinity of c = 0

$$I(z", c) = \sum_{n=0}^{\infty} (I^{(n)}(0, z")/n!)\ c_n . \qquad (4.13)$$

Confining our attention to the first two terms of the series, we obtain

$$I(z", c) = I(z", 0) + I(z", 0)c , \qquad (4.14)$$

where

$$I(z", 0) = I_1\ (z"),\ I'(z", 0) = (dI)/(dc)\Big|_{c=0} .$$

Hence

$$I'(z", 0) = - \int_0^{\pi} (\cos \phi \, d\phi)/(\exp(-2z \cos \phi)) = - I_1(2z") . \qquad (4.15)$$

Substituting I(z, 0) and I'(z, 0) in Eq. (4.14), we obtain

$$I(z", c) = I_1(z") - I_1(2z")c , \qquad (4.16)$$

where $I_1(z")$ and $I_1(2z")$ are modified first-order Bessel functions of the first kind.

For z" >> 1,

$$I_1(z) \approx (\exp(z"))/(\sqrt{2\pi z"}), \; I_1(2z") = (\exp(2z"))/(2\sqrt{\pi z"}) .$$

Finally, for the sliding velocity we obtain

$$v = \lambda_j /(a\tau_o \sqrt{2\pi z}) \; [\exp z" - (c \exp(\sqrt{2z"})/ \sqrt{2})] . \qquad (4.17)$$

This equation is correct for $2 \leq z" \leq 8$. For $z" \geq 8$, the Maclaurin series can never be cut off at two terms, and it is necessary to use electronic computers. Figure 4.13 shows a curve calculated on

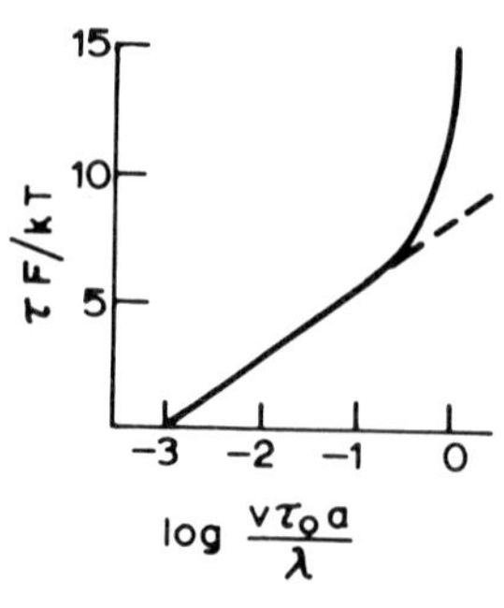

Fig. 4.13 Relation of friction force to sliding velocity on corrected semilogarithmic coordinates (continuous curve) (20).

the "Arrow" using values of the integral I(z, c) for z at intervals $\Delta z" = 0.2$. As shown, for low sliding velocities there is a linear dependence between F_f and ln v; at higher velocities F_f depends non-linearly upon ln v. These results also explain the data in Figs. 4.11 and 4.12.

Thus, the molecular-kinetic theory of friction (5) explains many experimental results. But still, it is not a general theory of friction, because it does not explain the frictional properties of rubbers at low temperatures (10). Also, Grosch (24), Bulgin (14), and others (25-27) have obtained experimental data showing $F_f(v)$ with a maximum. In other words, the molecular-kinetic theory is true only for high molecular-segment mobility. In this case one of its assumptions is clearly true, namely, that the number of kinetic units n_c or the molecular contact area was constant.

This hypothesis is completely correct when the viscous flow mechanism is applicable. However, with increase in the sliding velocity or decrease in temperature, especially in the range of friction-force maxima, it is impossible to take n_c = const. As a result of relaxation processes on the contact boundary, the contact-area formation kinetics cannot provide full (maximal) polymer segment contact with the hard friction surface. Consequently, though for low sliding velocities or high temperatures we can consider n_c = const., for increased velocities $n_c \neq$ const. Thus, generally it is necessary to know the effect of the time that the molecular segments are in contact with the solid. The relations F(v) and F(T) have been elucidated (13) for a wide range of velocities and temperatures. Let us examine some fundamental conclusions from this study.

Molecular chains or their segments can occur in two states: in state 1 the segment is in contact with a hard surface, whereas in state 2 it is not. Molecular segments can transfer from state 1 to state 2 and vice versa. Suppose the number of segments that are carried to a contact area unit in state 1 is equal to n_1, and in state 2 is equal to n_2. From general concepts it is possible to obtain the equations for transfer velocities from state 1 to state 2 and from state 2 to state 1

$$dn_1/dt = - n_1\nu_{12} + n_2\nu_{21} , \tag{4.18}$$

$$dn_2/dt = - n_2\nu_{21} + n_1\nu_{12} , \tag{4.19}$$

where ν_{12} and ν_{21} are transfer frequencies from state 1 to state 2 and from state 2 to state 1 respectively.

Taking into account that the number of segments is constant ($n_1 + n_2 = n_0$), in place of Eqs. (4.18) and (4.19) we obtain

$$dn_1/dt = - n_1\nu_1 + (n_0 - n_2)\nu_2 , \qquad (4.20)$$

where ν_1 is the bond-breaking frequency and ν_2 is the frequency of bond formation on the contact surface; ν_1 depends upon the tangential force, but ν_2 does not, because after the breaking of a segment only the normal pressure (if it is sufficiently large) can affect new contact formation.

The frequency of bond breaking, according to the molecular-kinetic theory (Eq. 4.4), is equal to

$$\nu_1 = 1/(\pi\tau_o) \int_0^{\pi} \exp(- (U - f_1\cos\phi\ (\lambda/2)/(kT)))d\phi = 1/(\tau_1(0))\ I_0(z') , \qquad (4.21)$$

where

$$1/(\tau_1(0)) = 1/(\tau_o)\ \exp(- u/kT) = \nu_1(0) ,$$

$I_0(z')$ is the zero-order Bessel function, and f_1 is the tangential force that bears on one bond ($f_1 = F_f/n_1$) (each segment has an adhesive junction).

Assuming further that the reciprocal surface transfer in the range of velocities under consideration does not affect the process of establishing equilibrium bonds (that is, once sliding has been established), from Eqs. (4.20) and (4.21) we obtain

$$n_1 = n_0\nu_2/(\nu_1 + \nu_2) = n_0/(\nu_1(0)\nu_2^{-1}I_0(z') + 1) = n_0/(AI_0(z') + 1) . \qquad (4.22)$$

Multiplying n_1 by the friction force f that bears on one bond, we obtain

$$F_f = n_1f_1 = n_0f_1/(AI_0(z') + 1) = (2kTn_0z')/(\lambda[AI_0(z') + 1] , \qquad (4.23)$$

where $A = \nu_2(0)/\nu_2$; $z' = \lambda f_1/2kT$, in which ν_2 does not depend upon f_1 and z'.

For sliding velocities, Eq. (4.5) remains applicable. Thus, the dependence of the friction force upon the sliding velocity is determined by the equations

$$F_f = [(2kTn_0)/\lambda_j][z'/(AI_0(z') - 1)] ,$$
$$v = [\lambda/ \tau_1(0)] I_1(z') . \tag{4.24}$$

For A = 1 and z << 1 we obtain

$$F_f = [(2v\tau_0 n_0 kT)/(\lambda^2)] \exp(U/kT) , \tag{4.25}$$

agreeing with Eq. (4.7) for $\Gamma = \lambda/2n_c$ and $n_c = n_0/2$.

Figure 4.14 shows the theoretical relation $F_f(v)$, according to Eq. (4.24), for two values of A, the ratio of the bond-breaking frequency at f = 0. With decrease in A the maximum shifts to the side of high velocities, but its width and height increase. The best correlation with experiment is observed at $A = 1.4 \times 10^{-5}$, that is, if the activation energy of the bond-forming process proves to be less than the activation energy of bond breaking, or approximately 7 kcal/mole.

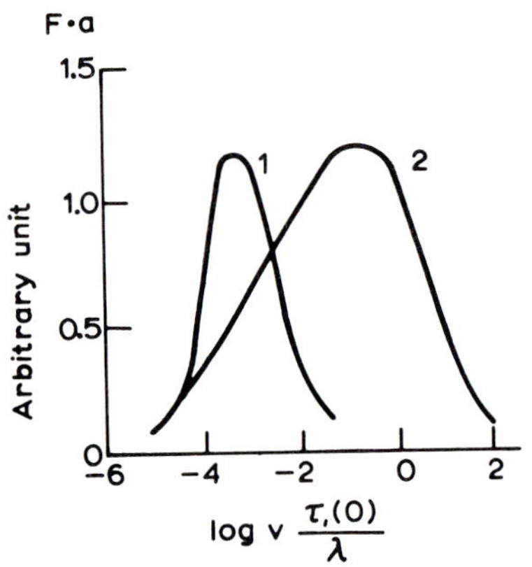

Fig. 4.14 Relation of the friction force to sliding velocity for a vulcanized rubber according to Eq. (4.24) (13): (1) $A_1 = 1$; (2) $A_2 = 1.41 \cdot 10^{-5}$. Here $a_1 = 20a_2$ (a = scale coefficient).

Figure 4.15 shows the theoretical relation of the friction force to temperature according to Eq. (4.24). The theoretical curve

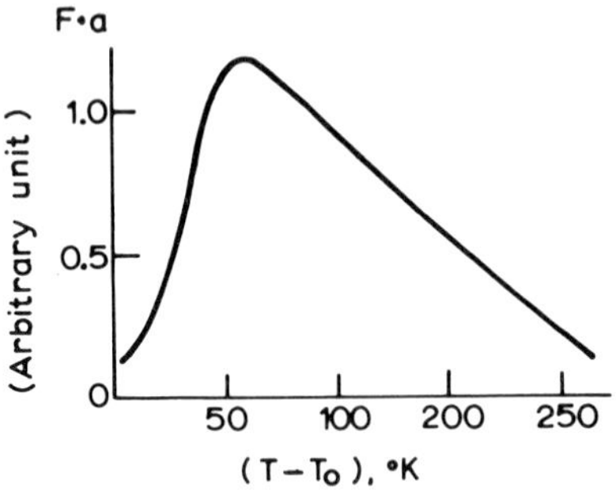

Fig. 4.15 Relation of the friction force to temperature for a vulcanized rubber according to Eq. (4.24) (13). T_o is practically equal to T_g.

agrees in form with experimental data (10) (Fig. 4.16). To illustrate the correlation of theory with data, Fig. 4.17 shows results calculated from Eqs. (4.24) and (4.25) (curves 1 and 2) and Grosch's (24) experimental data (curve 3). As seen in Fig. 4.17, the theoretical maximum is narrower than the experimental. This is fully understandable, because in the theory a single relaxation time was examined, whereas in a real polymer there is a range of relaxation times, whose contributions broaden the maximum. The theoretical curve leads to $F_f \to 0$ when $v \to 0$ or $v \to \infty$. For low velocities, this is corroborated by experiment, but for high velocities there is a divergence from experimental data because the

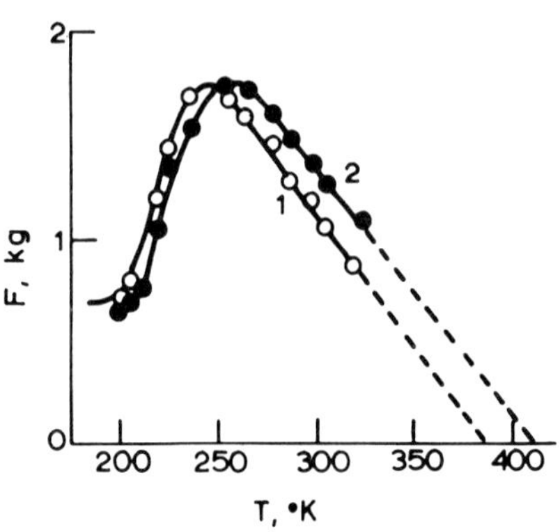

Fig. 4.16 Relation of the friction force to temperature for steel on spatially vulcanized butadiene-styrene rubber for p = 0.65 kg/cm^2 (10). Sliding velocities: (1) 0.2 mm/min; (2) 1 mm/min.

theory does not take into account new processes that become important at high rates (fracture of a surface, initial heating, etc.).

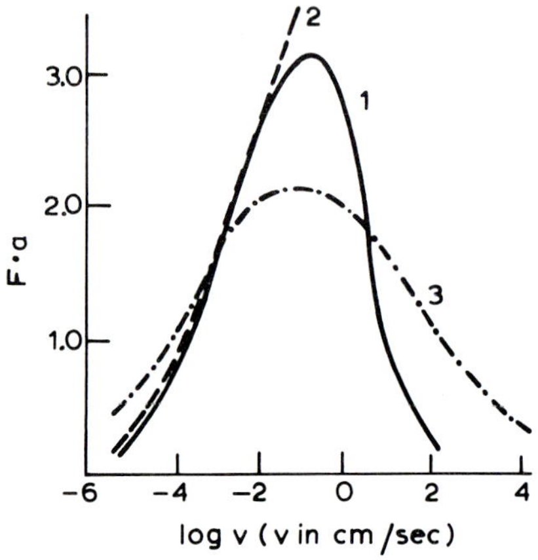

Fig. 4.17 Relation of the friction force to sliding velocity for a vulcanized rubber: (1) by Eq. (4.24); (2) by Eq. (4.5); (3) from experimental data (24).

The theory stated in this chapter permits not only the explanation of experimental data from a single point of view, but also the prediction of the effect of pressure on the functions $F_f(v)$ and $F(T)$. In fact, the integration of Eq. (4.20) for equilibrium conditions $dn_1/dt = 0$ leads to the equation

$$n_1 = n_0[\nu_2/(\nu_1 - \nu_2)] \ . \qquad (4.26)$$

The pressure effect can be accounted for as a change of the bond rupture frequency ν_2. Assuming to a first approximation that $\nu_2 = \alpha p$, we obtain

$$n_2 = n_0[\alpha p/(v_1 + \alpha p)] \ . \qquad (4.27)$$

This equation, as we shall see below, leads to the so-called Thirion's law. Furthermore, the theory of friction also allows prediction of the effect of temperature, polarity, density, the spatial network, and the nature of the hard surface upon the rate dependence of the friction force, and also the effect of velocity upon the temperature dependence of the friction force. Let us consider the effects of polarity and temperature on the functions $F_f(T)$ and $F_f(v)$.

Increase in the polarity of a natural rubber leads to a reduction of the constant A in Eq. (4.22) owing to a decrease in bond-rupture frequency ν_1 (0). At this point, the friction force shown in Fig. 4.18, verifies this conclusion. The figure shows the effect of rubber polarity (increase in volume of the polar styrene groups in the rubber chain) on function F(T). Temperature has a similar effect on the rate dependence of the friction force (Fig. 4.19). In this case, both the bond-breaking and bond-forming frequencies are changed; therefore, the effect is not so perspicuous. Calculation of the friction force by Eq. (4.24) shows that with rise in temperature the maximum $F_f(v)$ should shift to the high-velocity side and the curve of $F_f(v)$ should be steeper. The data in Fig. 4.19 confirm these conclusions.

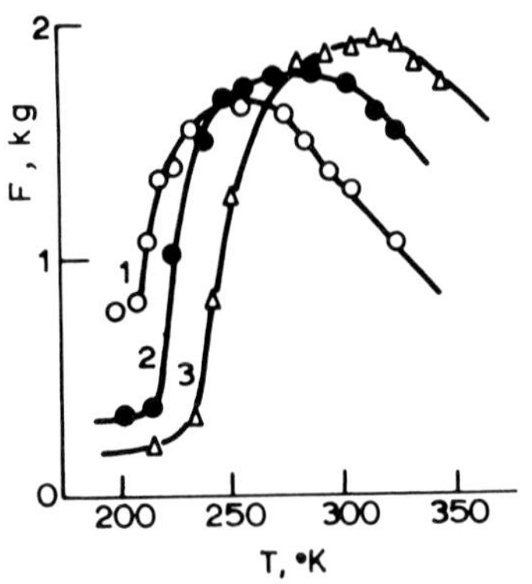

Fig. 4.18 Relation of the friction force to temperature for vulcanized styrene-butadiene rubbers on steel (15): $p = 0.66$ kg/cm^2; v = 1 mm/min. (1) Type 10; (2) Type 30; (3) Type 50.

The equation for the frequency when $U/kT >> 1$ should be written in the form*:

$$\nu = \nu_o [U/kT] \exp[- U/kT] . \qquad (4.28)$$

Using this equation, it is possible to obtain (64) a more general equation for the friction force:

*The equation $\nu = 1/\tau$ sometimes used for frequency is incorrect, because the mean value of the frequency $\nu = 1/\tau$ is not equal to $1/\tau$ in the general case.

$$F_f = [(2n_o kT)/\lambda_j][\bar{z}/(A[I_0(\bar{z}) - [(2kT)/(2U)]I_1(z')] + 1)] \quad (4.29)$$

and for the sliding velocity:

$$v = \lambda_j \nu_1(0)[I_1(\bar{z}') - [zkT/2U]I_0\bar{z} - [zkT/U]I_2(\bar{z})] ,$$

where $I_2(z)$ is a modified Bessel function of second order. When $\bar{z} << 1$, and $A = 1$, we have

$$F = [(2v\tau_o n_o kT)/\lambda_j^2] \exp[U/ kT] ,$$

which agrees with the equation obtained from the molecular-kinetic theory.

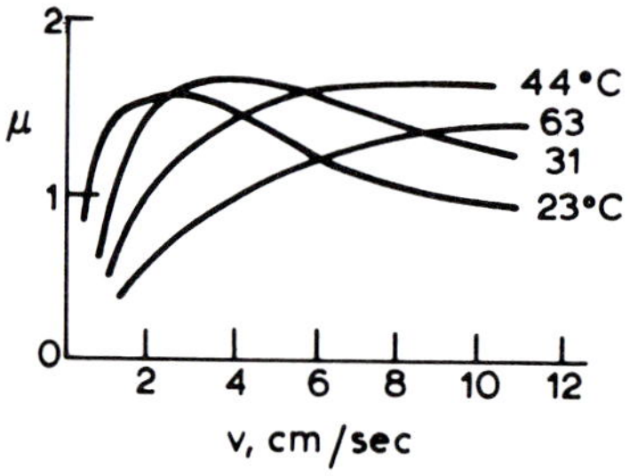

Fig. 4.19 Relation of the friction coefficient to sliding velocity for a vulcanized butyl rubber (14).

Experimental data demonstrate that for a wide range of temperatures and sliding velocities $\bar{z} >> 1$. Then, using the asymptotic form of the Bessel function for large arguments, we obtain for the friction constants c_f (or specific real friction force) the following equation:

$$F_r = c_f = [(1.1kTn_o)/\lambda_j][(1.4 + \ln v/v_o)/(v/v_1 + 1)] , \quad (4.30)$$

where $v_o = n_j\nu_1(0)$ and $v_1 = 2\nu_2$ (ν_1 is the bond-breaking frequency and ν_2 the bond-forming frequency).

The criterion of applicability of Eq. (4.30) is $v/v_o > 1$, which is the case when the sliding velocity v exceeds the bond-breaking frequency in the absence of shear force. Equation (4.30) for $v/v_1 << 1$ is similar to Eq. (4.9), which is derived from the molecular-kinetic theory of friction.

The maximum size of the specific actual friction force can be calculated from the equation

$$f_{r(max)} = c_{f(max)} = [(1.4n_o kT)/\lambda_j](v_1/v_{max}) . \qquad (4.31)$$

The sliding velocity corresponding to the value c_{max} is equal to

$$v_{max} = \frac{\lambda_j v(0)(U_2/(U_1 - U_2))}{1 - [(\alpha kT)/(U_1 - U_2)]} \exp[-(U_2/kT)] , \qquad (4.32)$$

where α is the molecular constant. With strong adhesion of segment to friction surface, $v_1/v_{max} \approx \ln (v_1/v_0)$, and then

$$f_{r(max)} = c_{f(max)} = 1.4n_o [(U_1 - U_2)/\lambda_j][1 - (\frac{kT \ln (U_1/U_2)}{U_1 - U_2})]. \qquad (4.33)$$

Calculation shows that the second factor in brackets has a small value ($\approx$0.1). Consequently, the maximum friction force depends only upon the magnitude of the interaction between a polymer chain and a hard surface, determined by the difference between U_1 and U_2. With a strong interaction, when $U_1 > U_2$, $f_{r(max)}$ does not depend upon temperature, but is determined only by the difference between the activation energy of breaking and reforming of bonds, the number of molecular contacts, and the distance between successive chain equilibrium positions on a hard surface. The equations obtained closely match our data and those of others (3,7,10,12,14,19,24). Especially good agreement has been observed between the theory and experiment in regard to the dependence of the friction force upon temperature.

The basic equation also includes the principle of time-temperature equivalence, whose applicability to polymer friction we have indicated (10,15,19,25). According to Eq. (4.32), there is a relation between the velocity, the corresponding friction-force maximum, and the temperature, which is expressed by the equation

$$\ln (v_{max}/\beta_5) = \alpha_1 T , \qquad (4.34)$$

where β_5 and α_1 are constants. It follows that to change the velocity v_{max} by an order of magnitude, it is necessary to change the temperature by an amount $\Delta T = T_2 - T_1$ such that $\Delta T = 2.3\alpha_1$.

The magnitude of α_1 ranges from 0.09 to 0.12. This corresponds to ΔT between 19 and 26°C. We may also calculate ΔT from the measured value of α_1, which depends upon the polymer and the hard surface. The theoretically calculated values nearly coincide with experimental data obtained by Grosch, Bartenev, and Yelkin, according to which ΔT ranges from 20 to 30°C.

Relaxation theories allow not only the explanation of fundamental data for rubbery-polymer sliding friction but also the prediction of the pressure effect on the position of $f_{r(max)}(v)$. The relaxation theory of friction (13) explains the fundamental data, unifies basic polymer-friction relations, and allows the prediction of new effects. From this theory, the asymmetry of the relations $F_f(v)$ and $F_f(T)$ becomes comprehensible, as well as the role of a polymer's relaxation spectrum. Thus, the relaxation theory correlates with and further develops the molecular-kinetic theory.

Effect of temperature and sliding velocity

Yelkin systematically studied the temperature dependence of the friction force for viscoelastic polymers (10,15,16,21,28) in the range of -200 to 200°C.

For low temperatures, the friction force increased with temperature. In a wide temperature range, the function F(T) was as shown in Fig. 4.20. The whole temperature range can be divided

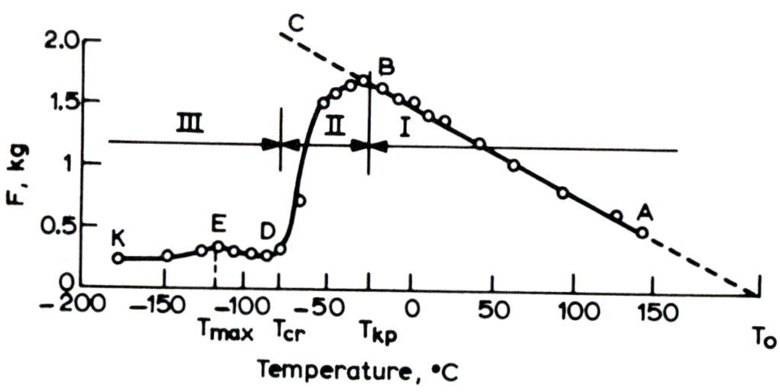

Fig. 4.20 Relation of the friction force to temperature for natural rubber on steel.

into three regions. In region I from point A to point B on the curve, a linear relation F(T) is observed according to the molecular-kinetic theory [see Eq. (4.10)]. In region II, a decrease in the friction force is observed with decreasing temperature, falling away from the linear relation predicted by the theory (dashed straight line CB). In region III, the friction force depends weakly upon temperature. The rubbery, transitional, and glassy states of a polymer correspond to these three regions. Two maxima are observed on the curve: the primary (point B) and the secondary, in the glassy-state zone (point E). According to some (10,15), the natures of the two maxima are different. The low-temperature maximum is related to mechanical-deformation losses through the polymer's bulk when sliding on a rough, hard surface. But the nature of the primary maximum cannot be explained by mechanical losses (24) or relaxation (13).

The presence of a high-temperature maximum (B) is caused by an increase in polymer chain mobility as the temperature increases in the zone to the right of the maximum (the high-temperature zone), which corresponds to a decrease in the friction force. On the other hand, in the zone to the left of the maximum, with a decrease in temperature a decrease in contact area through a change in the modulus E is assumed (15), and the friction force correspondingly decreases. The authors supposed (15) that the critical temperature T_{cr}, below which the polymer chains behave as rigid bodies, corresponds to point B in Fig. 4.20, and below T_{cr} the molecular-kinetic friction mechanism is not operative in its pure form.

It is well known that the position of the primary maximum depends upon the sliding velocity, and the curves shift symmetrically in relation to one another (Fig. 4.16). Further, for low temperatures the curves $F_f(T)$ shift analogously to the curves $F_f(v)$ in the high-temperature zone. This indicates that below T_{cr} the friction process is related to molecular-kinetic friction and the segments are in rather intense motion. The molecular-kinetic theory cannot explain this fact, but the relaxation theory can [Eq. (4.29)].

The change in the friction force or the friction coefficient in the temperature zone below T_{cr} is affected by the sliding velocity. As shown in Fig. 4.21, increase in the sliding velocity leads to a reduction in the slope of $\mu(T)$. Hence, it is possible to conclude that increase in sliding velocity leads to a weaker dependence of the friction force on temperature.

It is important to note (see Fig. 4.21) that the maximum coefficient of friction essentially does not depend upon the sliding velocity, which agrees with Eq. (4.33). According to some

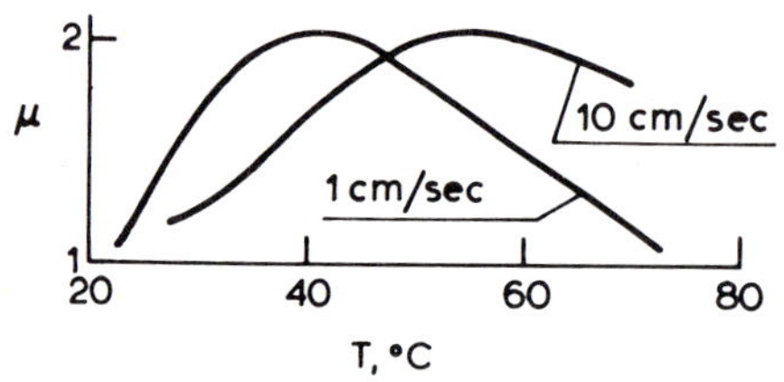

Fig. 4.21 Relation of the friction coefficient to temperature for a vulcanized rubber on steel (14).

data (14), a tenfold increase in the sliding velocity shifts the position of the maximum by 1-15°. These results show that for polymer friction we can apply the principle of time-temperature equivalence (19,25), expressed for rubbery-material friction by Eq. (4.34).

Grosch (24) systematically and thoroughly studied the relation of the friction coefficient to the sliding velocity; he applied the principle of time-temperature equivalence to a series of vulcanized rubbers. One of the "master" curves he obtained for μ vs a(T)v is shown in Fig. 4.22. The value of a(T) is equal to that measured in the relation between relaxation times of polymer

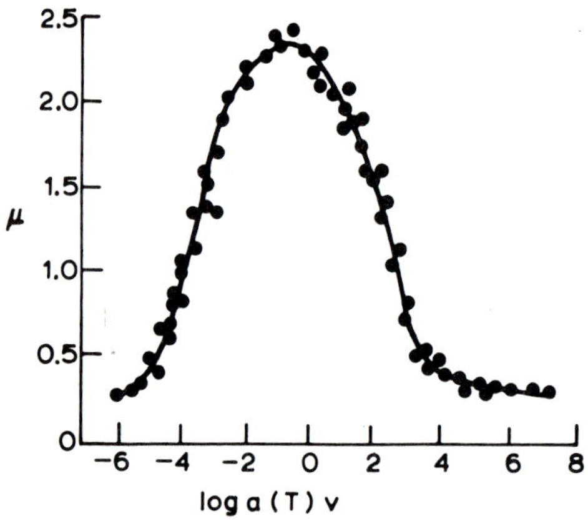

Fig. 4.22 Generalized relation of the friction force to log α(T)v for a vulcanized rubber on glass (23).

chains (see Chap. 1). As is evident, all the points for a particular polymer at different temperatures lie on one general curve. These results confirm the applicability to elastic materials of the principle of time-temperature equivalence and thereby indicate a single polymer-friction relaxation mechanism for a wide range of velocities and temperatures.

Applying the principle of time-temperature equivalence, Bartenev and Yelkin (19) constructed the function $F_f(v)$ (Fig. 4.23) from functions $F_f(T)$. Thus, the shift of the maximum in the friction

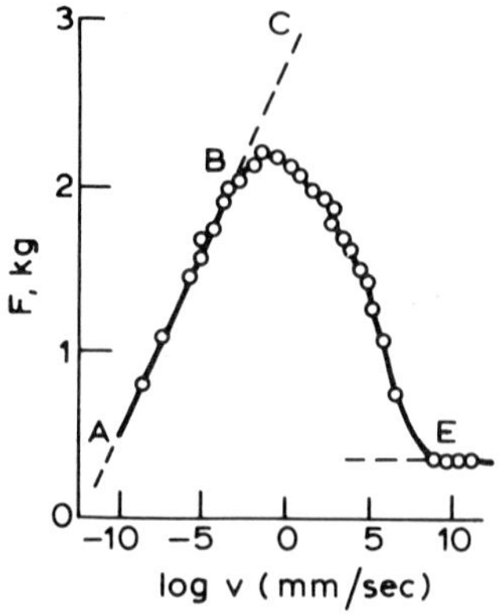

Fig. 4.23 Relation of the friction force to sliding velocity for vulcanized butadiene-acrylonitrite rubber SKN-50 (19). T = 40°C; N = 0.65 kg.

force as a function of the sliding velocity and temperature fully conforms to the principle of time-temperature equivalence. Actually, by Eq. (1.10) the relation between the relaxation time and the temperature is such that shorter relaxation times correspond to higher temperatures. But, as we have seen in the molecular-kinetic theory, the sliding velocity of polymers is controlled by the relaxation times for the transfer of polymer chains from one equilibrium position to another. Shorter relaxation times match higher sliding velocities. Consequently, according to Ferry's principle, the highest temperatures ought to match the highest sliding velocities. This conclusion fully agrees with the results shown in Figs. 4.21-4.23 and is described by Eq. (4.34). Thus, for both high and low temperatures, the friction force is determined by relaxation processes in the transfer of segments from one contact point to another under the action of thermal motion and the tangential force.

According to Eq. (4.34) and the data of Fig. 4.22, the sliding velocity, which matches the maximum friction coefficient, increases with temperature. This relation can be expressed quantitatively, according to Grosch (24), through the frequency ν at which the increase in the modulus of elasticity is a maximum, if we compare situations with identical temperatures. Here $v_{max}/\nu = \lambda = 6 \times 10^{-7}$ cm independent of the type of vulcanized rubber. This condition for the maximum can be written differently: $v_{max}\tau_{max} = \lambda$. The specified value of the magnitude λ agrees with one earlier obtained (5).

All the experimental data that we have examined above can be explained on the basis of the relaxation theory of friction.

Molecular-mechanical theories of friction

As an illustration of another theoretical approach to an explanation of the experimental relations $F_f(T)$ and $F_f(v)$, let us examine studies by Bulgin (14), Savkoor (27), and Schallamach (9). Bulgin (14) proposes that the polymer surface can be represented as a large quantity of micro-asperities that are in contact with a hard solid. Each asperity has dimensions on the order of 10^{-7} cm and forms an adhesion bond with the counter-body surface (Fig. 4.24). The bond can be torn by an applied tangential force f_s.

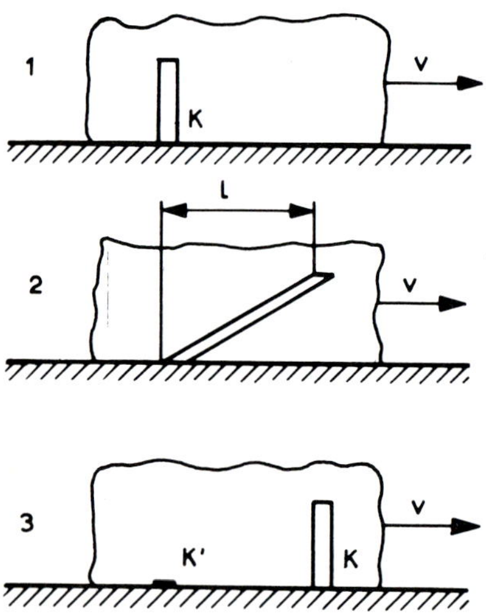

Fig. 4.24 Rupture of the friction bond according to Bulgin (14): (1) contact formation; (2) deformation contact; (3) rupture and renewal of contact.

It is assumed that the maximum force f_s, which ruptures the bond, does not depend upon the polymer's sliding velocity. For such a polymer, sliding micro-asperities that are held back by an adhesion bond will be deformed (Fig. 4.24, state 2). Therefore, the number of cycles (elongation and tear) will be a function of the sliding velocity and the polymer's properties. The tear energy is easy to calculate, taking into account the maximum stress σ_e on an element, the beginning cross-section of the element area is S_o, and its length L_o, and the strain ε. Then the tear energy can be expressed as

$$W_t = 1/2(\sigma_e S_o L_o).$$

The restored elastic energy W_r after tear and recovery in the element (that is, after one cycle) is equal to

$$W_r = 1/2\{\sigma_e S_o L_o \varepsilon (1 - \pi E''/E')\}, \tag{4.35}$$

where E' is the storage modulus, and E" is the loss modulus, which is determined by dynamic experiments (see Chap. 1). The losses during deformation are equal to

$$W_D = 1/2\{\sigma_e S_o L_o \varepsilon \pi (E''/E')\}. \tag{4.36}$$

If we take the number of elements per unit nominal surface as n, then the losses per unit friction surface are

$$W_{nD} = 1/2\{\sigma_e S_o L_o \varepsilon n \pi (E''/E')\}.$$

After one deformation cycle, an asperity slides a distance $\Delta L \approx L_o \varepsilon$. The work of friction forces for one cycle, $f_f \Delta L = f_f \varepsilon L_o$, is equal to the energy losses W; consequently

$$F_f \varepsilon = 1/2\{\sigma_e S_o \varepsilon n \pi (E''/E')\}, \tag{4.37}$$

from which

$$F_f = 1/2\{\sigma_e S_o n \pi (E''/E')\}. \tag{4.38}$$

The number of contacts, n, through which adhesion bonds are established depends upon the pressure P and hardness H. In the first approximation, $n \approx P/H$. For the friction force we obtain

$$F_f = \text{const}\ \sigma_e (P/H)(E''/E') \ . \qquad (4.39)$$

Thus, on the basis of Eq. (4.39) the friction force is proportional to the pressure, inversely proportional to the hardness, and proportional to the loss factor E"/E'. Equation (4.39) is confirmed qualitatively by experiment (Fig. 4.25).

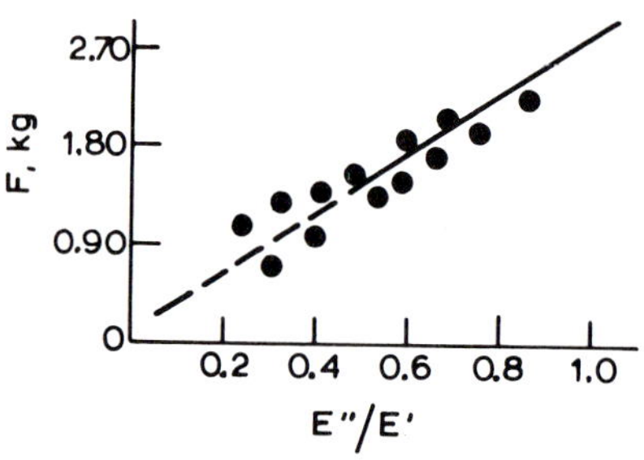

Fig. 4.25 The friction force as a function of E"/E' (14).

The relation of the friction force to the sliding velocity and temperature is determined, according to Buldin, by the frequency and temperature dependence of the loss factor E"/E'. From Eq. (4.39) and experimental data (Figs. 4.10 and 4.25), the corresponding deformation frequencies of a polymer micro-asperity during friction were calculated with the given values of E"/E'. Thus, a sliding velocity of 1 cm/sec corresponds to deformation frequencies from 10^5 to 10^7 cycle/sec. The mean jump length λ is $\approx$ 100 Å. This agrees with values obtained previously for λ (5,8).

The activation energy for the friction process calculated by relation ln v(1/T) is $\approx$ 10 kcal/mole. Because hysteresis losses at v = 0 are equal to 0, Bulgin thus concludes that the friction coefficient $\mu = 0$ for $v = 0$.

For rubbers the friction force does not depend upon the filler; but is determined only by the polymer's elastic properties. This confirms the purely adhesive nature of external friction in vulcanized rubbers. The filler principally affects the material's hardness.

Bulgin's theory is qualitative and contains crude assumptions. Thus, for example, it is assumed that the time of bond rupture, τ_r, is determined only by the bond strength Σ_{bs} and the sliding velocity v, that is, $\tau_r = \Delta L/v$. This assumption, especially for low sliding velocities, is incorrect, because it does not take into account the possibility of bond rupture by thermal motion. It is known (5) that the rupture time $\tau_r = \tau_o \exp (U/kT)$. Bulgin's theory does not take into account the effect of friction force on bond lifetime. Hence, the superposition of external and internal friction follows, which leads to a small value of the activation energy (10 kcal/mole). The latter is approximately one-half the activation energy in the friction process as obtained from other data (4,5,8,12). Furthermore, Bulgin's theory does not take into account losses in the bond rupture of the polymer-solid pair, because it considers only the deformation losses.

Bulgin's hypothesis about the proportionality between the contact area and pressure (especially for cutting) can be viewed as only a rough approximation. His theory does not take into account the relation of the contact area to the sliding velocity and temperature. For the maximum friction force, the real contact area depends significantly upon the sliding velocity (11,12). The maximum stress, σ_{max}, should depend upon temperature and the sliding velocity, but Bulgin's theory ignores this too. And finally, as has been shown (26), there is no foundation for using the loss factor in hysteresis, which is measured in a simple deformation cycle.

Schallamach also undertook a theoretical interpretation of the extreme dependence of the friction force upon the sliding velocity (9). Schallamach's theory is a development of Bartenev's molecular-kinetic theory. According to Bartenev, the transfer time of a kinetic unit of polymer on the surface of a solid is determined by the bond-breaking time for the polymer-solid pair. Earlier, we ignored the bond-forming time. Schallamach believes that generally it is impossible to ignore this quantity, and consequently the time of transfer $t = t_1 + t_2$, where t_1 is the bond-breaking time in the polymer-solid pair and t_2 is the bond-forming time in the same pair.

Furthermore, according to Schallamach, the number of bonds on a surface unit n depends upon time:

$$dn/dt = -n\nu_1 , \tag{4.40}$$

where ν is the bond-breaking frequency. The bond-forming frequency is $\nu_2 = \nu_0 \exp(-U/kT)$. The bond-breaking frequency ν_1, in the absence of ν_2, depends upon action on the kinetic element by the tangential force. The latter changes the potential barrier of the bond in the polymer-solid pair so that $U = U_o - \lambda f_i$, where λf_i is the elementary work done by the friction force, and U_o is the potential barrier of the bond in the polymer-solid pair. Further, using the law of additivity of elementary friction forces ($F = \Sigma f_i$), Schallamach concludes that $f_i = E\varepsilon i = Evt_i$, where E is the modulus of elasticity, ε_i is the deformation of the kinetic element (of a polymer chain), v is the sliding velocity, and t_i is the mean bond lifetime, equal to the bond-breaking time t_1. Then

$$\nu_1 = \nu_0 \exp(-U/kT)\exp[(\lambda_j Ev\tau_1)/kT] . \tag{4.41}$$

Substituting Eq. (4.41) into Eq. (4.40) and setting $\alpha = \lambda Ev$, we obtain

$$dn/dt = -n/\tau_1 \exp[(\alpha\tau_1)/kT] .$$

After integrating this equation, we obtain

$$\ln n/n_o = - [kT/(\alpha\tau_0)]\exp[\alpha\tau_1/kT] ,$$

or

$$n/n_o = \exp[-kT/\alpha\tau_1]\exp[\alpha\tau_1/kT] , \tag{4.42}$$

where n_o is the total number of bonds per unit of surface. This equation tells us the fraction of bonds that remain until time t. The mean bond lifetime is equal to

$$\bar{t} = \int_0^\infty (n/n_o)\,dt .$$

Consequently,

$$(\bar{t}/\tau_1) = (kT/\alpha\tau) \int_{kT/\alpha\tau}^{\infty} [(\exp(-y))/y]\, dy \, . \tag{4.43}$$

where $y = (kT/\alpha t)\exp(\alpha t/kT)$.

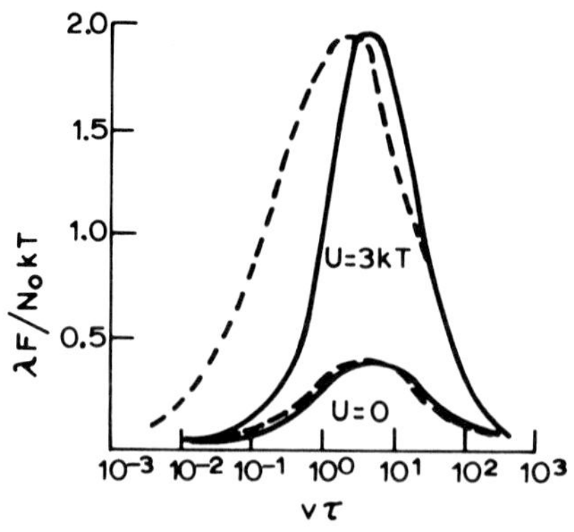

Fig. 4.26 Relation of the friction force, $\lambda F/N_o kT$, to vt (----) and $\phi(vt/\lambda)$ (——) (9).

Equation (4.43) cannot be expressed clearly in rational functions; its value was tabulated for small and large values of the arguments:

$kT/\alpha\tau_1 >> \bar{t}/\tau_1$, $kT/\alpha\tau_1 << \bar{t}/\tau_1$

The friction force can be written in the following form:

$$F_f = [n_0\alpha\tau_1/\lambda_j][\alpha\tau_1/kT] \, . \tag{4.44}$$

Figure 4.26 shows the theoretical relation $F_f(v)$. Schallamach's theory does not describe experimental data for very low and very high sliding velocities. His theory is semiphenomenological, because together with the constants of the molecular-kinetic process, the modulus of elasticity E ($\alpha = Ev\lambda$) enters into the expression for $F_v(v)$. In order to avoid this, the author assumes further that $E = E_\infty[1 - \exp(\bar{t}/t_1)]$. But this only complicates the general solution.

The fundamental equation (4.40) takes into account only the number of bonds broken over time and does not consider the process of bond formation [see Eq. (4.20)]. Here, it can be demonstrated that the relation of the number of tearing bonds to nontearing bonds depends upon the tangential force. Consequently, the tangential force affects not only the frequency of bond breaking, but the number of bonds being broken. Schallamach's theory does not consider this. Schallamach overlooks low sliding velocities, that is, the case when $\bar{t}$, determined by the sliding velocity, has no value for low velocities, because it is significantly longer than the thermal transfer time.

Savkoor (27) has proposed that large deformations, much larger than in the bulk, develop on a thin surface film because of adhesive bonds and the tangential force. The surface-film hysteresis properties determine the nature of polymer friction for elastic materials in particular. Further, taking one of the mechanical models of a polymer, Savkoor obtained a theoretical relation F(v), with a maximum. We believe that his theory contains a number of unfounded assumptions. Thus, his hypotheses of the proportionality of the real contact area to the logarithms of time and pressure are unfounded. As a result, the condition for the maximum of F(v) has no physical significance. His theory does not explain the temperature dependence of the friction force in the experimental data (15). Also, he assumes that the real contact area is independent of the sliding velocity (especially in the zone of the friction-force maximum) and of temperature. In other words, Savkoor's theory is a rather crude phenomenological theory of polymer external friction. Noteworthy, too, is the work of Boggs and Riemen (17), who showed that the friction force for crosslinked rubbers on rough surfaces depends upon the sliding velocity. In the Boggs-Riemen theory, for low velocities F is proportional to v, and the slope of $F_f(v)$ is proportional to hysteresis losses. With further increase in the sliding velocity, the friction force tends to diminish. To explain this relation, Boggs and Riemen employ an analogy between external friction and flow. They assume that for small values of v vulcanized rubber behaves as if it circumvented the roughness of a solid. This assumption corresponds to laminar flow of a liquid. The viscosity coefficient is analogous to hysteresis losses in a vulcanized rubber.

By analogy with hydrodynamics, in which laminar flow is characterized by the Reynolds number, Boggs and Riemen showed that the friction of a vulcanized rubber for low sliding velocities can be characterized qualitatively by the dimensionless number nv/E, where n is a measure that characterizes the hysteresis properties of a vulcanized rubber. There exists a certain critical value of this measure, below which the friction is "laminar" and above which it is "turbulent". In the latter case, polymer wear is observed. Savkoor and Boggs and Riemen's works are interesting in that phenomenologically they indicate an analogy between external and internal friction.

The theories of friction we have examined explain qualitatively the dependence of the friction force upon the sliding velocity and temperature. However, these theories are not useful because they are based on crude assumptions, and because their final expressions cannot be subjected to direct verification.

4.2 REAL AREA OF CONTACT

The real contact area and also the friction force can be examined from two points of view: the molecular-kinetic and the phenomenological (30). The phenomenological approach is significantly developed in the study of sliding friction (31). Both approaches lead to equivalent relations of the real contact area to the contact time [S(t)], the sliding velocity [$S_r(v)$], the pressure [$S_r(p)$], and the temperature [$S_r(T)$] (11-13,23,30-33).

Effect of contact time and sliding velocity

Let us examine from both points of view the actual contact area for constant values of temperature and normal load (30). From the molecular point of view, the number of chains, n_1, existing at the moment of time t in contact with a solid, can be determined by Eq. (4.20):

$$n_1 = [(n_o v_2)/(v_1 + v_2)] \{1 - \exp[- ((v_1 + v_2)/t)]\} +$$

$$n_1(0)\exp[- ((v_1 + v_2)/t)] , \qquad (4.45)$$

where n_o is the total number of chains, and $n_1(0)$ is the number of chains that are in contact at t = 0. With $t \to \infty$, from Eq. (4.45) it follows that $n_{1\infty} = (\nu_1 n_o)/(\nu_1 + \nu_2)$, that is, an equilibrium number of chains on the solid surface is determined by the correlation between frequencies at $\nu_1 = \nu_2$, $n_\infty = 1/2\ n_o$. As a result of adhesion at the boundary of the polymer-solid pair, $\nu_2 > \nu_1$;

therefore $n_{1\infty} > (1/2)\, n_o$, that is, most of molecules are linked to the surface.* For real contact area, an analogous relation had already been obtained phenomenologically by Kragelsky (33) for metals:

$$S(t) = S_\infty[1 - \exp(-\nu_2 t)] + S_o \exp(-\nu_2 t) \ , \qquad (4.46)$$

where S_∞ is the contact area at $t\to\infty$, and S_o is the contact area at $t = o$.

For small values of ν_1, Eq. (4.20) will take the form

$$dn/dt = (n_o - n)\nu_2 \ , \qquad (4.47)$$

where n is the number of molecular contacts for a polymer with a hard surface per unit nominal contact area. The value in brackets determines the free part of surface contact. Consequently, for strong adhesion ($\sqrt{\nu_2} >> \sqrt{\nu_1}$), the rate dn/dt of the contact area formation is proportional to the free contact area $n_o - n$ and to the "setting" frequency.

Now, let us examine the process of contact-area formation from the phenomenological point of view. We introduce the value of the relative contact area determined as $n/n_\infty = S_r/S_\infty$. The values n/n_o and S_r/S_H are often used, when $n_o = n_\infty$ or $S_\infty = S_n$. But generally $S_\infty \neq S_n$.

Under constant stress, the strain in the polymer $\varepsilon(t)$ may be described by Kelvin's equation (34):

$$\dot{\varepsilon} = \alpha(\varepsilon_\infty - \varepsilon) \ , \qquad (4.48)$$

where ε_∞ is the equilibrium value of deformation, and α is a constant Then, if $\varepsilon/\varepsilon_\infty \approx \phi/\phi_\infty$, Eq. (4.48) takes the form

$$\phi = \phi_\infty[1 - \exp(-\alpha t)] + \phi_o \exp(\alpha t) \ . \qquad (4.49)$$

It is easy to see that Eq. (4.49) is similar to Eq. (4.46), if we take $\alpha = \nu_2$. Equation (4.49) reflects the experimental data

---

* By a statistical method, it is possible to demonstrate that the condition $\nu_2 > \nu_1$ allows practically all free segments to be linked to the surface as long as there is the low pressure.

qualitatively, but has an important deficiency: the coefficient α depends upon the normal load and time.

A more exact relation was proposed by Rebinder and co-workers on the kinetics of the development of rubbery deformation in elastomers (35,36). They showed that the coefficient $\alpha = \sigma/\eta_\ell \varepsilon$, where η_ℓ is the constant of limiting viscosity and σ is the shear stress. At this point, Eq. (4.48) ought to be replaced by the following equation:

$$\eta_\ell \varepsilon \dot{\varepsilon} = (\varepsilon_\infty - \varepsilon)\sigma \ . \qquad (4.50)$$

Equation (4.50) is equal to

$$\varepsilon_\infty \ln[(\varepsilon_\infty - \varepsilon)/\varepsilon_\infty] + \varepsilon = \sigma t/\eta$$

or, with the condition $\varepsilon = \phi$, $\varepsilon_\infty = \phi_\infty$,

$$\ln(1 - \phi) + \phi = - \sigma t/\eta_\infty \ . \qquad (4.51)$$

It is possible to proceed directly from Eq. (4.48) to Eq. (4.50) if we assume that $\alpha = 1/t$ and $t = \eta/E$. Taking for small deformation $\sigma = E\varepsilon$, we obtain $\alpha = \sigma/\eta\varepsilon$. The calculation of this relation allows one to avoid a formal introduction of the relaxation-time spectrum and the necessity of integrating Eq. (4.48) for t.

Equation (4.51) describes the deformation of asperities in the contact zone with a constant stress σ. Actually, it can be supposed that the increase in the real contact area with time occurs because of an increase in the number of chains under constant stress. On the other hand, with an increase in the number of contacting chains, the real pressure P_ϕ in the zone of contact changes, proportionately to the ratio of the normal load N to the real contact area. This can be taken into account by the relation $\alpha = P/S_r n_\ell$ or $\alpha = p_n/\phi n_\ell$, where p_n is the nominal pressure that determines the normal load and the nominal contact area (p_n = constant during time of testing).

Substituting in Eq. (4.48) the relation $\alpha = p_n/\eta\phi$, we obtain an equation analagous to Eq. (4.50) and its integral (4.51). According to Eq. (4.51), experimental data on a plot of $\ln(1 - \phi) + \phi$ vs t should lie on a straight line. This is evident from Fig. 4.27. The following values of the viscosity η are determined from the slope:

p (kg/cm^2):	0.7	1.0	3.2	6.0
η (10^6 poise):	0.45	0.45	0.46	0.45

As we see, the magnitude η actually represents the deformation of the real contact area for polymers.

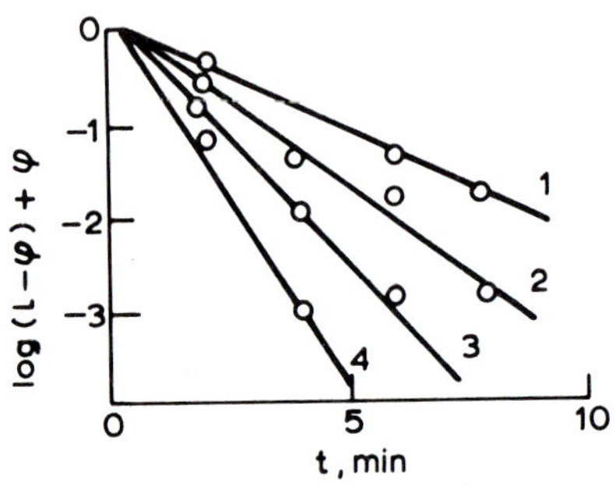

Fig. 4.27 Dependence of the relative real contact area upon time in a semilogarithmic plot, for a vulcanized butadiene-acrylonitrile rubber SKN-26 (30) during friction. Temperature 21°C; pressure: (1) 0.7 kg/cm^2; (2) 1.8 kg/cm^2; (3) 3.2 kg/cm^2; (4) 6 kg/cm^2.

Thus, the real contact area formation for polymers is determined by the law of rubbery deformation (4.48). It is interesting to compare the equivalent Eqs. (4.45) and (4.48). For $\alpha = \nu_1 + \nu_2$, there exists a close relation between the phenomenological and statistical approaches.

From the relation S(t), it is possible to go directly to the relation S(ν). According to Eq. (4.47), since the sliding velocity $v = dx/dt$, we note:

$$dn = (n_o - n)[(\nu dx)/v] \tag{4.52}$$

The frequency ν depends through the activation energy upon the distance between a section of a molecular chain and the surface of a solid. Integrating Eq. (4.52), we obtain

$$n = n_o[1 - \exp(v_o/v)] \quad (4.53)$$

where $v_o = \int \nu dx \approx$ const., while x varies from 0 to ℓ (ℓ is the mean distance between asperities of a hard surface; $v_o = \nu(\ell)$.

According to Eq. (4.53), for $v_o/v >> 0$, we have $n = n_o$, and for $v \to \infty$ we have $n \to \infty$: for low sliding velocities the actual contact area is constant, and for high sliding velocities it is equal to zero. It is possible to state, on the basis of Eq. (4.53), that the more mobile the polymer chains (the greater v_o), the wider the range of velocities at which n = const. These qualitative conclusions are confirmed by data. Figure 4.28 shows the relation of the real contact area S_v, measured at v = const., to the static contact area S_o.* Suppose $S_v/S_o = n/n_o$. Then it is possible to compare the experimental relation with the theoretical (4.53).

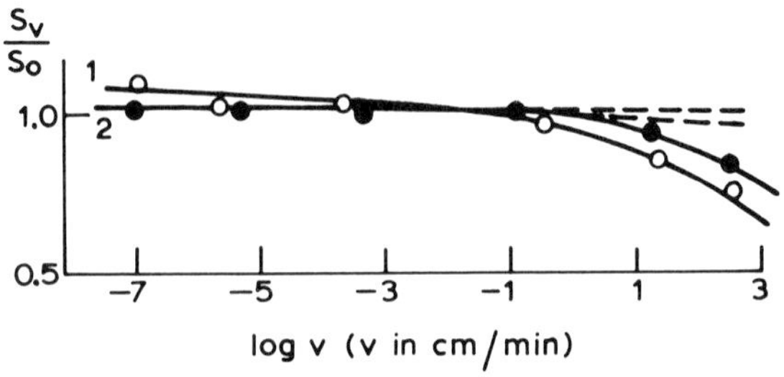

Fig. 4.28 Relation of the relative actual contact area, $\phi = S_v/S_o$, to the logarithm of the sliding velocity for vulcanized rubbers (11). S_v is the real contact equilibrium area for a given sliding velocity v; S_o is the actual contact (static) equilibrium area for a given specific load. $p = 5.5$ kg/cm^2; T = 20°C. Vulcanized rubbers: (1) SKN-18; (2) SKN-40.

As is evident, for low sliding velocities S_v/S_o = const. Consequently, although the polarity of a natural rubber is insignificant, nevertheless it affects the curve S(v). Here an increase of polarity leads to a reduction in segment mobility in the range at which S_v/S_o = const., so that the range becomes narrower (11). The relation $S_v(t)$ is of this kind and can be explained phenomenologically (22).

---

*Data obtained on an optical tribometer by Lavrentev (see Chap. 7).

The effect of normal load

As is well known, rubbery polymers are subjected to elastic deformation in the contact zone. According to experimental data (20), with rise in pressure the number of contacting molecules, or contact area, increases. Generally, Eq. (3.5) holds true:

$$\phi = 1 - (1 - \phi_0)\exp(- \gamma p/E) .$$

One of the serious deficiencies of this equation, which reflects the dependence of the real contact area upon pressure, consists in the fact that it contains the empirical constant γ. This does not allow the calculation of the real contact area of polymers.** Therefore, it is convenient to examine model surfaces, which can explain the physical significance of constant γ.

Archard (37) showed that for surfaces with a wide distribution of asperity sizes the dependence of the real contact area upon normal load is close to linear. For small pressures with $p/E << 1$, surface asperities quickly wear smooth and the law of change in the contact area is expressed by the exponential relation $S = p^{2/3}$ (9). Therefore, for the model of hemispherical asperities the contact area for one isolated asperity is

$$S = \pi[3/4[(1 - x^2)/E]\Omega' p]^{2/3} s^{2/3} , \qquad (4.54)$$

where x is Poisson's ratio (for vulcanized rubbers $x = 0.5$), and s is the surface area that presses on one irregularity. From Eq. (4.54 we obtain an equation for the relative contact area:

$$\phi = \pi(9/16\ r')^{2/3} s^{-1/3} (p/E)^{2/3} . \qquad (4.55)$$

A differential equation whose solution is Eq. (4.55) has the form

$$d\phi/dp = 2/3\ \pi(9/16\ r')^{2/3} s^{-1/3} (p/E)^{-1/3} \psi(S) , \qquad (4.56)$$

**This equation has since been successfully used to determine the relation between the magnitude of γ and the polymer surface contour.

where $\psi(S)$ is a function that takes into account the reciprocal effect of asperities during their deformation. Physically, the function $\psi(S)$ ought to satisfy the following conditions: for low pressures $\psi(S) = 1$; for high pressures $\phi \to 1$, $d\phi/dp \to 0$--that is, $\psi = 0$. The function $\psi(S) = 1 - \phi$ is the simplest function that satisfies these conditions. In this case, Eq. (4.56) has the form

$$\phi = 1 - \exp[- \gamma(p/E)^{2/3}] , \qquad (4.57)$$

where $\gamma = \pi(9/16)^{2/3}s^{-1/3}$. Keeping in mind that $s = n_o^{-1}$, where n_o is the number of asperities per unit specimen surface, we obtain for γ the equation

$$\gamma = 2.14(r')^{2/3}n_o^{1/3} . \qquad (4.58)$$

It has been shown (38) that the calculation of the actual contact area for vulcanized rubbers by Eq. (4.57) agrees satisfactorily with experimental data (20).

Figure 4.29 shows experimental dependence of the contact area upon pressure for these kinds of vulcanized rubber. With increase in vulcanized-rubber modulus, the maximum value S_∞ and rate (dS/dp) diminish somewhat according to Eq. (4.57). The slope of the straight line (Fig. 4.30) is equal to β/E by Eq. (3.5). The

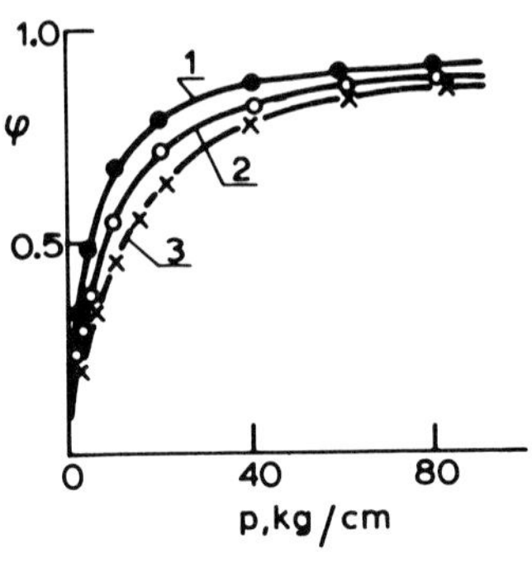

Fig. 4.29 Relation of the relative real contact area to specific load for vulcanized rubbers (20): (1) SKN-18; (2) SKN-26; (3) SKN-40.

vulcanized rubbers were obtained from one source; this gives a basis for assuming that β = const. SKN-40, SKN-26, and SKN-18 have a modulus in uniaxial compression respectively of 23, 20, and 17 kg/cm and the experimental values γ/E for these rubbers are equal to 0.54, 0.61, and 0.73. For given values of the modulus, values of β are calculated, which for the vulcanized rubbers examined are close to 1.24.

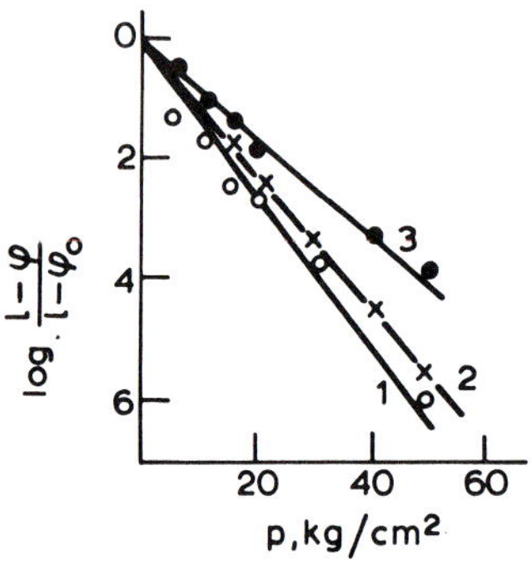

Fig. 4.30 Relation of the real contact area to pressure for vulcanized rubbers (20): (1) SKN-18; (2) SKN-26; (3) SKN-40.

Quite significant is the experimentally demonstrated fact that saturation of the real contact area is incomplete--that is, for $p \to \infty$, $S_r < S_N$. (S_N = nominal contact area.) Tolstoy pointed out the impossibility of complete contact for dispersed systems with rise in pressure and explained this by the presence of micropores and air cavities on the surface of the hard solid and similar phenomena. For increasing pressures, $S \to S_\infty$ up to some saturation pressure p_N. This means, according to the physical significance of Eqs. (3.5) and (4.56), that it is possible to substitute S_N for S_∞.

Effect of temperature

According to Eq. (4.45), under static conditions for $t \to \infty$ the number of molecular contacts is equal to

$$n = [n_0\nu_2/(\nu_1 + \nu_2)] \ . \qquad (4.59)$$

Because $\nu_1 = \nu_o \exp(-\ U_1/kT)$ and $\nu_2 = \nu_o \exp(-\ U_2/kT)$, we obtain

$$n = n_0/[1 + \exp(-(\Delta U/kT))] , \qquad (4.60)$$

where $\Delta U = (U_1 - U_2) > o$.

According to Eq. (4.60), with a rise in temperature the number of contacts falls. Here appears the molecular mechanism of thermo-elasticity for elastic materials. However, as Fig. 4.31 shows, with a rise in temperature the real contact area increases (12,22). This result is explained by the fact that the equilibrium position is practically unattainable because of the effect of long relaxation times. Equation (4.60) does not take this into account.

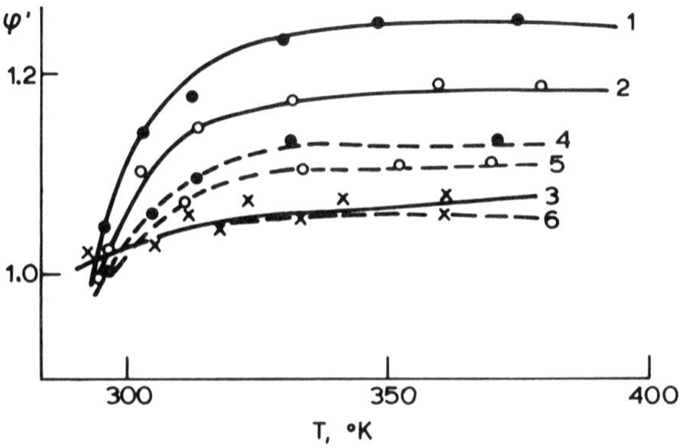

Fig. 4.31 Relation of the relative real contact area $\phi' = S_v/S_o$ to temperature for vulcanized butadiene acrylonitrile rubber SKN-40. S_v is the equilibrium real contact area at the temperature of testing (static or in motion); S_o--the settled static area of real contact at 20°C. — Solid lines: stationary contact; -- dashed lines: friction (ν = 0.35 cm/min). Specific loads: (1),(4) 0.4; (2),(5) 1.8; (3),(6) 6 kg/cm^2.

Inasmuch as in studying n(t) it is necessary to consider the process of nonequilibrium contact, it is impossible to disregard the exponential term in Eq. (4.45). Moreover, calculation of the increase in frequencies ν_1 and ν_2 with temperature ought to lead to an increase in the nonequilibrium real contact area. A theoretical calculation of the dependence of the nonequilibrium real contact area upon temperature is difficult. Therefore, to interpret the relation one uses Eq. (3.5) in Fig. 4.31. In this equation E is the modulus for a polymer, and is static or dynamic in accordance with experimental conditions. According to other data (40), the obtained values of the real contact area are fixed after 7 minutes of stationary contact. Therefore, in Eq. (3.5) it is necessary to substitute the following values of the nonequilibrium 7-minute static modulus for the vulcanized rubber studied:

T (°C):	20	40	60	80	100
E (kg/cm^2):	21.5	19.3	18.5	18.2	18.0

Comparison of the experimental and theoretical curves for the dependence of the real contact area upon temperature for an elastomer gives good results (22).

In using Eq. (4.59), we should make the following observation. According to this equation, the number of molecular contacts can also increase with temperature. All is determined by the correspondence between the activation energies U_1 and U_2. For $U_1 = U_2$, the number of molecular contacts generally does not depend upon temperature. Besides, the use in Eq. (4.59) of the expression for frequencies of the type $\nu = \nu_o \exp(- U/kT)$ for low temperatures is incorrect, because the frequencies are not proportional to the Boltzmann constant in $\exp(- U/kT)$. This is because the Boltzmann distribution generally is inapplicable to energies, and therefore it is better to examine molecular transfer from one position to another from a more general point of view, namely phonon statistics. The latter was developed for polymers by Holzmuller (41,42), who obtained for frequencies the equation

$$\nu = \nu_o \exp[(- \Delta U/kT)(\lambda_p/2\pi\alpha)] , \qquad (4.61)$$

where λ is the wavelength of a phonon and α is the distance between molecular segments.

Because the factor $\lambda_p/2\pi\alpha > 1$ for the transfer of a segment from one equilibrium position to another, there must be a kinetic energy that exceeds the magnitude ΔU. It is possible to present this conclusion differently by considering the temperature $T^* = T(2\pi\alpha/\lambda_p)$, which is less than the thermodynamic temperature T. In other words, as a result of intermolecular interactions, the thermal energy kT will be greater than the effective thermal energy kT* that the segments have. Such a modification in the equation for frequency leads to interesting and important consequences: it proves possible (43,44) to explain Ferry's equation (see Chap. 1) by Frenkel's theory (6). In view of the above, analysis of the temperature dependence of the molecular contact area is clearly difficult

The dependences of the real contact area upon time, velocity, pressure, and temperature that we examined are fairly well explained by phenomenological and molecular-kinetic concepts. The theoretical relations revealed are correct only to a first approximation, because they were obtained for simple models, and do not take into account the spectrum of relaxation times, supermolecular structures, and roughness of the hard surface. Comparison with experiment, however, allows us to conclude that the theoretical dependences reflect fundamental processes that determine the value of the real contact area for given external conditions. In the coming sections, it will be shown that the principles obtained can be taken as the basis of interpretation for dependence of the friction force upon pressure and contact time.

4.3 Friction Force

Let us examine the effect on the friction force of the following factors: the contact time and application rate of the tangential force; the nature of the polymer, the character of the surface, and the stress state of the polymer.

Effect of normal load

At a constant sliding velocity and temperature the fundamental equation for the molecular-kinetic theory of friction can be transformed into the form [see (4.1)]

$$F_f = n(2kT/\lambda j)I_1^*(\tau v/\lambda_j) \ , \qquad (4.62)$$

where I_1^* is the inverse Bessel function.

From Eq. (4.62) it is evident that the friction force is proportional to the number of contacts or the number of segments, n. The friction force depends strongly upon the normal load, while the energy barrier does so weakly. The energy barrier depends upon molecular forces of adhesion. Often, the adhesion forces surpass the strength of the polymers themselves, the molecular forces of cohesion being on the order of 10^4 kg/cm^2 (45). The specific molecular adhesion $P_M \approx 10^4$ kg/cm^2 or somewhat less.

Later, we shall propose that for polymers the effect of the normal load on the cohesion force, and consequently also on the potential barrier, is small. In consequence of this, $dU/dp_r = \omega$ can be considered a small magnitude, and higher derivatives can be ignored. Viewing U as a function of real pressure P_r and expanding Eq. (4.62) in series, the following linear approximation can be made if $P_r < P_M$:

$$F_f = n_c(2kT/\lambda_j)(I_1^*)_{p_r=0} + n_c\nu(2\tau_o/\lambda_j)\exp(U_o/kT)(I_1^*)_{p_r=0}\omega p_r \ , \quad (4.63)$$

where U_o is the potential barrier for $P_r = 0$. Substituting $P_r = P/n_c$, $n_c \approx S_r$, we write Eq. (4.63) in the form

$$F_f = c_1 S_r + c_2\omega N \ , \quad (4.64a)$$

or

$$F_f = (c_1 + c_2\omega p_r)S_r \ , \quad (4.64b)$$

where the coefficients c_1 and c_2 depend upon the sliding velocity, temperature, and molecular constants.

From Eq. (4.64), it follows that relation F(N) is determined for small values of ω by the dependence of the real contact area upon normal load. Furthermore, the load can act directly on the friction force. This is possible in two cases: when S_r = const., and when the second term in the right side of Eq. (4.64) is comparable to the first in magnitude. Equation (4.64) is close to Deryagin's law in physical significance; according to Deryagin's law, the friction force also depends directly upon p when S_r = const. However, the molecular constants in these two formulas are substantially different. It is particularly important to emphasize that c_1 and c_2 in Eq. (4.64) depend upon velocity and temperature, whereas in Deryagin's theory the friction force in essence does not depend upon velocity and temperature.

Figure 4.32 shows the function F(p) according to Eq. (4.64).

It has been shown (46,47) that for pressures from 1 to 150-200 kg/cm^2 the friction constant c_1 does not depend upon pressure. Here $c_2 = 0$. Consequently, over a fairly wide range of pressures the load dependence of the friction force for elastic materials is determined by the load dependence of the contact area.

It is known (47,48) that the real contact area in rubbers is satisfactorily described by an exponential Eq. (3.5); see Fig. 4.33. The law of friction can be expressed thus:

$$f = c[1 - (1 - \phi_0)\exp(- \alpha p)] \ . \tag{4.65}$$

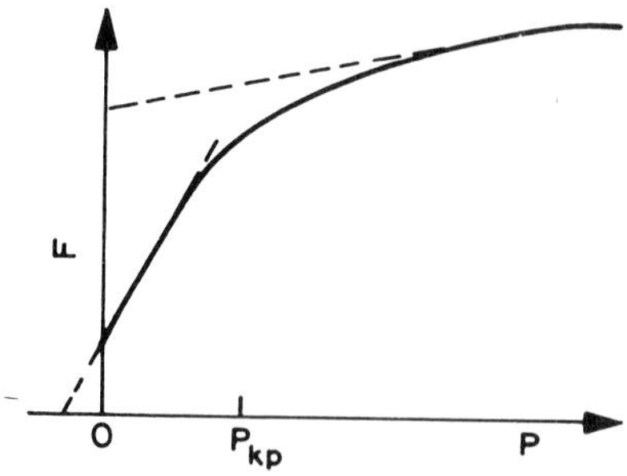

Fig. 4.32 Dependence of friction force upon normal pressure according to Eq. (4.64) (46).

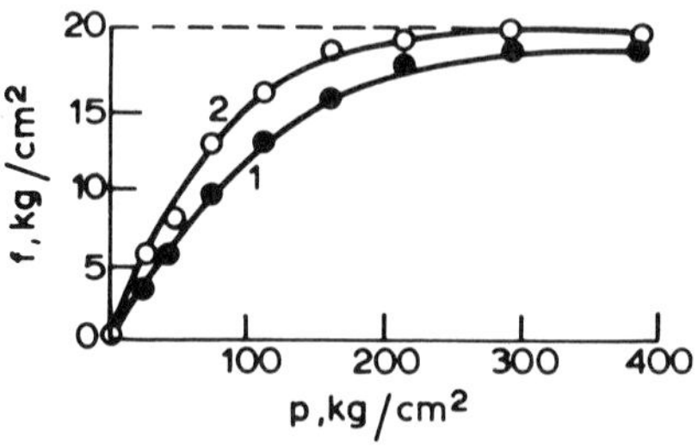

Fig. 4.33 Dependence of the specific friction force upon normal pressure for a vulcanized natural rubber on brass (48). (1) Initial friction (at the moment of shear); (2) steady friction (v = 0.02 mm/min).

For small values of α, we obtain the equation

$$f = c[\phi_0 + (1 - \phi_0)\alpha p] \ . \tag{4.66}$$

For $\phi_0 = 0$,

$$f = c\alpha p \ . \tag{4.67}$$

Equation (4.66) is analogous to Coulomb's law and is conformed for small p by many experimental data (2,3,7,49,50). Equation (4.67) is analogous to Amonton's law and is usually not confirmed for rubbery materials because $\phi_o \neq 0$. For more rigid polymers, however, Eq. (4.67) agrees rather well with the experimental data (3).

It is known that for rubber in the range of pressures 5-50 kg/cm^2 Thirion's empirical formula is applicable (56):

$$1/\mu = A + Bp \ , \tag{4.68}$$

where A and B are constants. For low pressures, this equation contradicts Coulomb's law and experimental data (49). Assuming, as before, $f = c\phi$, it is possible to write Eq. (4.68) as

$$\phi = [dp/(1 + \alpha p)] \ , \tag{4.69}$$

where $\alpha = 1/Ac$, $b = B/A$, $B = 1/c$. In some cases (20), the real contact area depends upon pressure in conformity with Eq. (4.69). The relaxation theory of friction [Eq. (4.27)] in this case reveals the physical significance of the constants a and b in Eq. (4.69).

Equation (4.69) is widely used in the study of polymer friction and wear. However, it has not been given a satisfactory physical explanation, since it is necessary to determine the constants by experiment for each specimen, a task that restricts the use of this equation for engineering calculations.

It would be interesting to derive an equation to describe the relation of the real contact area to the load for the range of values $P/E \gg 2$. An analysis of the contact process indicates that the surface layer of the specimen suffers a complicated stress, which is determined by the size of the load. Moreover, for small loads it is necessary to take into consideration the geommetry of the contacting surfaces (69).

From the change in the stress of the specimen's surface layer on contact, one can estimate the change of some macroscopic characteristics, for example, the modulus of elasticity. It is known that generally the moduli of the specimen's surface layer and of the bulk differ (Fig. 3.2).

With a calculation of these data, the original differential equation [Eq. (3.4)], which was used to derive Eq. (4.65), can be written in the form

$$d\phi/dP = \gamma(1 - \phi)(1/E(P)) \ .$$

Assuming for the parameter E(P) the linear relation to pressure

$E(P) = E_o + \alpha P$,

where P is the pressure and E_o is the modulus of elasticity for P = 0, we have

$d\phi/dP = \gamma(1 - \phi)[1/(E_o + \alpha P)]$.

Solving this equation, we obtain

$$\phi = [((E_o + \gamma P)^{\gamma/\alpha} - C)/(E_o + \gamma P)^{\gamma/\alpha}], \quad (4.70)$$

where C is the integration constant, determined from boundary conditions.

Equation (4.70) enters into Thirion's formula (6.49) for $\gamma/\alpha = 1$; here C = E. In this case, constant α in Eq. (4.69) is related to coefficient and the modulus E_o by the relation (see Eqs. (2.4) and (3.5)):

$$\alpha = \gamma/E_o . \quad (4.71)$$

The accuracy of the above assumption is confirmed by results obtained from experimental data.

Figure 4.34 shows the relation of coefficient α in Thirion's formula to the specimen's degree of roughness γ. The relation is linear, in complete correspondence with Eq. (4.71). From Eq. (4.71), it follows that the cotangent of the angle $\alpha = f(\gamma)$ with the axis of the abscissa determines the modulus. The calculated value of the modulus is 18.6 kg/cm^2, which is close to the actual modulus for the specimen's bulk. The relation of the contact area to pressure for a unfilled crosslinked rubber from SKN-40 rubber with a different degree of surface roughness is taken as a basis for calculation.

The relation of the coefficient α to $1/E_o$ is linear (Fig. 4.35); the coefficient γ, determined by the tangent of the angle $\alpha = \gamma(1/E_o)$ with the axis of the abscissa, is equal to 1.2. This magnitude agrees with a value determined by another method. Given in Fig. 4.35 are data obtained from the pressure dependence of the contact area for an unfilled crosslinked natural rubber with a different modulus (70).

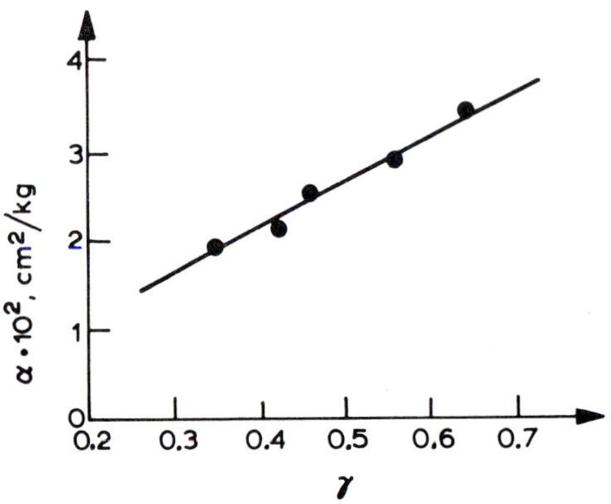

Fig. 4.34 Relation of coefficient α to the degree of roughness of the specimen, γ, for unfilled crosslinked SKN-40 rubber (60).

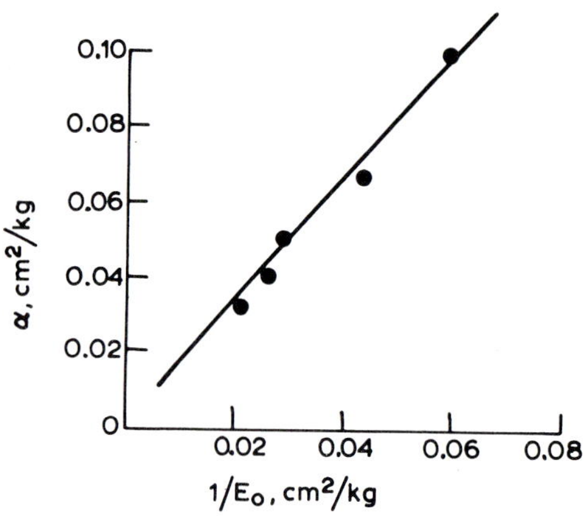

Fig. 4.35 Relation of coefficient α to $1/E_o$ for unfilled crosslinked natural rubber.

The first investigations into the dependence of the friction force upon pressure for elastic materials over a wide range of loads were published in our studies (47,48). Figure 4.35 shows some results, from which it is evident that for steady friction and loads that produce pressures greater than 250 kg/cm^2, the friction force attains its maximum value and then is practically independent of load. From this it follows that it is possible to disregard the second term in Eq. (4.64). The friction constant c here does not depend upon pressure.

For steady friction with loads above 250 kg/cm^2, the specific friction force $f_n = F/S_N$ is constant, because the real contact area is practically equal to the nominal contact area $S = S_N$. Here $f = c$.

Figure 4.36 shows data (39) for steady friction in a plot in which, according to Eq. (4.68), the relation of the inverse friction coefficient $1/\mu$ to the pressure $p = P/S_n$ should be expressed by a straight line. Divergences from this formula for small values of p are noted elsewhere (48,51). Divergences under large loads, related to attainment of the real contact area, occur earlier than follows from Thirion's formula. Here, the value of the maximum specific friction force will be constant, which agrees with the equation $f = c$ (Fig. 4.36). From the slope of straight line 2, it follows that the friction constant $c_f = 20.3$ kg/cm^2. Curve 3 is experimental.

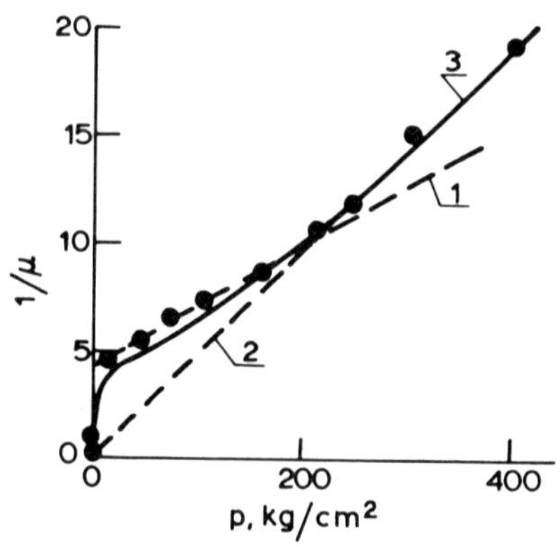

Fig. 4.36 Data of Fig. 4.33 plotted for comparison with Eq. (4.68) (39).

In all load ranges, the experimental data (Fig. 4.36), when plotted as $\log(cS_n - F)$ vs p (Fig. 4.37), satisfy Eq. (4.65). Consequently, Eq. (4.65) is a more exact form of the law of friction for elastomers. From the slope, it follows that the coefficient $\gamma = 0.17$. The coefficient's small value indicates that the friction surfaces had little roughness.

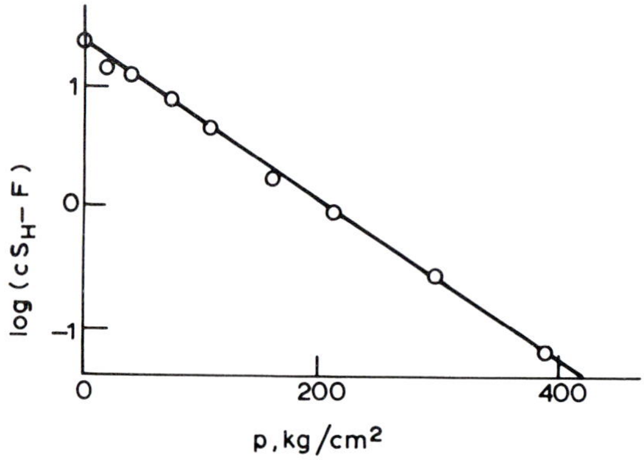

Fig. 4.37 Dependence of the friction force upon pressure (according to data of Fig. 4.36).

For pressures at which p >> Eγ, Deragin and Toporov (52) showed experimentally that the dependence of the friction force upon normal load is expressed by the binomial law of friction:

$$F = \mu_o(p + N_o) .$$

The polymer specimens took the form of films in thickness ≈ 100 μm. In this account (52), however, it was not indicated how change in film thickness with increase in pressure was evaluated, how static friction was measured after deduction of elastic deformation shear in the same polymer, etc. Furthermore, in spite of irrefutable experimental and theoretical proof that the relations S(N) and F(N) for loads up to 200-300 kg/cm^2 have the form shown in Fig. 4.33, results obtained by Deryagin and Toporov (Fig. 4.38) do not testify to this.*

These similar cases lead to the necessity of repeating Deryagin's experiment, using a method free from the deficiencies of the method used by Deryagin and Toporov. Relations F(P) were investigated for a series of vulcanized rubbers at loads from 50 to 1000 kg/cm^2 on an instrument for measuring volume compression (see Chap. 7). On this instrument, it is possible to study simultaneously the effects of pressure on the friction force and on the compressibility of crosslinked rubber. It is known (53) that in this range of

*In all probability, the formation of the contact area for thin films begins at low pressures.

pressures the compressibility of vulcanized rubbers is 2.6×10^5 cm^2/kg. The experimental results for two vulcanized rubbers and plastics are shown in Fig. 4.39. In the range 400-500 kg/cm^2 an inflection in the curves is observed, indicating a change in the relation. For pressure above 800 kg/cm^2, a rather sharp increase in the friction force for vulcanized rubbers is observed with rise in pressure.

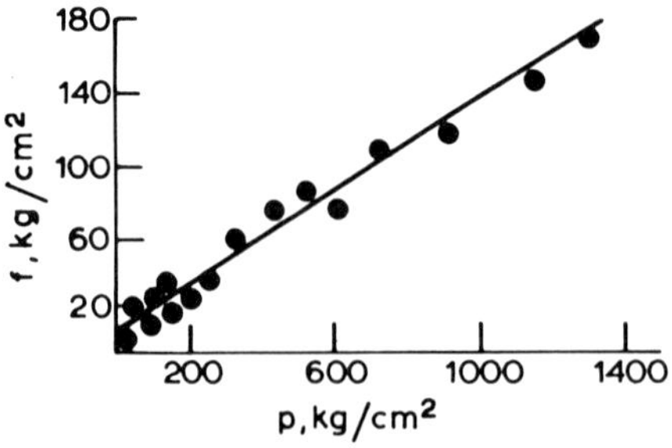

Fig. 4.38 Relation of the friction force to pressure for a nonfilled vulcanized rubber on steel (52).

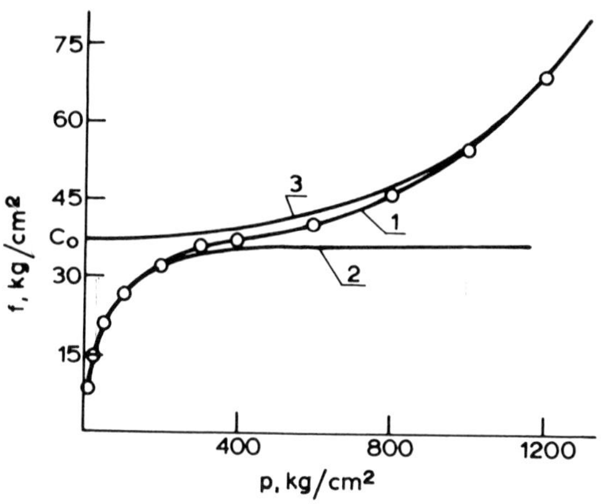

Fig. 4.39 Relation of the friction force to pressure for vulcanized SKN-40 on steel: (1) experimental curve; (2) theoretical curve from Eq. (4.65); (3) calculated curve for friction constant c.

For pressures of 10-200 kg/cm^2, change in the real contact area is the predominant factor; the curves have a tendency to saturate.

The direct effect of pressure on the friction force accompanies the saturation. Thus, for all loads the friction force is determined by Eq. (4.64). The direct effect of pressure on the friction force is related to Deryagin's law of friction. A direct correspondence is observed between the slope of the relation f(p) and the compressibility: the greater the compressibility, the steeper the slope for high pressures. The calculation of the pressure by the first approach, following Frenkel (6), can be accomplished in the following manner.

Change in the polymer volume ($V_o - V$) leads to an increase in the activation energy:

$$U = U_o + k(V_o - V) ,$$

where V_o is the polymer volume for p = o. Expressing the change in volume ($V_o - V$) in terms of the pressure and compressibility κ, we obtain $V_o - V = \kappa V_o p$; thus

$$U = U_o + k\kappa V_o p , \qquad (4.72)$$

where U_o is the activation energy for p = o. Then the constant $\omega = dU/dp$ from Eq. (4.64) is equal to $k\kappa V_o$. Consequently,

$$f = c_1\phi + c_2\kappa V_o p . \qquad (4.73)$$

Equation (4.73) can explain the relation f(p) in all ranges or pressures. But f(p) as expressed by this equation has more than theoretical significance. As a result of the sharply increase of pressures in systems with fillers (to 1000 kg/cm^2 and higher), it is necessary to take into account the significant increase in the friction force for high pressures. According to Eq. (4.73), if other conditions are equal, this is related to the compressibility coefficient for vulcanized rubbers. High compressibility (κ = 2.5 x 10^5 cm^2/kg) can also explain the fact that F(p) for higher pressures is greater in vulcanized rubbers than in rigid polymers (for PTFE κ = 1.6 x 10^{-5} cm^2/kg).

For small loads, the change in the friction force is determined by the increase in the actual contact area (Curve 2 in Fig. 4.37); for large loads the specific actual friction force or friction constant c is changed (Curve 3 in Fig. 4.39) (54).

Ratner and Sokolskaya (50) also investigated the law of friction for small loads. They gave the following formula for the law of friction:

$$F = \mu_o N_o + \mu_\infty p \ ,$$

or

$$\mu = \mu_\infty + [(\mu_o N_o)/\underline{p}] = \mu_\infty + (F_o/\underline{p}) \ , \qquad (4.74)$$

where F_o is the friction force for $\underline{p} = 0$, μ_∞ is the friction coefficient for $\underline{p} \to \infty$, and N_o is the intermolecular attractive force.

Ratner and Sokolskaya observe that their experimental data do not afford a basis* for writing the law of friction as Deryagin's formula, because $\mu_o \neq \mu_\infty$. Thus, for example, they demonstrated that during testing the value of F_o for the same vulcanized rubber remains constant on different surfaces (metals) when $\mu_o \neq \mu_\infty$.

Using μ_∞ and F_o from the vulcanized rubber's structure and its rigidity in Eq. (4.74), Ratner and Sokolskaya investigated the effect of the type of raw rubber and quantity of fillers on "static" friction. The type of rubber determines the intermolecular interaction, and the quantity of components affects the hardness of the polymer. The value of F_o depends upon the contact area, and consequently with increase in the modulus at constant pressure this value ought to fall. The modulus is easily controlled by varying the filler.

Later, in a study of soft unfilled crosslinked rubbers, Ratner and Sokolskaya showed that the relation

$$\mu = \mu_\infty + [B/(p^{1-h})] \ , \qquad (4.75)$$

where h is the exponent that characterizes hardness, better describes the experimental data.

Equations (4.74) and (4.75) are empirical and can be used only for a qualitative description of the law of friction for small loads. It has since been shown (49) that the "constant" μ_∞ is not a constant but depends upon load.

---

*This is fully understandable because in this range of pressures the actual contact area is unsteady.

Thus, the law of friction is determined by the pressure dependence of the actual contact area. For high pressures, when S = const., it is necessary to take into account the direct effect of pressure on the friction force (46,68). The law of friction in the form of a binomial function (4.73) is the most general and matches experimental data for a wide range of pressures.

Effect of contact time and the rate of application of the tangential force

The usual method of measuring the static friction force consists in determining the force at which the transfer of a body from rest to motion is observed. This method, related to the presence of absolute static friction, is employed not only in studying metal friction but also in studying vulcanized-rubber friction. However, as experiments reveal, for vulcanized rubbers the amount of static friction is not absolute, but depends upon many factors, above all experimental conditions (51). The determination of the static friction force for metals is not difficult, becuase usually it is larger than the kinetic friction force. For vulcanized rubbers, the kinetic friction force is larger than the static friction force; therefore, it is difficult to establish the moment of a specimen's transfer from the static state to motion.

Figure 4.40 shows the determination of the static friction force on a pendulum tribometer (see Chap. 7) at short contact times. As seen in the figure, the determination of static friction

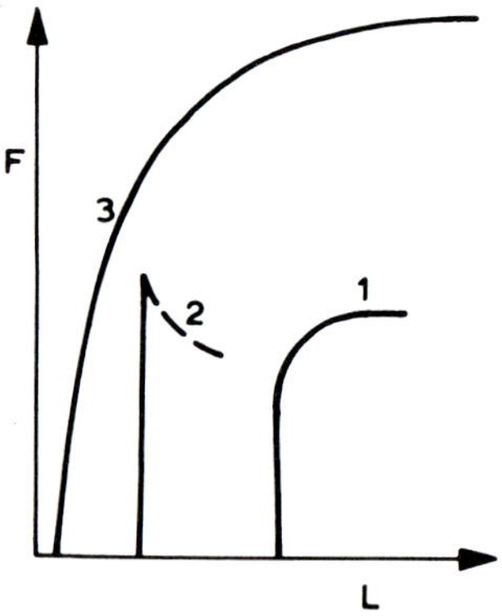

Fig. 4.40 Relation of the static friction force to distance of sliding: (1) vulcanized rubber under small loads, (2) steel, (3) vulcanized rubber under large loads (on steel).

force on steel is made rather simply: the maximum friction force is taken (curve 2). The determination of the static friction force of a vulcanized rubber is somewhat difficult. For small loads (curve 1) it is possible in practice. But for large normal loads, it is impossible to obtain a value for the transitional friction force, which corresponds to the beginning of the specimen's displacement, because for any tangential force sliding of the specimen is observed. The character of the curves also depends upon the stationary contact time.

It is necessary to note that difficulties in the determination of the static friction force are not merely due to some methods or instruments, but are inherent in the molecular situation. For vulcanized rubbers the static friction force always depends upon a number of factors. Figure 4.41 shows curves for the determination of the friction force on a pendulum tribometer by the usual method (curve OABC) or at the time when the tangential force ceases to grow at points A_1, A_2, A_3, A_4. The latter is achieved by shutting off the motor for a short time, while the tangential force is created only through the gravity force on the displaced pendulum.

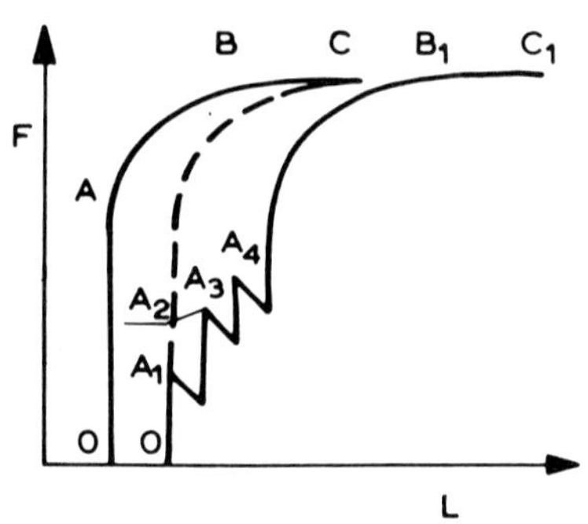

Fig. 4.41 Relation of the friction force to distance of sliding for vulcanized rubber on steel. At points A_1, A_2, A_3, A_4, the pulling force is discontinued.

Shutting off the motor at point A_1, we should obtain a static condition of the specimen mounted on a rigid block. However, sliding of the specimen is observed, because as the pulling force decreases, so does the slope of the pendulum. Restarting the motor again exactly after two minutes of rest, we obtain an increase in the tangential force to point A_2. Repeatedly switching off of the motor, we obtain the observed pattern of change in the friction force.

The experiment described reveals that the static friction force is not constant but depends upon the observation time. Point A (Fig. 4.30) does not correspond to the static friction force, but represents a conditional characteristic of initial friction, obtained by a fixed rate of application of tangential force. An identical experiment was performed for friction of steel on steel. No change in the friction force after cutting off the pulling force was noted for any observation time (from a second to several days).

Proceeding from the dependence of the initial friction upon observation time, it is possible to surmise the effect on the initial friction of the tangential force and its rate of application (that is, the increase in tangential force applied to the specimen for unit time). The rate of the application of the tangential force to the specimen depends not only upon the traction rate, but also upon the stiffness K of the dynamometer that measures the friction force. This stiffness is determined by the force (in kg) necessary for a given deformation of the dynamometer in mm. The relation between the rate of application of tangential force and the stiffness of the dynamometer is seen in the following observations. Usually, the friction force is determined by the extent of deformation of the spring or force arrangement. If between the specimen and the drum, which stretches the fiber when it rotates, there is a rigid bond (for example, steel wire), then the initial velocity of the specimen's displacement in practice is determined by only the drum's rotation velocity. The specimen at once passes from the static state to sliding at the given velocity. For such a bond, the transfer of the necessary tangential force to the specimen occurs almost instantaneously. When the dynamometer is placed between the specimen and the connecting rod, the traction force is applied to the specimen with a velocity that depends upon the stiffness of the dynamometer. If the stiffness is very low, a long time is needed to deform the spring sufficiently to produce a visible shear in the specimen.

Some results of an investigation into the effect of stiffness of the dynamometer on the initial friction force are shown in Fig. 4.42 (55). The preliminary specimen displacement, during which the initial friction force was measured, was set at 0.15 mm. The friction force of vulcanized SKS-30 butadiene-styrene rubber sliding on steel was measured, as was the mean rigidity of the dynamometer. An instrument of type TPM with various dynamometers and a pendulum tribometer with various loads were employed for measurement of the friction force. The pendulum tribometer's limits of stiffness are shown in Fig. 4.42 by arrows.

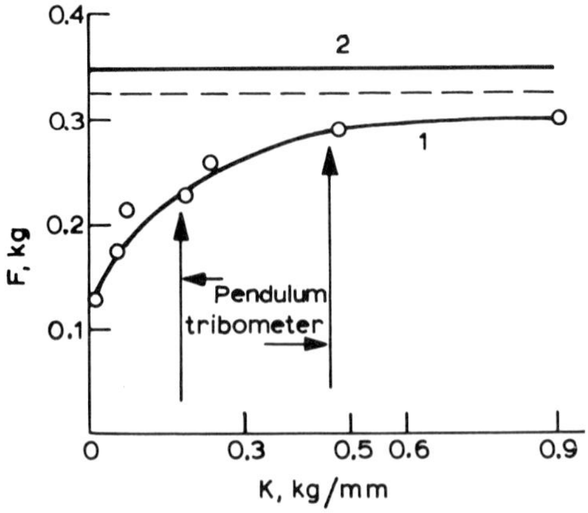

Fig. 4.42 Dependence of a vulcanized rubber's initial friction force upon stiffness of the dynamometer K´ (55): (1) initial friction, (2) kinetic friction.

As seen in the figure (curve 1) with increase in the dynamometer's stiffness the initial friction force increases and tends toward the value measured by the dynamometer of a tensimetric system which has very high stiffness. The friction force obtained with the aid of this dynamometer is shown in Fig. 4.42 by a dotted line.

The experiments discussed above indicate that the initial friction force depends upon the rate of application of the tangential force. It is interesting to note that the value of the friction force, measured by the tensimetric dynamometer, is close to the value of the kinetic friction force. The kinetic friction force (curve 2) does not depend upon the dynamometer's stiffness. This circumstance allows us to propose that for an absolutely stiff dynamometer and for the whole system generally, the magnitude of initial friction is extremely close to the kinetic friction at a steady sliding velocity.

The effect of stiffness on the magnitude of initial, or apparent, static friction is explained by the fact that from the moment of application of the tangential stress the specimen acquires a certain sliding velocity, which is very insignificant under small tangential stresses (55). Measurement of the initial friction force is based on the fixed friction force for a known displacement of the specimen; therefore, for exact (or even meaningful) measurement it is necessary to take into account the component of general displacement due to elastic displacement of the specimen (by means of elastic shear throughout the specimen). In some cases elastic shear can conceal the actual sliding of the specimen. Thirion has also noted this effect (56).

Figure 4.43 shows the friction force on a vulcanized SKN-18 butadiene-acrylonitrile rubber for various heights of a specimen projecting from the ring. The initial friction force was measured at a displacement equal to 0.15 mm. As seen in the figure, for h > 1 mm the initial friction force fell with increase of the specimen height (curve 1), but the kinetic friction force (curve 2) waş constant.

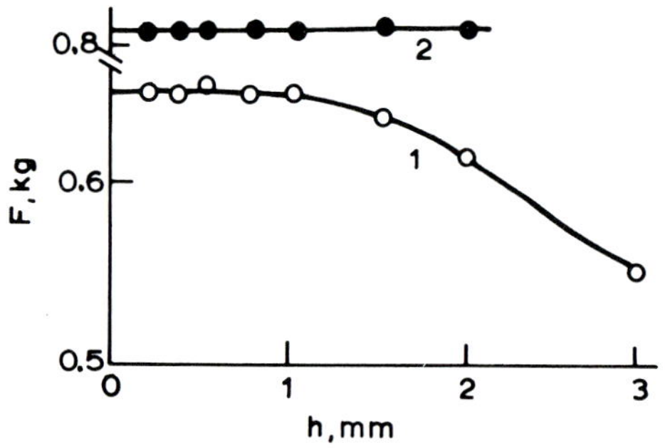

Fig. 4.43 Dependence of the friction force upon height of the projecting part of the specimen for SKN-18 a vulcanized butadiene-acrylonitrile rubber: (1) initial friction, (2) kinetic friction.

If a vulcanized rubber specimen is glued to a hard surface and a tangential force is applied to it, then the specimen will be subjected to elastic displacement only. With subsequent application and removal of tangential stress, this method allows the isolation, at a fixed initial friction, of a specimen's elastic and visco-elastic displacement from its sliding.

Figure 4.44 shows the friction force for specimens of a crosslinked natural rubber as a function of the rate of application of the tangential stress (55). Each curve consists of rectilinear and curvilinear sections. The rectilinear section characterizes the dependence of the tangential force upon the specimen's elastic shear, which is identical for all velocities. For values of the tangential stress at which divergence from the straight line is observed, it is evident that the initial friction force is different and depends upon the rate of application of the load. With increase in this rate, the initial friction force increases.

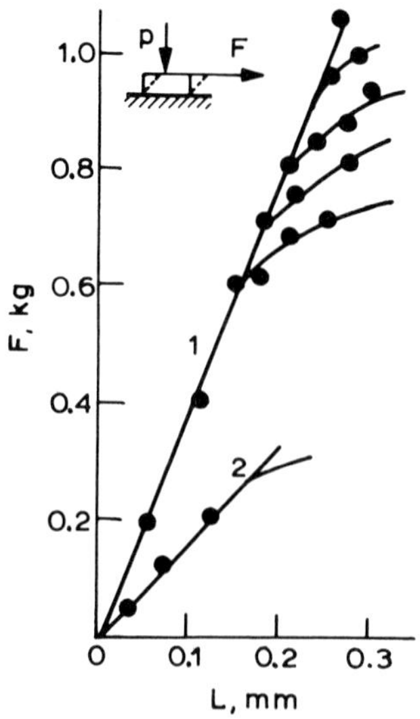

Fig. 4.44 Dependence of the initial friction force upon the shear distance L for a vulcanized natural rubber against steel at the pressures (1) 1.4 kg/cm^2; (2) 0.9 kg/cm^2.

If we plot the initial friction force against the logarithm of the rate of application of tangential stress, we obtain a straight line. This indicates the kinetic nature of static friction in visco-elastic polymers.

It is also possible to investigate the relation of contact time to friction force, if one studies the effect on the friction force of the so-called overload. Overload is the additional normal load acting on the specimen for a period of time and is not used in the determination of the friction force (the overload is removed just beforehand). For visco-elastic materials, overload plays a significant role and strongly affects the magnitude of the friction force.

Figure 4.45 shows the relation of the initial friction force to the extent of the overload (58). A polymer of average rigidity was studied (hardness of TM 2). The friction force was measured on a pendulum tribometer. As seen in the figure, the initial friction force increases with overload. In the case of small normal loads (curve 1) the friction force grows to a larger degree than for large normal loads (curve 2). For normal loads of 1.5 kg (curve 3), the friction force is practically independent of the extent of the overload.

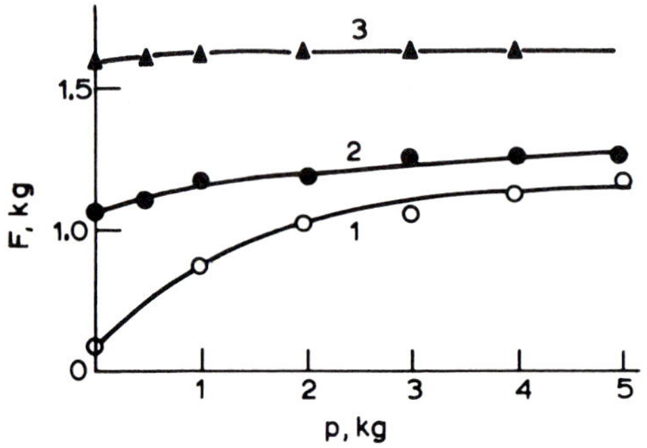

Fig. 4.45 Dependence of the initial friction force upon overload p' for a vulcanized rubber (64). Normal loads: (1) 0.4; (2) 0.9; (3) 1.5 kg. Time of preliminary contact, 1 min; load action time, 3 min.

Overload enlarges the contact area with time. For a rather long time after removal of the overload, its effect on the friction force is not noticed. This shows that the role of overload is based on relaxation processes that proceed on the surface of the specimen and affect the contact area. After a sufficient "rest", transient relaxation processes arise in the actual contact area.

Experiments also demonstrate that for the determination of the initial friction it is very important to take into account the preliminary contact time and for prolonged contact time to pay attention to the formation of stable chemical bonds (25).

In response to the preliminary contact time, the initial friction force can be smaller than the steady friction force (48). Further studies have demonstrated (16) that it is useful to distinguish the initial friction force F_o, the maximum unsteady (or irregular) friction force F_{max}, and the steady friction force F_∞ (Fig. 4.46). The numerical relationship among them depends upon the polymer's physical state and its temperature.

Studies have been conducted over a wide range of temperatures (from -100°C to +40°C) on initial and steady friction forces in two modes of operation (in the first mode the requisite temperature was

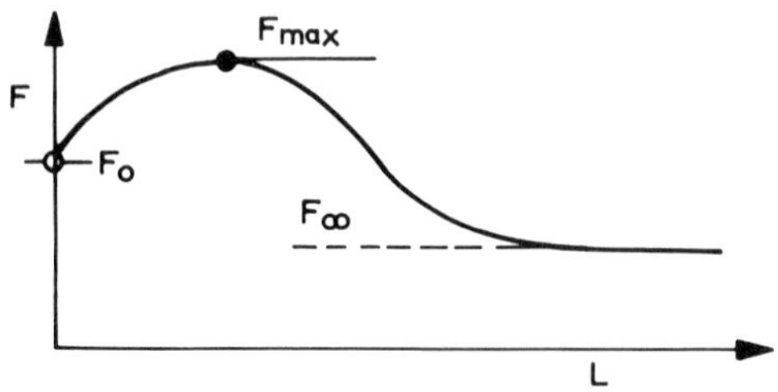

Fig. 4.46 Relation of the friction force to the sliding distance (16).

attained during a polymer's contact with a base, and in the second mode the temperature was attained before contact), showing the the friction force generally depends upon temperature as shown by a curve with a maximum (Fig. 4.47). In curves of $F_f(T)$ it is possible to note four characteristic sections: (1) the high-temperature section where the maximum and steady friction forces coincide; (2) the transitional one where all friction characteristics have different values; (3) the transitional one where initial and maximum friction forces coincide; (4) the broad low-temperature section where all three friction forces practically coincide. Analogous relations are observed during testing in the second mode, which produced a larger friction (16,19).

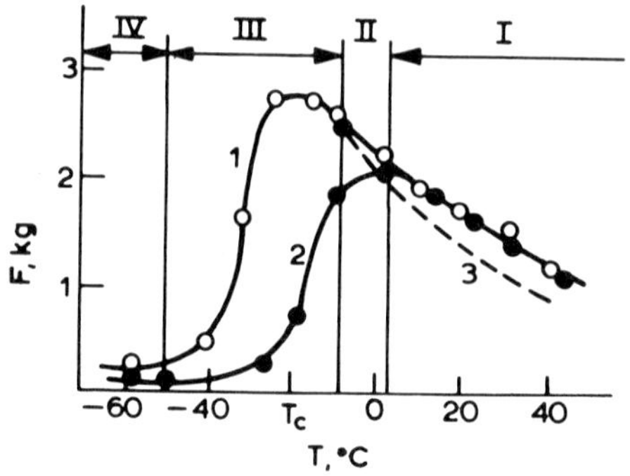

Fig. 4.47 Dependence of the friction force upon temperatures obtained in the first mode for a vulcanized SKN-40, butadiene-acrylonitrite rubber against steel: (1) maximum friction; (2) steady friction; (3) initial friction; v ≈ 1 mm/min; p = 0.65 kg/cm.

Increase in the steady friction force with decreasing temperature is connected with an increase in intermolecular interactions in the polymer-solid pair (second mode). For a temperature close to T_g

(first mode), as a result of the difference in the coefficients of linear expansion of the polymer and its base, the actual contact area separates, which leads to an abrupt decrease in the friction force. When one changes to polymers with lower T_g values, the temperature of the maximum friction force drops. For example, for butadiene-acrylonitrite copolymers SKN-40 ($T_g = -20°C$), SKN-26 ($T_g \leq -30°C$), and SKN-18 ($T_g = -50°C$), the largest value F_{max} is observed respectively for temperatures of -26, -35, and -53°C.

The maximum initial friction force also depends upon the rate of application of the tangential force (57). For friction of a vulcanizate on a vulcanizate, the maximum friction force increases with decreasing temperature (in the first mode). Thus, the initial friction force depends upon the mode of testing and the temperature and nature of the contacting bodies. The process of constant polymer deformation (ε = const.) is another extremely important factor in external friction (chiefly in elastic materials). In this case, that contact stress at the time of contact depends upon the polymer's modulus and the extent of strain.

The temperature dependence of the initial friction force for $\varepsilon = 20\%$ is characterized by a curve with a maximum. Here, with increase in the rubber's polarity, F_{max} shifts to the high-temperature side. A sharp decrease in the friction force for temperatures below T_g is explained by fracture in the actual contact area and decrease in normal load due to contraction stresses, which can be calculated by the equation (16,19)

$$\sigma = \int_{T}^{T_\eta} [(E(T))/(1 - x(T))] \ (\alpha_1 - \alpha_2) dT \ , \qquad (4.76)$$

where α_1 and α_2 are the coefficients of linear expansion for the polymer and its base, E is the modulus, and x is Poisson's ratio. The locus F_{max} does not depend upon the amount of deformation (Fig. 4.48), and is determined only by the polymer. The F_{max} occurs at the glass temperature of the polymer. For temperature above T_g an increase in the friction force with decreasing temperature is explained by the molecular-kinetic theory.

The special properties we have noted above reveal that the initial friction force for ε = const. has a molecular-kinetic nature.

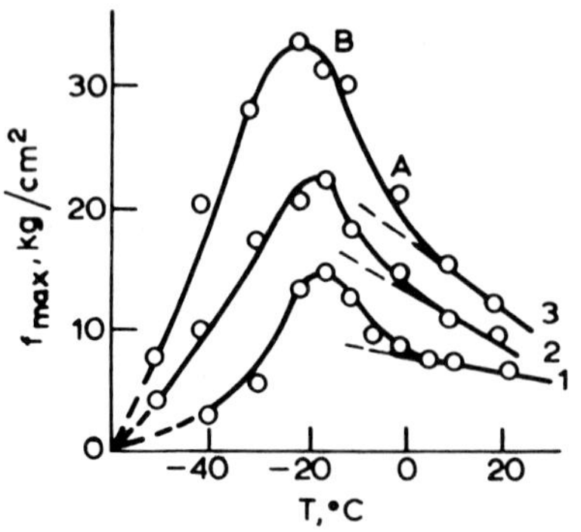

Fig. 4.48 Dependence of the initial friction force upon temperature for various deformation compressions of a vulcanized rubber: (1) ε = 10%; (2) ε = 20%; (3) ε = 30%.

Effect of the nature of the polymer and the characteristics of the surface

The contact area and the friction force are strongly influenced by the modulus of elasticity, which appears in Eq. (3.5) as a power of e. Figure 4.49 is an example of the dependence of the friction coefficient upon the inverse modulus. Furthermore, the value of the pressure P_N for which $S \approx S_N$ is, according to some data (20), proportional to the modulus: $p_N \approx 4E/\beta$. Given the modulus, this permits a fairly exact prediction of the value of p_N at which the friction force ceases to depend upon load.

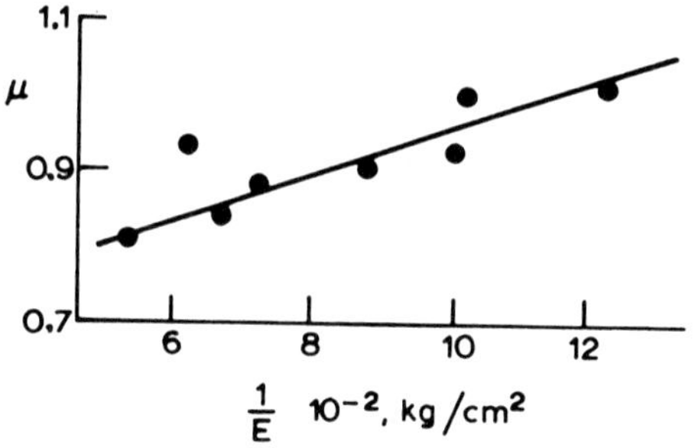

Fig. 4.49 Dependence of the friction coefficient upon the reciprocal of the modulus for vulcanized SKN-18 butadiene-acrylonitrite rubber.

The viscoelastic modulus E determines the dependence of the actual contact area upon time. According to Slonimsky and Rogovina (59)

$$E(t) = E_{\infty} + E_1(0)\exp[(\alpha t)^h] , \qquad (4.77)$$

where α and h are constants that characterize the polymer's relaxation properties, and $E_1(0)$ is the relaxation stage of the modulus at the initial moment t = 0. This relaxation is close to the experimental one (30).

As has been demonstrated (50), a polymer's modulus can be easily altered by the introduction of fillers. Studies on the effect of fillers on the friction properties of vulcanized rubbers indicate a close connection between polymer mechanical properties and the friction coefficient. The nature of the raw rubber determines the vulcanizate's modulus.

Table 4.1 shows values of the activation energy U for friction of various vulcanized rubbers (20). As shown in the table, the type of rubber and its polarity affect the activation energy. The largest values of the activation energy are observed in friction of a crosslinked rubber on polymers. This means that the polymer-polymer bond is more wear-resistant than a polymer-metal bond. Probably this is also explained by the large molecular contact area and interdiffusion of the molecules in the polymer and its base (60).

For friction on rough surfaces, mechanical losses are important, but in this case they are related to bulk deformation. Boggs and Riemen (61) note that between the friction force and hysteresis losses there is an approximately exponential relation.

Tabor (62) studied in detail the deformation component of the friction force in vulcanized rubbers on rough surfaces and demonstrated the important role of hysteresis losses throughout the polymer's bulk. In this case the coefficient of sliding friction with lubrication is proportional to the coefficient of rolling friction. The temperature and velocity dependences of mechanical losses determine these dependences for the friction force. Thus, a polymer's mechanical properties determine its friction properties indirectly, through their effect on the real contact area, and directly, through the same material's hysteresis properties. These effects result in a close relation between internal and external losses.

TABLE 4.1

Characteristic friction for pure crosslinked rubbers (based on standard formulas) on steel

Polymer	c (kg/cm^2)	U (kcal/mole)	Sources
Natural Rubber (NK)	19.0	18.5	8
SKS-10 (Butadiene-styrene rubber)		22.3	8
SKS-30 (Butadiene-styrene rubber)		27.2	8
SKBM (Butadiene rubber)		24.0	8
SKN-18 (Butadiene-acrylonitrile rubber)	14.0	24.0	20
SKN-26 (Butadiene-acrylonitrile rubber)	16.7	27.0	20
SKN-40 (Butadiene-acrylonitrile rubber)	18.0	33.0	20

Effect of internal stress in a polymer

The molecular-kinetic theory of external friction in rubbery polymers (5) proceeds from the hypothesis that the mechanism of polymer chain jumping on a hard surface is analogous to the thermal motion of molecules in liquids. The time between jumps which characterizes molecular mobility can be expressed as, according to Frenkel,

$$\tau = \tau_0 \exp(U/kT) \ .$$

If the polymer is in the stressed state, then time τ depends upon the stress σ on the polymer (63):

$$\tau = \tau_0 \exp[(U - \alpha\sigma)/kT] \ , \qquad (4.78)$$

where α is a constant related to the modulus and specific volume of the polymer.

For the action of the friction force,

$$\tau = \tau_0 \exp[(U - \alpha\sigma - \Gamma F)/kT] \ . \qquad (4.79)$$

It is convenient to reduce this equation to the form

$$F = A - B\sigma \ , \tag{4.80}$$

where

$$A = (kT/\Gamma)\ \ln\ (\tau_0/\tau) + U/\Gamma \ , \qquad B = \alpha/\Gamma \ .$$

The friction force, according to Eq. (4.80), ought to depend linearly upon the stress σ on a polymer. As will be shown below, experiment corroborates this relation (64).

For the derivation of Eq. (4.80) only the effect of stress on a polymer's activation energy during friction was considered; the effect of stress on the size of a chain's jumps, λ, was not taken into account. This size, which can be calculated from general concepts, increases along the direction of stretching and decreases in an opposite direction. It is possible to disregard this change in the first approximation.

Besides changing the potential barrier, the application of stress leads to the orientation of polymer chains throughout the bulk and on the surface of the specimen. The orientation effect can be measured (34) by the mean value of the cosine of the angle between the acting stress and the direction of a chain:

$$\cos\alpha = L(z) \approx \coth\ (1/z_4) - (1/z_4) \ , \tag{4.81}$$

where $z_4 = \alpha\sigma/kT$, and $L(z_4)$ is Langevin's function. For $z_4 < 1$, the function $L(z_4)$ is well approximated by the linear term: $L(z_4) \approx (1/3)z_4$. Hence we obtain

$$\cos\alpha = (1/3)(\alpha\sigma/kT) \ .$$

The chain orientation leads to a change in the overall friction force such that the friction force measured along the acting stress is larger than the crosswise force. The friction force can be expressed as the sum of two components, one characterizing the adhesion and the other the deformation. Applied external stress will affect not only the adhesion component, as shown above, but also the deformation component as a consequence of volume anisotropy.

For a quantitative study of the contribution of the two components to the total friction force, some experiments were carried out with smooth and very rough friction surfaces (64). Figures 4.50 and 4.51

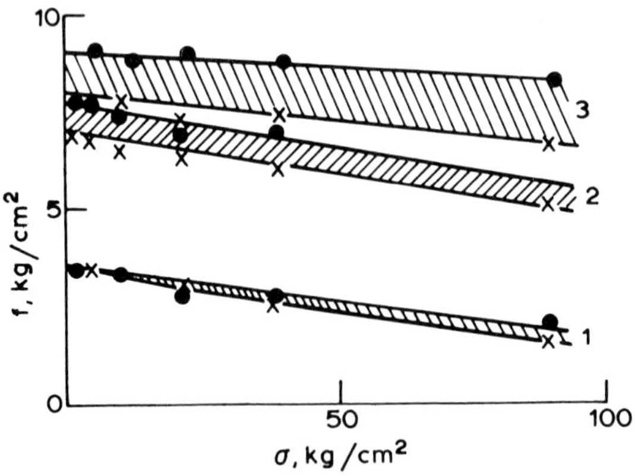

Fig. 4.50 Dependence of the specific friction force f upon real stress σ on a polymer for friction on a smooth steel surface (64): •, along stretched direction; x, perpendicular to the stretched direction. Normal load (in kg): (1) 0.2; (2) 0.4; (3) 0.8.

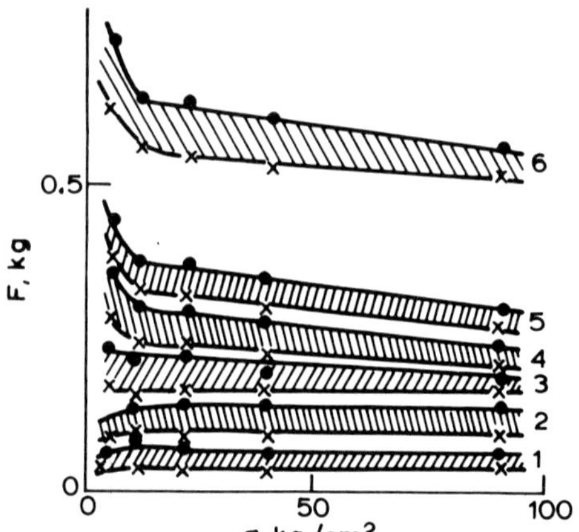

Fig. 4.51 Dependence of the friction force in spherical specimens upon stress on a polymer (64): •, parallel to the stretched direction; x, perpendicular to the stretched direction. Normal load (in kg): (1) 0.2; (2) 0.4; (3) 0.6; (4) 0.8; (5) 1.0; (6) 2.0.

show the results of these experiments. The friction force decreases with increase in load on the polymer. The larger the difference between the friction forces along and across the stretched direction, the larger is the normal load. This is true of friction on both smooth and rough surfaces. The friction force that is measured along the stretched direction is always larger than the friction force across the stretched direction.

For friction on smooth surfaces (Fig. 4.50) for all loads, the friction force depends upon load according to Eq. (4.80). For friction on rough surfaces under small loads, divergence from a straight line is observed (Fig. 4.51). Together with the factors noted above, moisture (55), the nominal contact area, cleanness of the surfaces (see Chap. 7), and surface abrasion affect the friction force in elastic materials, although only weakly.

It follows that despite great advances in the field of polymer friction, physics at the present time is still unable to predict the friction force. This is explained, mainly, by the fact that it is possible to determine experimentally only one of the fundamental characteristics of friction--the activation energy.

To facilitate an evaluation of the friction force when it cannot be determined by one or another principle, Tables 4.2-4.4 give values of the specific real friction force and the activation energy U for various friction pairs.

TABLE 4.2

Effect of the type of crosslinked rubber on the specific actual friction force (friction constant c) (20)

Unfilled crosslinked rubbers		
SKN-18 butadiene-acrylonitrile rubber	Steel Type ($\nabla 8$)	14.0, 27.0*
	Glass	17.5
	Polyvinylchloride	33.0
SKN-26 butadiene-acrylonitrile rubber	Steel ($\nabla 8$)	16.7, 27.0*
	Glass	18.4
	Polyvinylchloride	37.0
SKN-40 butadiene-acrylonitrile rubber	Steel ($\nabla 8$)	18.0
	Glass	10.4
	Polyvinylchloride	40.0

*For a filled crosslinked rubber.

Selection of friction pairs

Contemporary machine-building industries widely use various cured elastomers as friction and antifriction materials to operate in a wide range of sliding velocities, temperatures, and pressures.

TABLE 4.3

Effect of the nature of the friction surface on the specific real friction force for a natural rubber (7,8)

Friction surface	f_r (kg/cm^2)	Friction surface	f_r (kg/cm^2)
Steel (∇8)	18.5	Ebonite	24.5
Glass	21.0	Vulcanized natural rubber	35.0

TABLE 4.4

Effect of the equilibrium modulus on the activation energy of a vulcanized natural rubber for friction on steel (∇8) (7)

E (kg/cm^2)	U (kcal/mole)	E (kg/cm^2)	U (kcal/mole)
3.4	17.1	18.1	18.7
6.4	17.4	23.2	18.9
9.8	18.5	31.0	19.9

To select friction materials, it is necessary to take into account that for other equal conditions the following factors bring about an increase in the friction coefficient μ for vulcanized rubbers:

1. growth in the nominal contact area (p = const.);
2. reduction in the specimen thickness;
3. increase in the polymer polarity;
4. increase in the time of stationary contact, especially under atmospheric conditions;
5. deposit of polymer-film coatings on hard surfaces;
6. decrease in the modulus of elasticity;
7. increase in the sliding velocity and shear (to a fixed limit);
8. decrease of internal stress in the polymer;
9. decrease in roughness of the hard surface.

Many factors are interrelated and can act in a contrary way: for example, increase in polarity brings about an increase in the modulus of elasticity. The friction coefficient of vulcanized rubbers, as we have stated, is primarily determined by the type of rubber; fillers affect it much less (Table 4.5).

TABLE 4.5

Effect of the type and amount of filler on the friction coefficient for vulcanized rubbers (∇8) for v = 1.0 mm/min

Vulcanized rubbers	Filler	Amount of filler (wt %)	Friction coefficient p=1 kg/cm^2	Friction coefficient p=3 kg/cm^2
SKB butadiene rubber	Gas black	0	1.48	0.85
		40	1.55	0.89
		80	1.00	0.78
	Lampblack	0	1.48	0.85
		20	1.77	0.86
		40	1.80	0.86
		80	2.00	1.01
	Polyfluoroethylene TP-95	0	1.48	0.85
		10	1.66	0.87
		20	1.60	0.90
		40	1.66	0.92
		50	1.37	0.77
SKN-18 Butadiene-acrylonitrile rubber	Lampblack	0	1.73	0.89
		20	2.00	0.98
		40	1.87	1.02
		50	1.68	0.93
		80	1.37	0.92
	Polyfluoroethylene TP-95	0	1.73	0.89
		10	1.89	1.00
		20	2.00	0.98
		40	1.94	1.00
		50	1.68	0.88
SKN-26 Butadiene-acrylonitrile rubber	Lampblack	0	1.86	0.97
		20	2.00	1.06
		40	2.00	1.09
		50	1.74	1.04
		80	1.87	0.93
SKN-26	Oxide (magnetic)	0	1.86	0.97
		20	1.94	1.05
		40	1.94	1.18
		50	1.80	1.00
		80	1.94	1.13
	Chalk	0	1.86	0.97
		20	2.00	1.09
		40	1.86	1.06
		50	1.77	1.02
		80	1.80	0.95
	Polyfluoroethylene TP-95	0	1.86	0.94
		10	1.93	1.10
		20	2.10	1.08
		40	2.00	1.08
		50	1.77	1.02
SKS-30	Gas black	0	1.56	0.87
		40	1.96	1.03
		80	1.46	0.86
	Lampblack	0	1.56	0.87
		20	1.86	1.00
		40	1.95	1.03
		80	1.46	0.86

	Polyfluoro-ethylene	0	1.56	0.87
		20	2.06	1.04
		40	2.00	1.02
		50	1.86	0.93
KHAIPALON	Gas black	0	2.00	1.11
		40	1.47	0.94
		80	1.40	0.95
		100	0.93	0.82
	Lampblack	0	2.08	1.11
		40	1.87	1.11
		80	1.36	1.03
		100	0.93	0.86
	Chalk	0	2.00	1.11
		40	1.92	1.10
		80	1.66	1.02
		100	1.60	1.02

To increase the friction coefficient for load p = const. (65), it is necessary to increase the real contact area or to increase the nominal contact area or to change friction force from rough to smooth surfaces. Generally speaking, the friction coefficient (the ratio of the friction force to the normal load) does not take into account the role of the nominal geometric contact area: For p = const., decrease in the specimen's thickness and increase in the friction coefficient. With changes in external conditions, it is necessary to remember that a decrease in temperature increases the friction coefficient only to a certain limit, and moisture has but a weak effect on it. For increase in the friction coefficient, it follows that one must consider the nature and contour of the hard surface. According to the data in Table 4.5 and using Eq. (4.65), it is possible to calculate friction coefficients in each range of normal loads (up to 200-300 kg/cm^2).

This list of conditions for an increase in μ is not exhaustive. Thus, often it is necessary to take into account mechanical losses, the coefficient of overlapping surfaces, conditions of thermal conductivity for the polymer-metal pair, and so on. It follows that to select antifriction materials one must take into account the fact that a decrease in the adhesion of the polymer-metal pair generally leads to a decrease in the friction coefficient. To this end, modification of friction surfaces is employed. For example, fluorination of a vulcanized rubber surface does not change its bulk and structural properties, but strongly reduces adhesion.

Lubrication also effectively reduces the friction coefficient, but except for friction between hard metal bodies its role is significantly less. We note specific problems with the application of lubrication to a crosslinked-rubber-metal pair. The friction coefficient depends upon pressure. For friction without lubrication, it changes according to the relation $1/\mu = a + bp$ (66), but only up to a certain value of pressure, above which the relation reverses. This means that the friction coefficient for small pressures falls with a rise in pressure, and for large pressures it rises. A film of lubricant collapses significantly more readily during friction of a polymer-metal pair than during friction of a metal-metal pair. According to data obtained by White and Denny (67), for pressures 10-20 kg/cm^2 with various lubricants, μ does not depend upon pressure.

In conclusion, we should note that in a short time significant progress has been made in understanding the nature of vulcanized-rubber friction. The time is near when, with the aid of known physico-chemical parameters for polymers, it will be possible to calculate the friction force.

REFERENCES

1 R. Ariano, Rubber Chem. Technol., 3(1930)287.
2 E.L. Roth, R.L. Driscoll and W.L. Holt, Rubber Chem. Technol., 16(1943)155.
3 J. Harry and J. Prock, India Rubber World, 128(1953)619.
4 A. Schallamach, Proc. Phys. Soc., B 66(1953)386.
5 G.M. Bartenev, Dokl. Akad, Nauk. SSSR, 96(1954)1161.
6 Ya.I. Frenkel, Kineticheskaya Teoriya Zhidkostei (Kinetic Theory of Liquids), AN SSSR, Moscow, 1945.
7 G.M. Bartenev and Z.Ye. Styran, Dokl. Akad. Nauk. SSSR, 121(1958)87.
8 G.M. Bartenev and Z.Ye. Styran, Vysokomol Soedin., 1(No. 7)(1956)978.
9 A. Schallamach, Wear, 6(No. 5)(1953); J. Polym. Sci., 9(1952)385.
10 G.M. Bartenev and A.I. Yelkin, Dokl. Akad. Nauk. SSSR, 151(No. 2)(1963)320.
11 G.M. Bartenev, V.V. Lavrentev and N.A. Konstantinova, Mekh. Polim., (No. 7)(1967)726.
12 G.M. Bartenev, V.V. Lavrentev and N.A. Konstantinova, Mekh. Polim., (No. 4)(1965)123; Wear, 18(1971)439.
13 V.V. Lavrentev and K.K. Ostryeiki, Mekh. Polim., (No. 6)(1967) 1125-1139; Plaste Kautsch, 10(1968)729.
14 D. Bulgin, Publ. Plast. Weekly, 143(No. 16)(1962)636.
15 G.M. Bartenev and A.I. Yelkin, Mekh. Polim., (No. 1)(1967)123.
16 G.M. Bartenev and A.I. Yelkin, Vysokomol Soedin, 7(No. 6)(1965)992; 10A(1968)808.
17 F. Boggs and J. Riemen, Rubber Age, 81(No. 4)(1957).
18 Z.Ye. Styran, dissertation, MITKLT, 1962.
19 G.M. Bartenev and A.I. Yelkin, Wear, 9(1965)8; J. Polym. Sci., C, (No. 16)(1967)1973; Wear, 11(1968)393.
20 N.A. Konstantinova, dissertation, MGPI, 1967.
21 A.I. Yelkin and V.N. Nikolaev, Dokl. Akad. Nauk. SSSR, 173(No. 6)(1967)1302.
22 G.M. Bartenev, V.V. Lavrentev and N.A. Konstantinova, Mekh. Polim., (No. 3)(1967)309.
23 V.V. Lavrentev, Vysokomol Soedin., 4(No. 8)(1962)1151.
24 K.A. Grosch, Proc. Roy. Soc., A 274(No. 1356)(1963)21.
25 V.V. Lavrentev, Plaste Kautsch., (No. 6)(1962)282.
26 J.A. Greenwood, H. Mischel and D. Tabor, Proc. Roy. Soc., A 259(No. 1229)(1961)480.
27 A. Savkoor, Wear, 8(No. 3)(1965)221.
28 A.I. Yelkin, dissertation, MGPI, 1970.
29 V.N. Nikolayev, dissertation, MGPI, 1968.
30 V.V. Lavrentev, Dokl. Akad. Nauk. SSSR, 175(No. 1)(1967)125.
31 I.V. Kragelsky, Treniye i Iznos (Friction and Wear), Mashgiz, 1952.
32 N.S. Korenevskaya, V.V. Lavrentev, S.M. Yagnatinskaya, V.G. Rayevsky and S.S. Boyutskii, Vysokomol. Soedin., 8(1960)1247.
33 I.V. Kragelsky, Trudy I Vsyesoyuznoi Konferentsii Potreniyu i Iznosu v Mashinakh, (Papers I of All-Soviet Conference on Friction and Wear in Machines), T. I., AN SSSR, Moscow, 1939.
34 T. Alfrey, Jr., High Polymers, No. 6, Mechanical Properties of High Polymers, Interscience, New York, 1947.

35 L.V. Ivanova-Chumakova and P.A. Rebinder, Dokl. Akad. Nauk. SSSR, 81(1951)219; Kolloidn. Zh., 18(1956)429, 450, 682.
36 A.V. Ivanova, V.F. Chuvayev and P.A. Rebinder, Dokl. Akad. Nauk. SSSR, 139(1961)83.
37 J.F. Archard, J. Appl. Phys., 24(1953)981; Proc. Roy. Soc., A, 243(1957)190.
38 K.K. Ostreiko, dissertation, Nl Khl im. Karpova, 1970.
39 D.M. Tolstoy, Dokl. Akad. Nauk. SSSR, 90(No. 5)(1953); in Razritie Teorii Trenia i Iznosa, AN SSSR, Moscow, (1957)170-172.
40 V.V. Lavrentev, Doctoral dissertation, NtlFKhcl im. Karpova, 1970.
41 W. Holzmuller, Physik der Kunststoffe, Berlin, 1961.
42 W. Holzmuller, Kolloid Z., 203(1965)7.
43 G.M. Bartenev, I.V. Razumovskaya and D.S. Sanditov, Vysokomol., Soedin., 10B (1968)745.
44 D.S. Sanditov, dissertation, MGPI, 1970.
45 P.P. Kobeko, Amorfnyye Veshchestva, (Amorphous Substances), AN SSSR, Moscow, 1952.
46 G.M. Bartenev, Dokl. Akad. Nauk. SSSR, 103(1955)1017.
47 G.M. Bartenev and V.V. Lavrentev, Wear, 4(1961)154-160; 18(1971)439-449.
48 V.V. Lavrentev, Dokl. Akad. Nauk. SSSR, 115(No. 4)(1957)717.
49 S.B. Ratner and V.V. Lavrentev, Sh. Tekh. Fiz., 26(No. 4)(1956)85.
50 S.B. Ratner and V.D. Sokolskaya, Dokl. Akad, Nauk. SSSR, 86(No. 1)(1952); 99(No. 3)(1954)431; Khim. Prom., (No. 1)(1955)27.
51 V.V. Lavrentev, Kolloidn. Zh., 19(No. 4)(1957)522.
52 B.V. Deryagin and Yu.P. Toporov, Dokl. Akad. Nauk. SSSR, 146(No. 6)(1962)1356.
53 C.E. Weir, J. Res. Nat. Bur. Stand., 50(1953)311.
54 G.M. Bartenev, V.V. Lavrentev et al., Vysokomol. Soedin., A, 13(No. 10)(1971)2354.
55 G.M. Bartenev and V.V. Lavrentev, Sysokomol. Soedin., 2(No. 2)(1960)238.
56 P. Thirion, Rubber Chem. Technol., 21(1948)505.
57 G.M. Bartenev, A.I. Yelkin and M.V. Voyevodskaya (Eds.), Teoriya Treniya i Iznosa (Theory of Friction and Wear), Nauka, Moscow, 19
58 A.I. Vekselman, Mekh. Polim., (No. 3)(1967)533.
59 G.L. Slonimsky and L.B. Rogovina, Vysokomol. Soedin., 6(1964)620.
60 S.S. Voyutsky, Adyeziija i Autoreziya (Autoadhesion of Polymers and Adhesion), translated by S. Kaganoff, Interscience, New York, 1963.
61 F. Boggs and W. Riemen, Rubber Age, 81(1957)613.
62 D. Tabor, Rubber Chem. Technol., 33(No. 1)(1960)142.
63 A.P. Aleksandrov, Trudy I and II Konferentsii ps Vysokomolekulyar Coyedineniyam (Papers I and II of the Conference on High-molecula compounds), AN SSSR, Moscow, 1945, p. 49.
64 V.V. Lavrentev (Ed.), Teoriya Treniya i Iznosa (Theory of Frictio and Wear), Nauka, Moscow, 1965, p. 17.
65 G.M. Bartenev, V.V. Lavrentev and A.I. Yelkin, Kauch. Rezina, (No. 3)(1963)20.
66 D.F. Denny, Proc. Phys. Soc., B, 66(1953)721.
67 C.M. White and D.F. Denny, The Healing Mechanism of Flexible Packing, Great Britain, Ministry of Supply, Sci. and Tech. Memo No. 3-47, London, 1948.
68 G.M. Bartenev, V.V. Lavrentev and V.S. Voyevodsky, Mekh. Polim., (No. 1)(1971)140.
69 V.V. Lavrentev, N.A. Konstantinova and K.K. Ostreiko, Kautsch. Gummi., (No. 11)(1973)136.
70 G.M. Bartenev and V.V. Lavrentev, Vysokomol. Soedin., A8(No. 10)(1971)2354.

Chapter 5

SLIDING AND FRICTION OF POLYMERS IN THE VISCOUS STATE

5.1 MECHANISMS OF BOUNDARY SLIDING IN VISCOUS AND DISPERSED SYSTEMS

In the preceding chapters, we have examined the fundamental principles of friction of polymers in the glassy and rubbery states. Any polymer can also exist in the viscous state. Nonvulcanized natural rubbers, blended with polymers and stock based on them, can serve as an example. The frictional properties of such materials have a number of special characteristics. This field of external friction has been studied very little, primarily because of experimental difficulties.

As a result of easy deformability in the viscous state during contact with a solid, polymers form a real contact area comparable to the nominal one. Here, even in the case of weak adhesion of the polymer to the hard surface, tangential stresses, sometimes surpassing the maximum shear stresses throughout the bulk, arise on the contact boundary during shear. Therefore, under these conditions, as a rule, pure external friction does not develop. One of the fundamental experimental difficulties in studying external-friction systems in the viscous state is related to this lack of pure external friction.

From general physical considerations, it is clear that with a rise in temperature a polymer loses dimensional stability and its molecular cohesion with the hard surface decreases (Table 5.1). This leads to localization of the shear formation throughout the bulk, that is, to a transition from external to internal friction. The transition process is diffuse, and often it seems impossible to make a distinction between external and internal friction. But fundamental principles are common to both. One distinctive characteristic is that in external friction, the sliding of polymer molecules proceeds more readily on the hard surface, whereas in internal friction, the molecules slide more readily on one another. Quantitatively, this is expressed in different values of the activation energy for the two processes. In external friction, too, the geometry of the contact surface is important, and consequently it is necessary to

TABLE 5.1
Maximum adhesion values for polyethylene and polystyrene on steel

Polymer	Temperature (°C)	Adhesion (kg/cm)	Limit of Strength During Stretching (kg/cm^2)
Polyethylene	22	1.94	1.41
	-24	3.45	2.82
	-60	4.44	5.70
Polystyrene	82	2.11	2.68
	64	2.71	3.00
	45	3.38	3.35
	25	4.12	3.77
	-24	5.60	5.07

calculate the real contact area and the effect of normal load on the friction force. Among other conditions for realization of external friction, it is necessary that the specific shear stresses be less than the polymer's limit of plasticity and larger than the maximum tangential stresses on the boundary of the polymer-solid pair. From the molecular point of view it is necessary that the bonding energy of a polymer chain on a hard surface be lower than that of intermolecular bonds in the polymer's bulk. For multicompone or dispersed systems, sliding phenomena on the boundary of the division are more complicated, and generally one can speak of the phenomenon of boundary sliding.*

Despite the fact that with increasing temperature the specific adhesion decreases, external friction of polymer in the viscous state is characterized by rather high friction coefficients. Thus, according to data obtained by Lukomskaya and others (2), the friction coefficients for natural rubber mix and vulcanized SKS-30 butadiene-styrene rubber at a temperature of 60°C are equal respectively to 5.8 and 5.5. The larger friction coefficients for polymers are due to a large area of molecular contact as a result of larger values of molecular mobility. With decreasing temperature (Table 5.1), adhesion is increased, as is to be expected. But simultaneously the limit of fluidity and strength of the molecular network increase even more. Friction remains purely external.

In contrast to sliding of vulcanized rubber mixes and of polymer melts, boundary sliding in dispersed systems has been fairly well studied. The results obtained in this field have been applied to

*Although here the generally accepted term "boundary sliding" is used, we shall examine the case of pure sliding or external frictio of dispersed systems.

polymer boundary sliding. The mechanism of sliding in dispersed systems was first studied comprehensively by Tolstoy (3, 4), who examined five possible cases of boundary sliding:

1. Sliding occurs on the geometric boundary between the hard wall and the dispersion medium or between the hard boundary monolayer of a surface-active substance and the dispersion medium.
2. Between the wall and particles, there is a layer of dispersion medium with bulk properties. Generally sliding is confined to the internal-friction layer.
3. Sliding is confined to the internal friction of the diffused, adsorbed layer of the dispersion medium on the wall, or of the diffused, adsorbed layers close to the particle wall, or simultaneously of those and other layers.
4. Sliding proceeds on the geometric boundary between boundary-adsorbed layers of the dispersion medium and the particles, close to the wall, or between these boundary layers and boundary-adsorbed layers on these particles. This mechanism is analogous to one of the sliding mechanisms studied by Akhmatov (5) in hard aliphatic acids deposited from a solution in a volatile solvent.
5. There is a sliding mechanism similar to the motion of small balls in a rolling bearing. This concept was advanced by Bloom (6) and also by Velikovski (7) as a possible mechanism of friction with boundary lubrication. Here, the small balls act as dispersed particles with their own adsorbed layers.

Theoretically, it is possible to combine the cases noted above. For melts, rubbers and vulcanizate stocks at ordinary conversion temperatures, the most probable sliding mechanisms are the third and fifth. For highly viscous melts the first alternative is also possible.

In some cases, for example, for lubricating greases (8), the stock or melt is extruded out of tubes in the form of hard extrudates. Sometimes flow with slipping is observed in the wall (9). Vinogradov and co-workers (10, 11) have published a series of studies on sliding in dispersed systems. In particular, they noted the circulation of dispersed particles in the boundary layer.

According to Rebinder (12), with increasing dispersed-phase concentration, the role of boundary sliding increases. This also relates to sliding in melts and raw stocks.

Below, we shall examine only those theories of possible sliding mechanisms that can be used for an analysis of viscous melts, sliding rubbers, and vulcanizate stocks on hard, smooth surfaces.

5.2 THE THEORY OF BOUNDARY SLIDING

Tolstoy (13) developed the theory of boundary sliding for liquids on hard surfaces.

Let us examine a monomolecular layer of a liquid on the boundary with a solid. The sliding velocity of this layer is $v = (d\nu/dn)\lambda_m$, where λ_m is the median distance between molecules that are sliding under thermal motion from one equilibrium position to another (14). The mobility of particles in the bulk, according to Frenkel's theory, can be expressed as

$$q = [\lambda_m^2/(6kT\tau_0)]\exp[-(U)/(kT)], \qquad (5.1)$$

$$q = [(dv/dn)\lambda_m]/F = 1/(\eta\lambda_m), \qquad (5.2)$$

where F is the force acting on one molecule, and η is the viscosity.

An analogous equation can be written for liquid molecules adjoining a solid boundary [taking $(t_o)_b = t_o$]. The ratio of mobilities is equal to

$$q_b/q = \exp[(U - U_b)/(kT)] \qquad (5.3)$$

According to Frenkel, the activation energy is defined as the energy necessary for the formation throughout a liquid volume of micro-cavities and holes in which the molecule can fit. This energy can be calculated as the surface energy of micro-cavities:

$$U = S\gamma_{\ell v},$$

where $\gamma_{\ell v}$ is the surface tension on the boundary of a liquid with a vacuum and $S = \pi\lambda_m$ is the effective surface of a micro-cavity (with effective diameter λ_m).

The formation energy of a boundary micro-cavity is:

$$U_b = \alpha S(\gamma_S - \gamma_{\ell v}) + (1 - \alpha)S\gamma_{\ell v},$$

where α is the surface part of a molecular micro-cavity of a hard surface. Then on the basis of Eq. (5.3) we have

$$q_b/q = \exp[\alpha S(W_c - W_a)/(kT)] = \exp[(\alpha S\gamma_{\ell v})(1 - \cos\theta_\alpha)/(kT)], \qquad (5.4)$$

where W_c and W_a are the work done against cohesion and adhesion respectively, and θ_α is the contact angle of wetting.

According to Eq. (5.4), liquid sliding depends upon the wettability of a hard surface and is determined by the difference between the work of cohesion and of adhesion. For complete wetting of a solid surface by a liquid, $\theta_\alpha = 0$ or K = A, and $U_b = U$. Thus, for complete wetting, the molecular mobility on the dividing boundary is equal to the mobility in the bulk.

The ratio U_b/U is related to the phenomenological sliding coefficient (Fig. 5.1):

$$q_b/q = (dv/dn)_b/(dv/dn) = [(W_aB)/\lambda_m]/[(W_aB)/(\varepsilon + \lambda_m)] = \varepsilon/\lambda_m + 1. \quad (5.5)$$

Hence

$$\varepsilon = \lambda_m[(q_b/q) - 1), \quad (5.6)$$

where ε is the sliding coefficient.

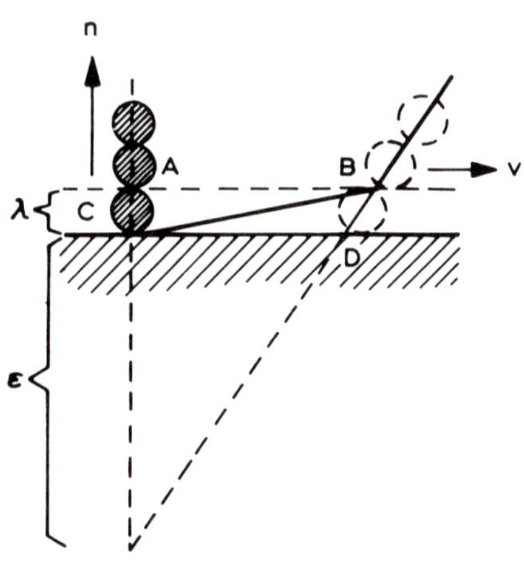

Fig. 5.1 Hypothetical extension of flow beyond the limit of a hard wall, equivalent to sliding (13): (1) displacement of molecules proportional to the sliding velocity; λ_m = distance between molecules; ε = sliding coefficient.

The above theory and Eqs. (5.4) and (5.5) allow quantitative evaluations of the effect of boundary sliding for the study of polymer rheological properties. Tolstoy (3) showed that, for example, the relative error in the discharge volume or fluidity of the crosslinked system, determined by capillary or shear viscosimeters (without a calculation of boundary sliding) ranges from several units to tens of units. The relative error of measurement for the discharge volume Q of a capillary is

$$(Q - Q_o)/Q = [(4\lambda_m)/r']((q_b/q) - 1), \qquad (5.7)$$

where Q is the discharge volume per unit time with sliding, Q_o is the same without sliding, and r′ is the radius of the capillary.

Equations (5.6) and (5.7) can be used to evaluate the intensity of boundary sliding in a liquid capillary. Thus, for example, for lubricating grease with $\eta = 10^6$ poise, with small velocity gradients, and with a boundary layer of thickness $h_o = 10^{-6}$ cm, the relative error $(Q - Q_o)/Q = 4$. This corresponds to a fivefold increase in apparent fluidity.

As follows from Eq. (5.7), the intensity of boundary sliding grows with increase of λ and decrease of wettability or with increase in the difference between work of cohesion and work of adhesion. Thus, Tolstoy's theory allows quantitative evaluation of the effect of boundary sliding. Here, it is important that for calculation it is possible to use the values of the adhesion and cohesion energies, which are rather well known (15,16) in the literature.

For adsorption of liquid molecules on the surface of a solid,

$$W_a = W_c - \Delta F,$$

where ΔF is change in the free surface energy of the solid. Because adsorption is an exothermic process and ΔF is a negative value, the wc of adhesion exceeds work of cohesion. Hence $q_b < q$, that is, in this case the sliding mechanism is related to the internal friction of an absorbed liquid film and to adjoining boundary layers of the liquid. For some sliding melts, for example polyethylene, complete wetting and low adsorption energy are not observed. Consequently, for poor adhesion and little change in surface energy of the solid, the theory predicts liquid sliding on the hard surface.

Tolstoy's theory, which was developed for liquids, does not take into account the fact that stresses σ, which arise on the interface of the polymer-hard-solid pair, ought to lead to a reduction in the effective activation energy of the sliding process. Consequently, Eq. (5.1) ought to be written in the form

$$q = \lambda_m^2/(6kT\tau_0)\ \exp(-(U - \gamma\sigma)/(kT)), \tag{5.8}$$

In other words, a polymer's viscosity depends upon the stress, as has been demonstrated experimentally (14).

For pure boundary sliding U_b will be expressed not by Eq. (5.1), but by Eq. (5.8). This means that tangential stresses that arise on the sliding boundary will promote an increase in the sliding velocity. In this case, we have pure external friction of visco-elastic polymers against hard smooth surfaces. Thus, the theory of external friction for rigid polymers interrelates with the theory of boundary sliding. For further development of the theory, it is necessary to consider the molecular-kinetic theory of polymer friction (see Chapter 4).

For an evaluation of the effect of polymer boundary sliding, it is useful to present data on the contact angle for various liquids and plastics (Table 5.2), and also values of the maximum work of

TABLE 5.2

Values of the contact angle for various liquids and polymers (17)

	Contact Angle of Liquid (Degrees)			
Polymer	Distilled Water	Tap Water	Autol 4, a Lubricating Oil	Oil D-1
Capron (nylon)	65	76	50	25
Resin AK-7	64	74	44	42
Polyamide-68	73	79	46	41
Polyethylene (linear)	86	85	52	25
Polytetrafluoro-ethylene (PTFE)	105	95	75	75

adhesion $(W_a)_{max}$ which is required for the detachment of liquid from the solid, on whose surface remains an equilibrium adsorbed film (Table 5.3). This work can be expressed in the following form:

$$(W_a)_{max} = \gamma_{\ell v}(1 + \cos\theta_\alpha), \tag{5.9}$$

It follows from Tables 5.2 and 5.3 that PTFE and polyethylene should have the greatest effect. This is confirmed in the use of these plastics (15).

TABLE 5.3

Values of the maximum work of adhesion $(W_a)_{max}$ for various solid polymers at 20°C (16)

Polymer	$(W_a)_{max}$ (erg/cm^2)	Polymer	$(W_a)_{max}$ (erg/cm^2)
Polytetrafluoro-ethylene	58.0	Polyvinylidene-flouride	88.5
Polytrifluoro-ethylene	72.0	Polyvinyl-fluoride	92.0
Polystyrene	78.0	Polyvinyl-chloride	92.0
Polyethylene	82.0	Polyvinylidene-chloride	92.0
Polyethylene-terephthalate	88.0	Polyhexamethylene-ethylenediamide	97.0

5.3 SOME EXPERIMENTAL RESULTS AND BASIC PRINCIPLES

According to Rebinder (17) and Volarovich (18), boundary sliding is inherent in dispersed systems that possess shearing strength. Vulcanizate stocks, clay suspensions, polymer systems and such materials belong to this kind of system. Moreover, from Tolstoy's theory, for $W_c - W_a > 0$ polymer melts have boundary sliding. Thus, it is possible to speak of boundary sliding for a considerable number of materials (11,17-21). Let us examine the principal results of some studies on this topic. One of the fundamental conclusions is that boundary sliding in dispersed systems begins only at a certain critical tangential stress, corresponding to the boundary layer's limit of fluidity, which does not exceed that in the bulk of the system.

Effect of normal load

There exist systems for which the boundary sliding force does not depend upon the normal load or depends upon it linearly. In the case of a binomial relation such as Deryagin's law:

$$F_o = \mu_o(P + N_o),$$

the real contact area, as Tolstoy demonstrated, ought to be constant (4). Furthermore, for dispersed systems it is necessary to take into account the role of a layer in contact with a hard solid. If we designate the real boundary contact area as S_o', and the real supporting area as S, then the function F(P) will be expressed by the equation

$$F_o = \mu_o(S_o/S)(P + p_oS_o), \tag{5.10}$$

If we examine sliding of a dispersed medium in a boundary layer of thickness h_o, then the general boundary friction force is equal to

$$F = F_o + F_\eta, \tag{5.11}$$

where F_o is the pure external friction force; F_η is the friction force that is conditioned on the viscous resistance layer and does not depend upon the normal load.

Figure 5.2 shows the effect of pressure (4) on the friction force (tangential stress) for paraffin paste sliding on polished steel. As shown, the function F(N) obeys Deryagin's binomial law. The sliding velocity affects the coefficient $\mu_o(S_o'/S)$, which increases with velocity. Because, according to Deryagin's theory, μ_o does not depend upon the sliding velocity, it is necessary to assume that the real boundary contact area depends upon velocity. Consequently, the portion of liquid phase on the contact boundary decreases with increase in velocity.

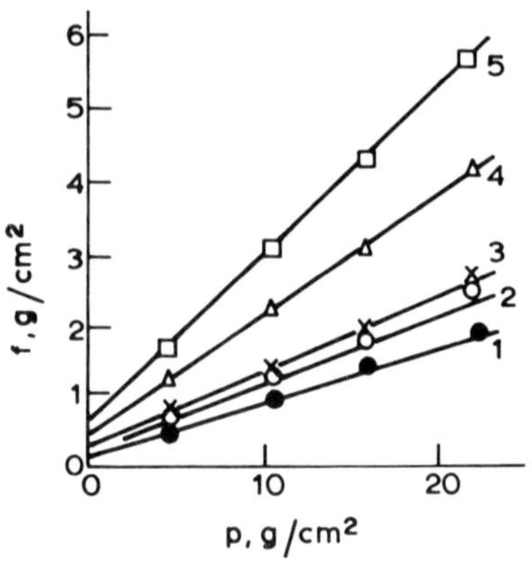

Fig. 5.2 Dependence of mean tangential stress upon mean normal load for 75% paraffin paste on nonpolar liquid petrolatum during sliding on clean polished steel (4). Sliding velocities: (1) 0; (2) 1.34 x 10^3 cm/sec; (3) 2.25 x 10^3 cm/sec; (4) 3.25 x 10^{-2} cm/sec; (5) 6.87 x 10^{-2} cm/sec.

For paste Al_2O_3, the function f(p) is such that $\mu(S_o'/S)$ = const in the range of velocities studied (Fig. 5.3).

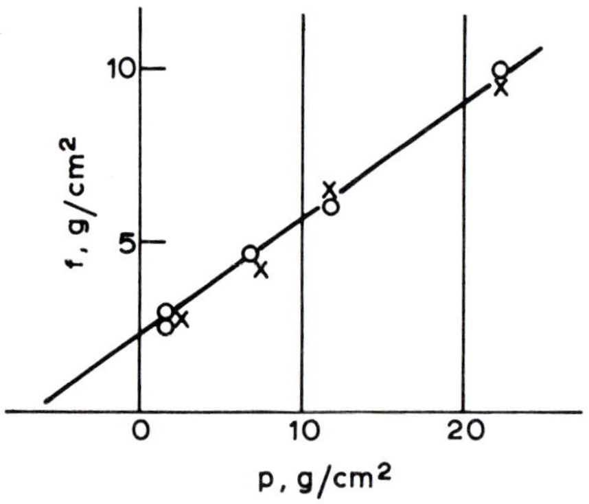

Fig. 5.3 Dependence of the mean tangential stress upon mean normal load for 53.6% paste Al_2O_3 on nonpolar liquid petrolatum during sliding on clean polished steel. Sliding velocities: 0, 1.34 x 10^{-3} cm/sec; X, 6.87 x 10^{-2} cm/sec.

The velocity of the raw rubber yuroprene against clean polished steel was studied on an STL-2 tribometer (see Chap. 7) for a wide range of pressures and temperatures. As seen in Fig. 5.4, the functions f(p) are linear and satisfy Eq. (5.10). The change in the slope of the straight lines indicates that with a rise in temperature the friction coefficient μ_o decreases. A decrease in the friction coefficient with increasing temperature is also observed for vulcanized rubbers (see Chap. 4).

The nature of the function $f_f(T)$ during sliding of nonvulcanized yuroprene has also been studied. Figure 5.5 shows functions $f_f(T)$ for various normal loads. As seen in the figure, satisfactory straight lines are observed for these functions. The linear functions $f_f(T)$ can be explained theoretically by Eq. (4.9), which results from the molecular-kinetic theory of friction. The relation of the slope of the straight lines $F_f(T)$ indicates a change in the polymer's molecular contact with a hard surface, produced by change in pressure.

A very interesting special characteristic of nonvulcanized rubber friction is its dependence upon pressure, which is analogous to that of a vulcanized rubber under large pressures, when S = const. But the temperature dependences of vulcanized and nonvulcanized rubbers are identical. Consequently, for rubber friction Eq. (5.10) holds true, where μ_o = const, but with a rise in temperature the relation S_o'/S changes, in a way that is determined by the development of molecular contact in the polymer chains. Actually, with rise in temperature, polymer-chain mobility is increased and new molecular

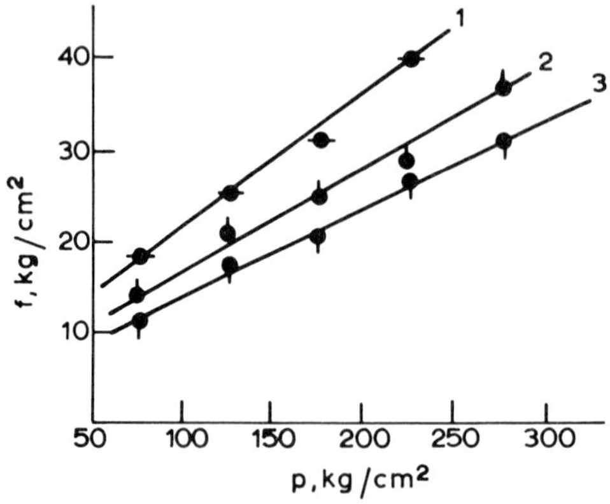

Fig. 5.4 Dependence of the specific friction force (tangential stress) upon pressure during sliding of yuroprene against steel at temperatures: (1) 22°C; (2) 50°C; (3) 75°C.

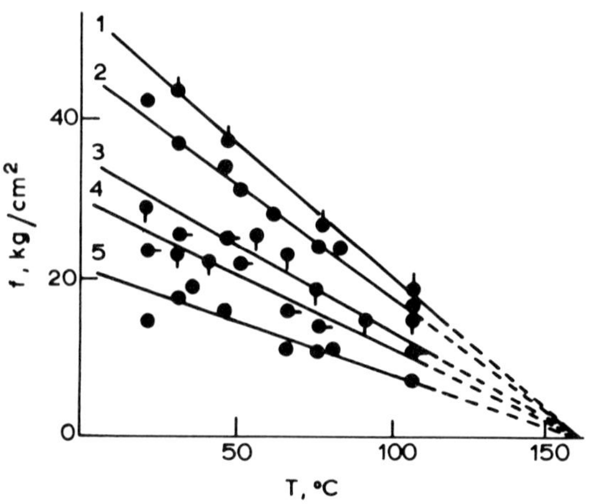

Fig. 5.5 Dependence of the specific friction force of yuroprene against polished steel upon temperature at v = 0.44 cm/min. Normal loads: (1) 275; (2) 225; (3) 175; (4) 125; (5) 75 kg/cm^2.

conformations are realized, which promote the formation of new points of real contact. From the data in Fig. 5.5 and Eq. (4.1), it is possible to calculate the effective activation energy. The terminal temperature at which the straight lines intersect is equal to $T_o = U_{ef}/[k/n(\nu_o/\nu)]$. The activation energy thus determined is equal to 10 Kcal/mole. This value is close to that for viscous flow in nonvulcanized rubber.

Lukovskii and co-workers showed (2) that for crude rubber stocks, sliding on hard surfaces, a correlation is observed between the external friction coefficient and the friction coefficient determined on Mooney's viscometer.

Effect of sliding velocity

The study of the effect of the sliding velocity on the friction force has central significance for understanding the friction mechanism. However, we have not succeeded in arriving at quantitative results. Available studies (4,9) indicate that for the relation P(v) a power law should be used. According to Tolstoy (4),

$$F = F_o + mv^n \qquad (5.12)$$

where F_o is the tangential force at v = 0; m and n are constants.

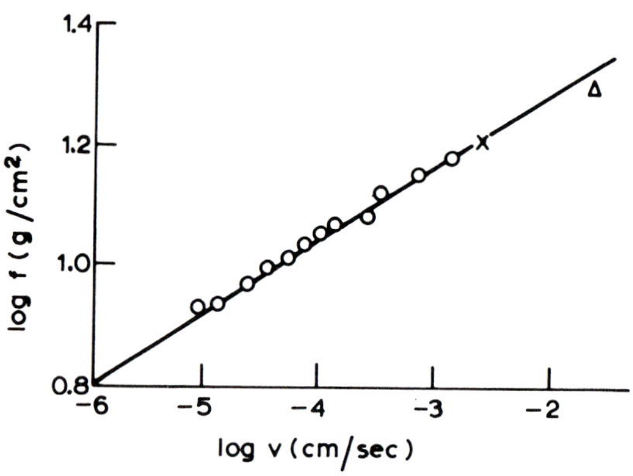

Fig. 5.6 Dependence on the sliding velocity of resistance to sliding of non-polar resinized lubricant on refined oil MPV against polished steel ($f = 1.8 + 31.1\ v^{0.282}$) #(4): 0, p = 1.5 g/cm^2, v = 1.34 x 10^{-3} cm/sec; X, p = 1.5 g/cm^2, v = 2.2 x 10^{-3} cm/sec; Δ, p = 1.5 g/cm^2 v = 2.5 x 10^{-2} cm/sec.

Mooney (9) and others (2) have used the relation

$$v \propto s^n \qquad (5.13)$$

where n is constant, and s is the shear stress.

On the other hand, according to Eq. (5.2) an inverse proportionality exists between the sliding velocity and the stress, which in turn is proportional to the force F. Equation (5.3) is a particular case of the more general Eq. (5.8), which under small shear converts into Eq. (5.2). One would therefore expect, according to the theory of boundary sliding, to be able to apply Eq. (5.2); however, experimental results obtained at low stresses and velocities, confirm Eq. (5.8). Thus, the theory of boundary sliding does not always explain the experimental results. Equation (5.13) can be applied to some rubber stocks (2,9). However, experimental results do not support this assertion. Thus, a theory of boundary sliding applicable to polymer friction is required to compensate for the lack of reliable and exhaustive experimental data.

Using power-law equation (5.12) and the equation for the total shear force (5.11), we obtain the general equation

$$f = f_o + f_\eta = \mu_o(S_o/S)(p + p_o(S/S_n)) + (kv^n)/S_n, \tag{5.14}$$

where f_o and f_η are specific shear stresses.

We can draw the following conclusions from the above:

1. The relation of the boundary friction force to pressure is expressed by Deryagin's binomial law.
2. The specific boundary friction force decreases linearly with increasing in temperature according to the molecular-kinetic theory of friction and the theory of boundary sliding.
3. Boundary sliding is inherent in a wide class of materials: low-molecular-weight liquids, dispersed systems, rubber stocks, rubbers, and polymer melts.
4. The effect of boundary sliding is determined by the difference in the values of cohesion and adhesion according to Eq. (5.4). In increases with decreasing temperature, and also with decreasing cross-section of the flow and with increasing difference between the works of cohesion and adhesion.
5. Three types of dispersed systems exist, whose resistance to sliding depends: (a) upon the sliding velocity and pressure, (b) upon velocity alone, (c) upon pressure alone. The most prevalent systems are of Type a (dispersed systems and polymers).
6. As with external friction, the states and treatment of the surfaces exert a large effect on the boundary friction force.

5.4 THE SIGNIFICANCE OF THE EFFECT OF BOUNDARY SLIDING

Boundary sliding has a special significance for technological processing of polymer materials. The processes of extrusion, calendering, molding, etc. cannot occur without polymer sliding in the viscoelastic state on solid surfaces. In several cases (9,15), the absence of boundary sliding has led to disruption of the technological process. Some procedures for increasing of boundary sliding (for example, the introduction of substances that reduce the adhesion of polymers to hard surfaces) are widely used in the manufacture of most polymer goods. In large measure, the application of lubricant also depends upon the degree of boundary sliding. The effect of boundary sliding is particularly important for the use of lubricating greases and lubricants on the base of polymer materials. Unfortunatel quantitative theories of polymer treatment do not take into account the effect of boundary sliding.

It is necessary to consider boundary sliding for the determination of rheological constants. Thus, for example, the determination of viscosity coefficients can lead to large errors (4) without an evaluation of the effect of boundary sliding. A calculation of this parameter is especially important for a determination of the viscosity in highly viscous systems. In some cases, thanks to bounda sliding, the yield stress determined by rotary viscometers can be less than the real stress (7).

REFERENCES

1 G. Kraus and J.E. Manson, J. Polym. Sci., 6(1952)629.
2 A.I. Lukomskaya, L.F. Gudkova, S.B. Merezbrannyi, P.N. Orlovskii and M.M. Reznikovekii, Kauch. Rezina, (No. 4)(1962)21.
3 D.M. Tolstoy, Kolloidn. Zh., 9(No.6)(1947)450.
4 D.M. Tolstoy, Sbornik Trudov, Posvyashchepurykh Pamyati Akademika P.P. Lazarev (Symposium of Works Dedicated to the Memory of the Scholar P.P. Lazarev), AN SSSR, Moscow, 1956.
5 A.S. Akhmatov, Molekulyarnaya Fizika Granichnogo Treniya (Molecular Physics of Boundary Friction), Fizmatgiz, 1963.
6 Bloom, Pet. Z., 25(1921)46.
7 M.P. Velikovskii, Konsistantnyye Smazki (Lubricating Greases), Gostoptekhizdat, 1945.
8 M.P. Volarovich, D.M. Tolstoy and E.P. Loshakova, Trudy II Vsyesoyuznoi Konferentsii po Treniyi i Iznosu v Mashinakh (Papers II of the All-Soviet Conference on Friction and Wear in Machines), Issue 1, 1949, p. 130.
9 M. Mooney, Proc. Int. Rubber Conf., Washington, November, 1969, pp. 8-13.
10 G.V. Vinogradov and V.P. Pavlov, Dokl. Akad. Nauk SSSR, 61(1948) 1049; 71(1950)307, 1069; 114(1957)977; 84(1952)309.
11 W. Pavlov, G. Vinogradov, V. Sinizyn and J. Deinega, Rheol. Acta, 1(No. 4-6)(1961)470-490.
12 P.A. Rebinder (Ed.) Viyazkost Shidkostei i Kolloidnykh Pastvorov (Viscosity of Liquids and Colloidal Solutions), AN SSSR, Moscow, 1941, p. 364.
13 D.M. Tolstoy, Dokl. Akad. Nauk SSSR, 85(No. 5)(1952)1089.
14 Ya. I. Frenkel, Kineticheskaya Teoriya Zhidkostei (Kinetic Theory of Liquids), AN SSSR, Moscow, 1945.
15 D.M. MacKelvey, Pererabotka Polimerov (Treatment of Polymers), Khimiya, 1965.
16 Anon., Collection, Konstruktsionnye Svoistva Plastmass (Structural Properties of Plastics), Khimiya, 1965.
17 P.A. Rebinder, Fiziko-Khimicheskaya Mekhanika (Physico-Chemical Mechanics), Znamiyd, 1958.
18 M.P. Volarovich, Zh. Fiz. Khim., 4(No. 6)(1935)807.
19 B.V. Deryagin and N.A. Krylov (Ed.), Vyazkost Zhidkostei i Kolloidnykh Pastvorov (Viscosity of Liquids and Colloidal Solutions), AN SSSR, Moscow, 1941.
20 G.V. Vinogradov and V.P. Pavlov, Dokl. Akad. Nauk SSSR, 61(No. 6), (1948)1049.
21 G.I. Fuks, Dokl. Akad. Nauk SSSR, 66(1949)1125.

Chapter 6

WEAR OF POLYMERS

In this chapter, we shall examine the phenomenon of polymer wear during external friction. Studies of the mechanism and basic principles of wear are essential to the solution of the fundamental problem, i.e., to maximize the life of polymer products. Wear is significantly more complicated than external friction; it is the result of a combination of physico-chemical processes that take place on polymer friction surfaces and boundary layers.

By the nature of its basic process, polymer wear can be tentatively divided into fatigue and abrasive (micro-cutting). Rubbery polymers also wear out as a result of the process of roll formation. For friction on hard surfaces with little heat release, the main mechanism is fatigue wear, the general theory of which was developed by Kragelsk (1-3).

Quantitatively, wear of materials is expressed by the volume of the worn material, V (in cm^3); or the weight of the worn material, q (in g); or the thickness (height) of the worn layer, h (in cm). It may also be expressed in normalized form, as in the normalized weight $J_q = q/S_nL$ (g/cm^3), linear magnitude $J_h = h/L$, or volume per unit energy of wear $J_W = V/A$ (cm^2/kg). Finally, it may be expressed as specific weight of wear, $i_q = q/S_r\ell$; or specific linear wear, $i_h = V_\ell/S_rd$. Besides the quantities mentioned above, the customary coefficients of wear resistance β_q, β_h, and β_W in their general form are given by $\beta = i^{-1}$. Here, L is the sliding distance; A = FL, the friction work; S_r is the real contact area; S_n is the fracture area (usually the nominal contact area); ℓ is the friction track length, equal to the diameter of a contact point; d is the diameter of a single contact point; h is the thickness of a layer removed from the real contact area during a single cycle; F is the friction force; and V is the volume of separated material transferred during sliding a distance equal to the diameter of a contact point, ℓ = d. Here $Vd = V_\ell L$.

J_h, J_W, and i_h are the most prevalent criteria, and among them definite relations are known (Table 6.1). The relationship among

TABLE 6.1

The relationship among criteria of wear

Formula in terms of	Linear Intensity of Wear, J_h	Specific Linear Wear, i_h	Energy of Wear, J_W
Physical quantities	$J_h = \frac{h}{L} = \frac{V}{S_n L}$	$i_h = \frac{V_\ell}{S_r d} = \frac{h}{d}$	$J_W = \frac{V}{A} = \frac{V}{FL}$
Linear intensity of wear	-----	$i_h = J_h(\frac{S_n}{S_r})$	$J_W = J_h(\frac{S_n}{F})$
Specific intensity of wear	$J_h = i_h(\frac{S_r}{S_n})$	-----	$J_W = i_h(\frac{S_r}{F})$
Energy of wear	$J_h = J_W(\frac{F}{S_r})$	$i_h = J_W(\frac{F}{S_r})$	-----

wear criteria, physico-mechanical indicators, and external parameters is determined by the wear mechanism within the framework of the established theory of wear. The criteria of wear we have noted are relative characteristics, because they do not take into account the effects of time, velocity, pressure and other parameters on the magnitude of wear.

6.1 RELATIONSHIP BETWEEN FRICTION AND WEAR

A direct relationship exists between the phenomena of friction and wear. This relationship is already discernible in the definition of the energy criterion of wear and in the role of the real contact area in the specific intensity of wear. The real contact area can be expressed for a wide range of pressures by the Bartenev-Lavrentev exponential equation (5). The equation for the wear intensity is

$$i_h = J_h[1 - (1 - \phi_0)\exp(-\alpha p)]^{-1}.$$

For large loads ϕ = const; therefore, the wear intensity, for other conditions fixed, should not depend upon pressure. By definition, the coefficient of energy wear resistance is equal to

$$\beta_W = A/V = (FL)/(S_H h) = \beta_h f, \tag{6.1}$$

where $f = F/S$. The specific friction force is proportional to the specific real contact area $f = c_6\phi$; consequently

$$\beta_W = c_6 \phi \beta_h. \tag{6.2}$$

Equation (6.2) shows that the coefficient β_W is proportional to the friction constant [i.e., on the specific actual (real) friction force (see Chap. 4)] and does not depend upon pressure. This means that the relation of β_W to pressure is determined entirely by the relation of ϕ to pressure:

$$\phi = 1 - (1 - \phi_0)\exp(-\alpha p),$$

where $\alpha = \gamma/E$; here γ is the roughness coefficient, and E is Young's modulus.

Experiment demonstrates that $\phi_0 << 1$; therefore for small pressures

$$\beta_W = \beta_h c\alpha p = \beta_h c_6 \gamma\, p/E, \tag{6.3}$$

that is, the coefficient of wear resistance is proportional to the modulus E. This has been confirmed elsewhere (6-8). The relation β_W/β_h for p = const is a characteristic of the material. For large values for p, when $\phi \approx 1$, we have $\beta_W = c_6 \beta_h$.

In Chap. 4, we showed that the specific real friction force is

$$c_f = \frac{2}{\lambda S_k} (U - kT \ln \frac{B}{V})$$

(c_f may be equal to c_6). Using this equation, it is possible to analyze the temperature and velocity relations of β_W for small sliding velocities. The specific real friction force determines the intensity of tangential stresses that arise on sections of real contact. Quantitatively, the tangential stress τ' under conditions of elastic contact is determined by the relation

$$\tau' = kf = k\mu p_r, \tag{6.4}$$

where k is the proportionality coefficient and p_r is the actual normal pressure. When the stress τ' exceeds the strength of the material, τ'_{cr} (that is, for $\tau' > \tau'_{cr}$), failure of the material occurs. For a constant value of p_r, the stress τ' is determined by the value of the friction coefficient. One may therefore introduce (8) the concept of the critical friction coefficient μ_{cr}, equal to τ'_{cr}/kp_r. Therefore, for $\mu \geq \tau'_{cr}/kp_r$ cracks on the polymer surface are observed, but for $\mu < \tau'_{cr}/kp_r$ the fatigue form of wear occurs. Thus, the value of the friction coefficient determines the wear mechanism of the material and is one of the most important characteristics of the wear process. If $\mu > \mu_{cr}$, the magnitude of wear in stabilized vulcanizates grows by 1-2 orders (8).

For a constant friction coefficient, at some critical pressure P_{cr} a sharp increase in the coefficient of wear resistance β_W occurs. When $P > P_{cr}$, a change in the wear surface is observed, resulting from an increase in the particle size of the detached material. For friction on smooth surfaces, μ is determined by adhesion interactions; therefore, lubrication sharply reduces friction and wear too (8).

According to Ratner and Melmikova (9), the wear of a vulcanized rubber on an abrasive paper can be expressed by the empirical equation

$$V = k\mu\eta/\underline{\sigma}, \qquad (6.5)$$

where k is the proportionality coefficient, μ the friction coefficient, η the viscosity that characterizes the elasticity of a vulcanized rubber (GOST, All-Soviet State Standard 6950-54), and $\underline{\sigma}$ the resistance to crack by All-Union Standard 270-64. Equation (6.5) has been verified by the authors through much experimentation.

Despite the fact that Eq. (6.5) agrees satisfactorily with the experimental data, it is applicable only over a narrow range of variation in the friction coefficient. Fundamentally, this equation expresses an empirical relation and is useful for the comparison of wear resistance for various elastic materials. A strong dependence of wear upon the friction coefficient is observed for plastics. Repeated experimentation confirms the conclusion that for otherwise equal conditions, the lower μ is, the less wear there is (10-17). More generally, increase in the friction coefficient cannot lead unequivocally to a rise or a reduction of wear resistance. In many cases, the friction coefficient exerts only an indirect effect on wear. However, for clarification of this problem it is necessary to examine in more detail various aspects of wear. It will become clear that in several cases polymer friction properties are not related to the mechanism and intensity of wear.

6.2 FATIGUE WEAR

Usually, wear of polymer materials occurs as a result of the fatigue mechanism of failure. This mechanism is primarily defined by the character of the friction contact: multiple deformation of the polymer occurs during the external friction process at separate points of real contact, and leads to fracture and subsequent tearing of the material. The degree and frequency of deformation depend upon the surface geometry and properties, and also upon the sliding velocity, pressure, and temperature.

Kragelsky first analyzed theoretically the fatigue mechanism of wear (1-4). The theory of fatigue wear of materials has gained wide acceptance both in the Soviet Union (6, 8-10, 17) and abroad (18, 19).

Theoretical studies

In Kragelsky's theory, one criterion (1) is the measure of specific wear i_q, which is the weight of worn material, q, over an area equal to the real contact area S_r, due to sliding a distance equal to the mean diameter of a contact point, d:

$$i_q = q/(S_r d). \tag{6.6}$$

In general the amount of worn material $q = q_1 m$, where $m = L/d$ is the ratio of the friction path length to the mean diameter of a contact point ($d = \ell$; $m = L/\ell$).

The gravimetric intensity of wear by definition (see the first page of this chapter) is equal to $J_q = q/S_n L$. Then

$$J_q = i_q(S_r/S_n) = i_q(p_n/p_r), \tag{6.7}$$

where p_n is the nominal pressure, and p_r is the real pressure. Equation (6.7) shows that for a constant value of i_q the intensity of wear depends upon the relative actual contact area $S_r/S_n = \phi$.

For plastic contact, to a first approximation, one has

$$S_r = N/(\bar{c}\Sigma_T).$$

Consequently, from Eq. (6.7) we obtain

$$J_q = i_q\left(\frac{N}{S_n \bar{c}\Sigma_T}\right) = i_q\left(\frac{p_n}{\bar{c}\Sigma_T}\right). \tag{6.8}$$

For spherical micro-asperities $C \approx 3$, and the intensity of wear will be equal to

$$J_q = i_q\ (p_n/HB), \tag{6.9}$$

where $HB = \bar{c}\Sigma_T$, the hardness of the material.

Babichev (15) showed experimentally that for a large class of polymers relative wear resistance is proportional to hardness. In the theory of fatigue wear, we have considered that the above fracture of the material occurs as a result of repeated deformations of the bulk for both plastic and elastic contacts. Let us examine Eq. (6.6) in more detail. If in this equation we substitute $q = Q/n$ (where n is the number of interactions in the deformed bulk of the material), then we obtain

$$i_q = \sigma/(S_r dn) = (v\rho g)/(S_r dn), \tag{6.10}$$

where ρ is the density, and g the acceleration of gravity. The volume of the worn material, $V = S_n h_{rm}$ (where h_{rm} is the relative height of the removed material), is determined by the geometry of the deformed surfaces, which can be characterized by the bearing-surface curvature (see Chap. 2). On this basis, we can write

$$h_{rm} = h_{max} \int_0^{\varepsilon} b\varepsilon^{\nu} d\varepsilon = \frac{b\varepsilon^{\nu + 1}}{\nu + 1} h_{max}. \tag{6.11}$$

The relative real contact area ϕ can be expressed by Demkin's equation (see Chap. 2):

$$\phi = b\varepsilon^{\nu}. \tag{6.12}$$

Substituting Eqs. (6.11) and (6.12) into Eq. (6.10), we obtain for the specific intensity of wear the equation

$$i_q = \frac{\varepsilon h_{max} \rho g}{(\nu + 1)nd}, \tag{6.13}$$

or

$$i_h = \frac{a}{(\nu + 1)nd}, \tag{6.14}$$

where a is the absolute approach, h_{max} is the maximum height of microasperities, and ν is the bearing-surface curvature parameter. In Eq. (6.14) a, ν, and d are determined by the surface geometry, but d can depend in addition upon experimental conditions: velocity, temperature, and properties of the polymer. Between these quantities and n there exists a not yet established functional relation (1). If d is proportional to a, then $i_h \approx 2n^{-1}$ does not depend upon the mechanical properties of the material that can be measured at one value of the stress.

It is also possible to examine the specific wear resistance that follows from a determination of the linear wear intensity:

$$J_h = V/(S_n L).$$

Considering that generally $S_r < S_n$, the volume of worn material will be equal to $V = Sh_{cr}$. Since $S_r < S_n$, the effective friction track amounts to $L \approx L_\phi = nd$. Then

$$J_h = (S_r h_{cr})/(S_n nd) = i_h \phi . \tag{6.15}$$

Here $i_h = h_{cr}/nd$, the specific linear wear intensity on the real contact area. If the relative thickness of the deformed zone, h_{cr}, is proportional to the mean diameter of a contact point, d, then $i_h \approx h^{-1}$.

According to Eq. (6.14), the fatigue friction is characterized by the number n of interactions that lead to a separation of layer material. For elastic contact, n is between 10^5 and 10^8, and for plastic contact it is between 10^2 and 10^4. Micro-cutting and deep digging (see Fig. 2.1) lead to fracture of the material at n = 1.

The maximum number of deformation cycles of surface layers, n, depends upon the stress σ on the material through the well-known formula (20)

$$n_m = A\sigma^{-t}. \tag{6.16}$$

Later, it was demonstrated (6) that $A \approx \Sigma_o^{+}$ (where Σ_o is the strength for a single loading). In view of the fact that Eq. (6.16) holds true for bulk fatigue, and wear is characterized by contact fatigue, Kragelsky and Nyepomnyashchy (4) constructed equipment for determining n under contact-fatigue conditions.

When adhesion is large (welding, setting) and friction depends on the deformation component, the value of contact stress σ entering into Eq. (6.16) can be expressed (21) through the specific real friction force f_r, which is equal to the shear stress on the polymer surface layer:

$$\sigma = 3f_r = 3\mu\ (\Sigma_o P)/S_r. \qquad (6.17)$$

The coefficient t of dynamic endurance in Eq. (6.16) is determined experimentally on a plot of $\ell n\ \sigma$ against $\ell n\ n_m$. By extrapolation, it is found to equal Σ_o. It is interesting to note that value Σ_o is approximately equal to the polymer's strength, determined by All-Soviet (GOST) State Standard 270-64 (22).

By using in Eq. (6.16) the expression for contact strength (6.17), we obtain

$$n = \left(\frac{\Sigma_o}{\sigma}\right)^t = \left(\frac{\sigma_o \phi}{3\mu p}\right)^t . \qquad (6.18)$$

Because the exponent t usually is larger than one, the value of n depends rather strongly upon the pressure p, the friction coefficient μ, and the relative real contact area ($p = P/S_n$).

Having substituted in Eq. (6.18) the value ϕ for elastic contact according to Demkin (23), from Eq. (6.14) we then obtain for the wear intensity (see also Chap. 2)

$$J_h = \frac{c_1 p^{1 + \beta t}}{\lambda} \left(\frac{E}{1 - x^2}\right)^{(t - \beta t - 1)} \left(\frac{k_1 \mu}{c_2 \sigma_o}\right)^t . \qquad (6.19)$$

Here

$$c_1 = \frac{0.6\sqrt{\nu}}{K_2 (\nu + 1)\nu};\quad \beta = (2\nu + 1)^{-1};$$

$$c_2 = \left(\frac{1}{2}\right)^{\frac{1}{(2\nu + 1)}} \left(\frac{0.75\pi}{k_2 \nu (\nu + 1)}\right)^{\frac{2\nu}{(2\nu + 1)}} \left(\frac{rb^{1/\nu}}{h_{max}}\right)^{\frac{\nu}{(2\nu + 1)}}$$

where k_1, k_2 are coefficients.

Kragelsky and co-workers (24) have presented values of all parameters and recommendations for calculation of the wear intensity by Eq. (6.19). We should note that Eq. (6.19) can be utilized to determine J_h at low sliding velocities and pressures, because, first, the real contact area (see Chap. 3) cannot be expressed for all pressures by the exponential equation (6.2) and, second, μ depends upon the sliding velocity.

For plastic contact, the wear resistance is equal to:

$$J_h = \frac{c_3}{\varepsilon_0^t}\left(\frac{p}{HB}\right)^{\gamma}\left(\frac{\sigma_\tau + K\tau'}{\sigma - K\tau'}\right)^{t/2}, \qquad (6.20)$$

where

$$\gamma = 1 + (t + 1)/(2v); \; c_3 = (h_{max}/rb^{1/\nu})^{(1 + t)/2}(\nu^{1/2}/2^{3/2}),$$

where h_{max} is the height of surface irregularities, and r is the mean radius of their asperities.

On the basis of Eqs. (6.19) and (6.20), it is possible to make a judgment on the relation of wear resistance to a number of friction parameters:

1. wear resistance depends upon pressure according to a power law (with exponent > 1);
2. with an increase in the friction coefficient, J_h rises sharply
3. with a rise in the modulus, J_h increases (for σ_o = const);
4. J_h decreases with increasing HB for plastic contact, but it is important to take into account the magnitude of the deformation ε_o;
5. if we vary the surface geometry (roughness), then the wear resistance depends upon the ratio h_{max}/r.

Fatigue fracture was postulated on the basis of studies on vulcanized rubber wear resistance by Reznikovsky, who examined a model "polymer deformed through sliding of hard indenters" (Fig. 6.1), and calculated the coefficient of energy wear resistance, β_W. If we assume that the friction track is proportional to the number of cycles n that fracture the layer of polymer material, and A_q is the work of fracture for one cycle, then

$$\beta_W = (A_q n_m)/(hS_n).$$

Using Eq. (6.18), by Rezinkovsky's formula it is possible to obtai

$$\beta_W = k\mu(\sigma_o/k_1)^t E^{(2/3)(1 - t)} p_n^{(1 - t)/3}(z/r)^{(5 - 2t)/3}, \qquad (6.21)$$

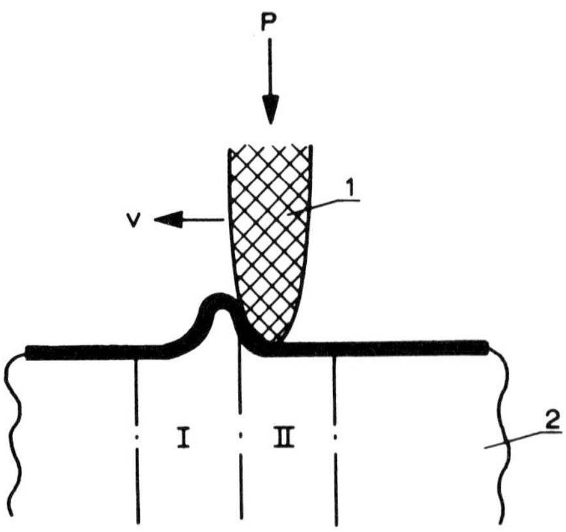

Fig. 6.1 Illustration of the friction wear mechanism which leads to the formation of transverse bands (25): (I) Compression Zone, (II) Stretching Zone; (1) hard, blunt indentor, (2) worn specimen.

where k, k_1 are coefficients, z is pitch of asperities, and r is the radius of asperities. The use of Eq. (6.21) for practical calculations is very much hampered by the fact that it is necessary to determine independently the coefficients k and k_1, and also surface parameters z and r. Therefore, this equation has not been accepted widely.

Schallamach (7,26,27) first quantitatively described the so-called "characteristic" wear, which arises during friction on sufficiently smooth polymer surfaces against a solid as a result of overstrain on parts of the surface layer. The zone of the largest stress develops against a hard protuberance. Depending upon the material's properties and the friction force during one or another traverse, ruptures appear in the surface layer. Transverse bands are formed on the polymer surface, the so-called "Schallamach's bands" (Fig. 6.2, a and b). Cross bands are formed during friction on smooth surfaces of all kinds of elastomers. A similar wear pattern is observed during sliding of vulcanized rubber on a (diffraction) grating. During friction on a grating, some plastics also form cross bands (25,28).

There is a tendency to form logitudinal fracture bands (Fig. 6.2), with increase in the sharpness of the asperities and the rigidity of the polymer. The formation of such bands is an indication of transition to the abrasive wear mechanism. In the glassy state, the fracture mechanism is basically abrasive [see (6.3)].

Schallamach showed that the volume of a worn elastomer is equal to

$$V = \text{const}\ r(P/E), \tag{6.22}$$

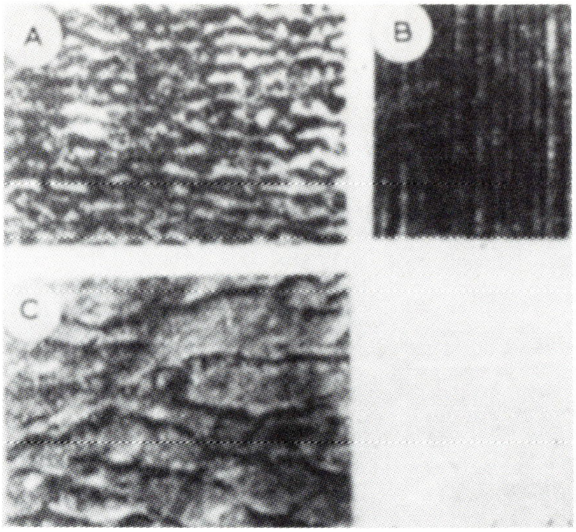

Fig. 6.2 Photographs of the surfaces of polymer specimens after wear against various hard surfaces. Vertical sliding (25). (a) polyamide on a grating, (b) vulcanized rubber on a grating, (c) vulcanized rubber on an abrasive paper.

where r is the radius of an indenter or an asperity. Reasoning from the fact that during friction of soft vulcanized rubbers against an abrasive paper cross bars are observed, Schallamach (27) explained the linear relation of wear to pressure by abrasive wear. Ratner (9) thinks that the explanation given by Schallamach is inconsistent because the abrasive paper is a system of sharp protuberances that cut the material, but Eq. (6.2) was derived for systems of dull protuberances. However, there is no doubt of the occurrence of cross bands during wear by an abrasive paper, an effect which, according to Ratner, characterizes friction wear. Schallamach's explanation of the wear mechanism for vulcanized rubbers against an abrasive paper as the tearing of the polymer surface seems to us to be correct. The point is that often it is impossible to characterize the wear mechanism by the type of wear surface (abrasive paper gives us abrasive wear). For wear against abrasive paper, Eq. (6.22) has been confirmed by others (8,9,10,15,29,30).

In the case of wear against a grating, the pressure exponent [see Eq. (6.19)], according to the theory of fatigue wear, is larger than one.

Ratner has demonstrated (25) that increase in the material's elasticity and decrease in the roughness of the hard surface promote the formation of cross bands. Diminution in the material's volume

V can be represented in the following form:

$$V = \text{const}\ Z^3, \tag{6.23}$$

where Z is the distance between cross bands.

Basic experimental results

Let us examine the effect of basic factors on polymer wear. According to Eq. (6.19), the fatigue exponent t, strength σ, modulus E, and friction coefficient μ exert the largest effect on wear. Surface geometry (the ratio h_{max}/r) and pressure p affect wear the least. To analyze the relation of wear to these variables, it is convenient to express wear resistance in the following form [see Eq. (6.19)]:

$$J = J_h = c'\sigma_o^{-t}[E^{(t - \beta t - 1)}][p_n^{(1 + \beta t)}][\mu^t] , \tag{6.24}$$

where c' is a constant.

If we determine wear intensity by the equation

$$J = J_1 p_n^{\alpha} \tag{6.25}$$

(where J_1 is the wear intensity for p = 1), then the exponent α depends only on the nature of the polymer and its surface geometry. Under ordinary conditions, $\alpha > 1$ for fatigue wear, and $\alpha = 1$ for abrasive wear (10,29). Thus, α characterizes the type of wear and depends upon only the polymer's properties (through t) and the surface geometry (through β). For vulcanized rubbers, α ranges from 1 to 5 and t ranges from 3 to 56 (10). For otherwise equal conditions, α and t obey a linear relation which, according to Fig. 6.3, holds true for a series of vulcanized rubbers that have been obtained from various rubbers and ingredients. The introduction of an antifatigue agent raises t and α simultaneously.

For fatigue wear, the polymer wear resistance is rather large, and the determination of the coefficient of wear resistance takes a long time (especially for friction on smooth surfaces). For an evaluation of fatigue wear resistance, researchers use metal gratings (30,32-34), considering that friction on a grating can simulate the process of fatigue wear for friction on smooth surfaces. By this method, the time of testing is considerably reduced. A fixed correlation exists between experiments on gratings and smooth metal surfaces.

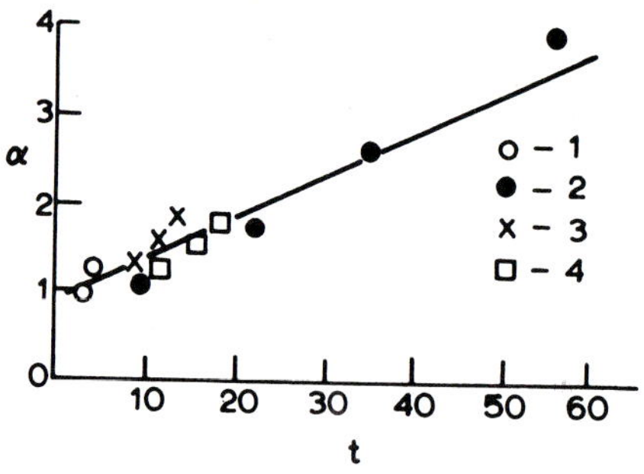

Fig. 6.3 Correlation of values α and t for vulcanized rubbers (10): (1) unfilled SKB (butadiene rubber); (2) filled SKB; (3) filled SKB with antifatigue agent; (4) unfilled Nairit chloroprene rubber equal neoprene) with DBF.

Testing on gratings produces a sensitive indication of the degree of fatigue resistance for polymers and vulcanized rubbers (35). Figure 6.4 presents as an example the relation of the rate of wear

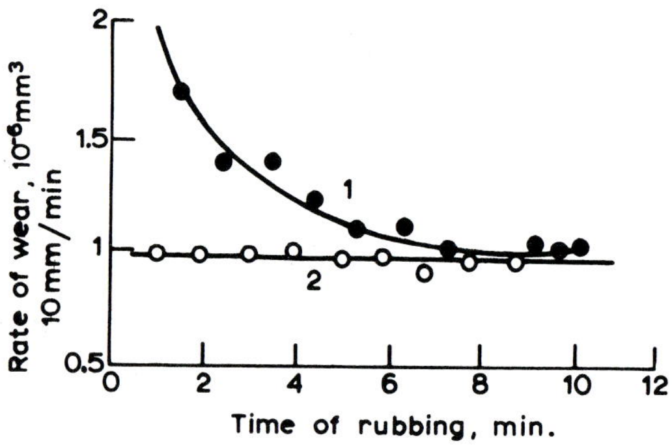

Fig. 6.4 Relation of the rate of wear of a vulcanized rubber against grit to duration of wear (35): (1) preliminary fatigued specimen; (2) specimen not subjected to fatigue.

to the sliding time for two vulcanized rubbers, one preliminarily fatigued (curve 1) and the other not subjected to fatigue (curve 2). As seen in the figure, the rate of wear of a preliminarily fatigued specimen is twice as high as that of the nonfatigued specimen in the first minute of the test. With passage of time, as a result of wear of the preliminarily fatigued layer, the rates become practically equal. Analogous results have been obtained for surface fatigue of polymethyl methacrylate (35).

The question arises, is there fatigue of a latent form in wear? The results of experiments that are given in Fig. 6.4 allow us to answer this question in the affirmative. But then the number of cycles that lead to fracture of a layer should be related not only to the tangential stresses τ', but also to the normal stresses τ''. Possible cases when $\tau'' >> \tau'$, for example, occur in the friction of rough surfaces with lubrication. Then the number of cycles, n, will characterize only the fatiguability of the material in bulk. These difficulties in explaining some facts by the theory have been noted previously (10). Thus, the basic validity of the theory of fatigue wear requires that the tangential stress be the fracturing stress, which is true only for friction materials, that is, when $\tau' >> \tau''$.

Now let us examine the effect of various factors on polymer wear:

EFFECT OF PRESSURE (4,7,8,17,24,32,36,37). Polymer wear resistance depends in a rather complicated way upon pressure, which determines the real contact area, the size of the deformed zone, the mean diameter of a contact point, and the friction force. Increase in pressure at a constant sliding velocity indirectly (through a rise in temperature) affects wear resistance and polymer fatigue properties. But the most important factor is the real contact area, which (in contrast with metals) depends strongly upon pressure.

According to calculation, the linear intensity of wear is equal to $J_h \approx i_h\phi$. For vulcanized rubbers, the intensity of specific wear, i_h, depends weakly upon pressure (4,35,37). Consequently, $J_h \approx \phi$. It is known (see Chap. 4) that ϕ depends exponentially upon pressure for vulcanized rubbers and some plastics; here surface asperities exert a significant effect on the value of ϕ. For smooth surface (∇9-10 class of treatment fineness and higher) at values of p/E of order 0.1, the relative contact area $\phi \approx 1$. Thus, wear of vulcanized rubbers should depend weakly upon load. This conclusion is confirmed by Rybalov's experimental data (37), shown in Fig. 6.5. Some increases in J due to increase in p probably result from initial heating of the polymer surface and the effect of wear particles on the formation of the actual contact area. For friction of soft, unfilled, crosslinked rubbers on smooth metal surfaces, pressure can significantly affect the measure of wear resistance. This is explained by the transition from fatigue wear to roll formation. (See below.)

For wear of vulcanized rubbers on a grating, values of α in Eq. (6.25) change from 2 to 7 and higher, and for wear against steel,

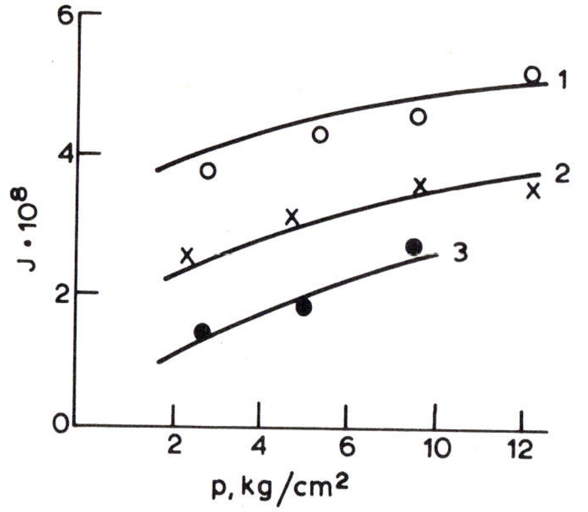

Fig. 6.5 Relation of wear resistance to pressure for vulcanized rubbers during friction against a smooth steel surface (37): (1) vulcanized rubber SKEP; (2) vulcanized rubber SKD-32; (3) vulcanized rubber SKN-18 + SKN-40.

from 1.1 to 2.7 (32,35). Thus, wear on a grating depends strongly upon pressure. It has bee noted that values of α against a grating and on steel correlate (Fig. 6.6). The quantity α depends upon the

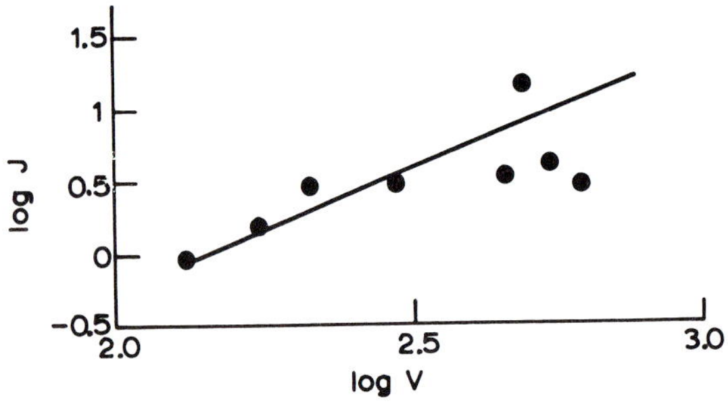

Fig. 6.6 Correlation of wear for various vulcanized rubbers on steel and grating on a logarithmic plot (log J vs log v_3) (32): J in mm^3, v in cm/min.

nature of the polymer and varies from 1.02 for a natural rubber to 3.84 for a butadiene-nitrile rubber (SKN-40) (32). It follows that the data given in Fig. 6.6 barely reflect the correlation betwee wear on steel and on a grating. Most experimental points lie on a straight line parallel to the abcissa - that is, for a series of vulcanized rubbers there is no correlation between wear on steel and on a grating. Bilik has indicated a similar lack of correlation for other polymers (plastics) (28). Ratner and Farberove (16) have demonstrated that for wear on a grating straight lines are obtained for a large number of plastics on a logarithmic plot (log v vs log p). This means that $J \propto p^{\alpha}$ in accordance with Eq. (6.25). It is necessary to note that work of this kind has only qualitative significance, because the exponent α depends in a complicated way upon the properties of the material and on surface friction.

For plastics, the relation of wear to pressure appears more clearly than for vulcanized rubbers. Thus, according to Mitrovich (38), the linear intensity of wear, J_h, depends upon pressure (Fig. 6.7). Nonlinearity can be explained by the initial heating of surface friction and by the adhering of nylon 6 (polycaprolactam) to a steel

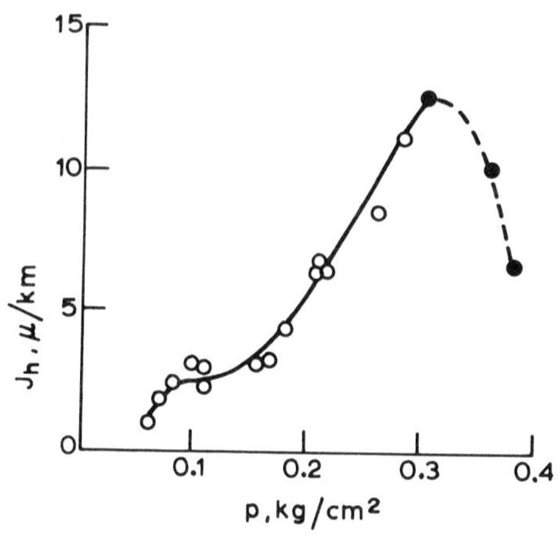

Fig. 6.7 Relation of linear magnitude of wear J_h to pressure p for nylon 6 (polycaprolactam) at a sliding velocity of 13.2 m/min on a smooth steel surface (38).

surface. Pruzhansky (39) observed a linear dependence of gravimetric wear upon pressure. As Ratner, Klitenik, and Lure (10) have shown, the exponent α in Eq. (6.25) depends upon polymer fatigue properties in complete accordance with Kragelsky's theory. To a first approximation, $\alpha = 1 + 3 \log n_{1/2}$, where $n_{1/2}$ is the number of cycles required for a reduction in strength in accordance with Eq. (6.18).

Let us observe that the friction coefficient also depends upon pressure [see Eq. (6.24)]; therefore, there cannot be a single interpretation of the effect of pressure on J_h. (Only in the range of small pressures S does μ = const.)

EFFECT OF FRICTION COEFFICIENT. The fatigue theory of friction allows a calculation of the effect of the friction coefficient on the magnitude of wear. In Sec. 6.1 we have already discussed the relation between friction and wear in polymers. Let us examine this question in more detail, proceeding from the fatigue wear.

According to Eq. (6.24), the wear can be expressed in the following form:

$$J = c''\mu^t p^{(1 + \beta t)} = c'_h \mu^t p^{\alpha}, \tag{6.26}$$

where c" is constant. Generally speaking, the relation between wear and friction becomes clearer if we describe the specific wear in accordance with Eqs. (6.10) and (6.18):

$$i_q = (V\rho g \cdot 3^t f_r^{\tau})/(sd\Sigma_o^t), \tag{6.27}$$

where f_r is the specific real friction force. For sliding of vulcanized rubbers on smooth surfaces, $S_r \approx S_n$. Then

$$i_q = (V\rho g \cdot 3^t f_r^t)/(Sd\Sigma_o^t) = [(h\rho g \cdot 3^t)/(d\Sigma_o^t)](f_r^t), \tag{6.28}$$

or

$$i_q = c'_2 f_r^t,$$

where

$$c'_2 = (h\rho g \cdot 3^t)/(d\Sigma_o^t).$$

Consequently, according to Eq. (6.28), specific gravimetric wear i_q is proportional to f_r^t. In turn (see Chap. 4), $f_r = c_f$ (where c_f is the friction constant) for vulcanized rubbers in the rubbery range of high elasticity, according to Bartenev's theory. The friction constant c_f varies from 20-60 kg/cm^2 for vulcanized rubber (5,40), and the fatigue exponent t is equal to 3.3-3.8 (41). Keeping in mind that c'_2 in Eq. (6.28) is small because for vulcanized rubbers Σ_o is on the order of 1500 kg/cm^2, it is expected that with increase in the specific real friction force f_r for small values of f_r wear will be practically constant if other conditions are equal. With increasing f_r, for some values of f_r, the wear ought to grow sharply, inasmuch as f_r^t changes rather rapidly with f_r. Such a relation of

i_q to the friction coefficient is actually observed* (Fig. 6.8).

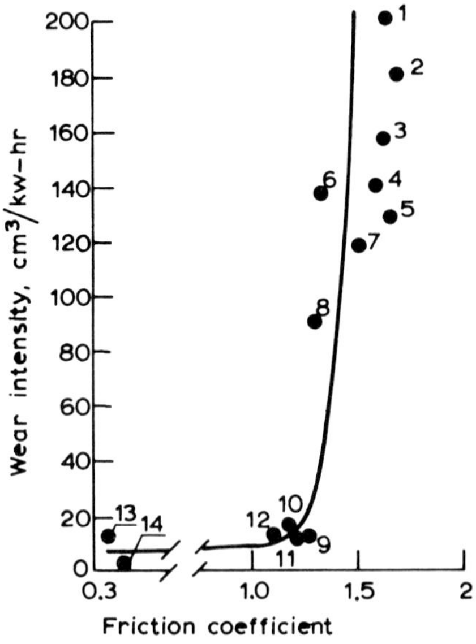

Fig. 6.8 Dependence of wear intensity upon the friction coefficient (tested on a grooved metal surface for p = 0.63 kg/cm^2) (8): (1) yuroprene 1712 (butadiene-styrene, oil-filled); (2) SKB (butadiene rubber); (3) butyl rubber; (4) combination of yuroprene 1500 and yuroprene 1712 (1:1); (5) natural rubber; (6) SKS-30 (butadiene-styrene rubber); (7) Nairit, chloroprene rubber (equivalent to neoprene); (8) yuroprene 1500; (9) SKN-26; (10) SKS-25-MVP-5; (11) KSK-30-1; (12) SKMVP-15; (13) natural rubber (water lubricant); (14) yuroprene 1500 (water lubricant).

Yevstratov, Reznikovsky, and others (8), for a series of vulcanized rubbers sliding on hard rough surfaces, presented so-called "critical values" of the friction force, above which wear catastrophically increased. It is significant that in this range there occurs a transition from fatigue wear to roll formation; consequently, fatigue wear is not realized for large friction forces or for good adhesion of polymer-solid pairs. However, until now the limits of applicability of the fatigue theory of friction have not been clearly determined.

Unfortunately, in polymer-wear research the specific real friction force is not determined; instead, wear is investigated in relation to the friction force or to the friction coefficient although these cannot serve as polymer characteristics (in comparison with friction

*One must make use of the correlation between f_r and μ.

constant c_f). In this case, nevertheless, it is preferable to use the friction force and not the friction coefficient. Figure 6.9 shows results of research on the wear of vulcanized SKN-18 and SKN-26 sliding on a smooth steel surface. As seen in the figure, for values

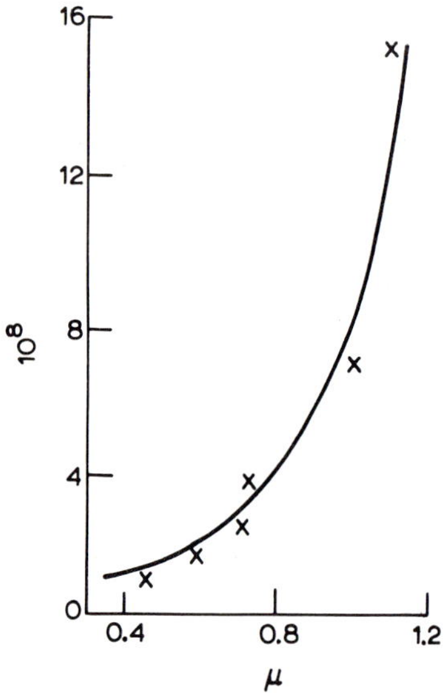

Fig. 6.9 Dependence of specific wear upon the friction coefficient for vulcanized SKN-18 and SKN-26 against a smooth steel surface (37).

of $\mu \approx 1$ wear sharply increases. With increase in the friction coefficient and other conditions equal, the contact temperature changes, which leads to a change in the friction mode. In this connection, there are very interesting studies by Rybalov (37) on wear of various vulcanized rubbers at a given temperature and mode of friction. Rybalov showed that in a range of temperatures from 50 to 250°C vulcanized rubbers that possess a lower friction coefficient have less wear. In order to exclude the effect of bulk properties, the friction coefficient was changed with the aid of fluorination of the very same specimen's surface. These experiments also indicate that polymer wear is approximately independent of the friction force.

For plastic wear, the effect of μ is difficult to discern; with change in the friction coefficient either the nature of the polymer's friction or its physico-mechanical properties change. Ratner, Klitenik, and Lure (10) applied lubrication in order to exclude these effects. They demonstrated that with reduction of μ, wear is reduced significantly, in accordance with Eq. (6.26). However, these studies cannot promote further application of the theory of fatigue wear to plastics, because they are not quantitative. For plastics we do not know the value of the exponent α, which characterizes a polymer's fatigue properties.

EFFECT OF TEMPERATURE. Temperature is one of the basic factors that control wear (3,8,28,37,42-50). The clarification of the action of temperature on wear is a task that has not been completed yet, because of the many factors upon which temperature has an effect. These factors can be tentatively broken down into friction and bulk effects. Friction heating produces an effect upon the friction force and the actual contact area, and bulk heating changes physico-chemical properties of the polymer.

From studies on friction, it is known (see Chaps. 3 and 4) that in the rubbery state the real contact area changes little with increasing temperature, but the friction constant C decreases rather significantly. In the glassy state, both the friction force and the contact area S depend weakly upon temperature. A significant change in S takes place in the transitional zone between the rubbery and glassy states.

Of the factors that usually effect wear--see Eq. (6.19)--let us consider the strength Σ_o, the modulus E, and the fatigue exponent t. For a vulcanized rubber, Fig. 6.10 presents the dependences upon temperature of these three quantities as well as the friction coefficient μ. The modulus practically does not change; therefore, the wear, according to Eq. (6.24), is proportional to the factor $(\mu/\Sigma_o)^t$, whose value increases as the temperature rises from 50 to 150°C as a result of the reduction in Σ_o and growth in μ. Once the values of all parameters have been determined (24), it is possible to calculate the influence of temperature on wear by the theory of fatigue [see Eq. (6.24)] (51). As shown in Table 6.2, rather good correlation is observed between the theoretical and experimental results.

It is necessary to note that the temperature during wear, by contrast with friction, affects the polymer's structure. During severe wear, thermal-oxidative processes are observed, which often lead to degradation of the polymer. Degradation as a thermal-activation process grows sharply in a certain narrow range of temperatures (8,17,37,49), which are called critical. Above this range of

TABLE 6.2

Calculated and experimental values of wear i (37)

	$10^8 i$	
Temperature (°C)	Calc.	Expt.
50	0.08	--
100	0.715	0.715
150	2.6	2.0
200	2.8	3.9
250	--	6.2

temperatures (for vulcanized rubbers from 150 to 250°C) the mechanism of wear changes. Wear takes on a catastrophic character [see (6.4)]. Rybalov showed (37) that temperature is the determining factor for wear. All other factors (sliding velocity, pressure, and friction coefficient) basically affect catastrophic wear through change in temperature. This conclusion can be drawn from studies conducted on wear for fixed thermal modes (T = const) and for various pressures and sliding velocities.

Here, it is not essential for a rise in temperature that there be intensive wear in the critical range of temperatures (Fig. 6.11). The regulation of friction temperature allows a reduction in wear with a rise in sliding velocity or pressure. Figure 6.11 shows three

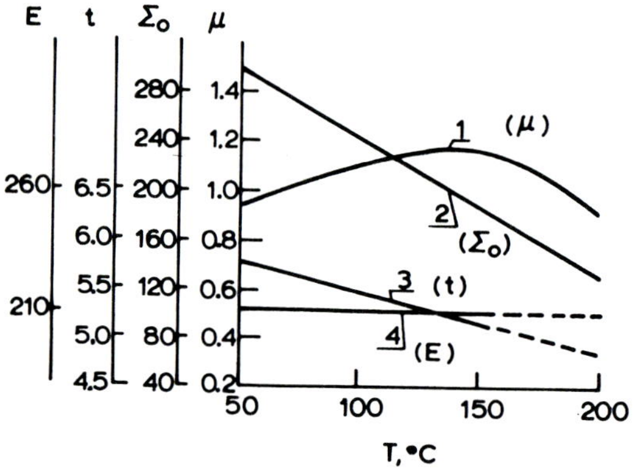

Fig. 6.10 Effect of temperature on the friction coefficient of a vulcanized rubber, μ (curve 1), strength Σ_o (curve 2), fatigue exponent t (curve 3), and modulus E (curve 4); E and Σ_o are in kg/cm^2.

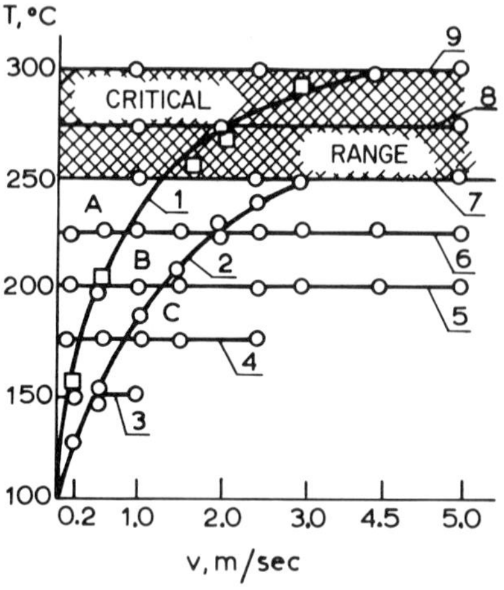

Fig. 6.11 Effect of friction mode on wear for vulcanized butadiene-nitrile rubbers (51): (1,2) friction process without thermo-regulation for pressures of 5 and 2.5 kg/cm^2 respectively; (3-5) with thermo-regulation at pressures of 2.5; (6-9) with thermoregulation at pressure of 5 kg/cm^2; A, B, C, zones of testing modes.

zones corresponding to modes of testing. For zone A, points above curve 1 are obtained during heating of the surfaces; in zone C, points below curve 2 are obtained during cooling of a friction surface; and in zone B, points are obtained during both heating and cooling of a friction surface. Without thermo-regulation, intensive wear sets in at a sliding velocity of 3 cm/sec (curve 2); but with thermo-regulation, at the same pressure, at a sliding velocity > 5 m/sec (straight line 7), wear is not observed. These very important results demonstrate that wear of smooth surfaces is determined by the temperature on the friction surface. It seems to us that this kind of research promotes a deeper understanding of the polymer wear mechanism.

In connection with the large effect temperature has on wear, the study of temperature fields that emerge during friction is of great significance (3, 42 and also references for Chap. 3). The task of studying and calculating temperatures in the contact zone is important for the processing and utilization of friction materials. Many studies have dealt with this area of friction and wear, in particular a monograph by Chichinadze (52). Once again (as in Chap. 3) we note the important effect of the period of mutual contact on the measurement of the friction temperature.

According to Eq. (6.24), temperature affects wear indirectly--through change in the strength Σ_o, the modulus E, the friction coefficient μ, and the exponent t. In the rubbery state, with increasing temperature (see Fig. 6.10), t and Σ_o decrease, E is constant, and the friction coefficient μ passes through a maximum, whose position depends upon the sliding velocity (see Chap. 4). From a calculation of the dependence of wear upon the friction coefficient, one might be led to predict a reduction in wear in the range of temperatures in which μ decreases. However, this is not observed, and wear grows catastrophically (Fig. 6.11). Evidently the reduction in strength Σ_o and thermo-chemical processes have a stronger effect on wear (53). According to data gathered by Schallamach (54) and Ratner (55), the surface temperature is higher by several degrees than the temperature of the environment. However, elementary wear events take place at contact points at locally different temperatures. Chemical processes that arise on the surface layer indicate this--for example, the kinetics of the addition of sulfur to raw rubber during the wear process. To clarify the wear mechanism it is not enough to comprehend changes in overall parameters (E, Σ_o, μ, t) and the wear mode (T, v); it is also important to take into account the change in the structure of the material as a result of physico-chemical processes.

Fatigue wear in plastics has been little studied. In the glassy state, plastics are basically characterized by the abrasive mechanism of wear for sliding, both on abrasive paper and on hard rough surface (56). Crystalline polymers which have a rubbery component wear out similarly to vulcanized rubbers. The effect of temperature on plasti wear resistance can be examined through the change of constant α in Eq. (6.25). Ratner, Lure, and Farberova (16,56-59) have shown that for fatigue wear $\alpha > 1$. Because α is characterized by the number of deformation cycles that fracture materials and because α increases with temperature, during the polymer's transition from the brittle to the ductile state, the rise in temperature leads to an increase in fatigue-wear resistance and in general wear resistance for plastics. It has also been noted that, at high enough temperatures, wear by abrasive paper is fatigue wear. Deriving from the molecular mechanism, fatigue wear is related to a material's durability. Ratner has suggested that the wear mechanism has a thermo-activation nature for fracture and is characterized by the relation

$$J = 1/\tau \ . \qquad (6.29)$$

For cyclic loads, the formula for durability in vulcanized rubbers has the form

$$\tau = C \exp [(U' - \gamma\sigma)/kT], \tag{6.30}$$

where C is a constant. The activation energy U' for fracture in the dynamic mode is less than the activation energy in the static mode.

Having substituted Eq. (6.17) for the strength in Eq. (6.30), we then obtain from Eq. (6.29)

$$J = J_o \exp [-(U' - \underline{\lambda} p\gamma)/kT] , \tag{6.31}$$

where J_o is a constant, and $\underline{\lambda} = 3v\mu$ is a constant which depends upon the material's structure and the friction coefficient.

We can make the following observations about Eq. (6.31). Undoubtedly, the process of fatigue wear is related to the thermo-fluctuating mechanism of durability. However, Eq. (6.31) holds true for static equilibrium conditions, whereas as wear is a dynamic process. The contribution of the friction coefficient to the exponent cannot be interpreted physically, because it depends upon the sliding velocity and temperature (see Chaps. 3 and 4). Therefore, Eq. (6.31) loses significance, since the temperature exponent becomes complicated. Nevertheless, especially for plastics, it is possible to assume that $\underline{\lambda}$ in Eq. (6.31) is a constant.

It is noteworthy that according to Eq. (6.31), for a friction force equal to zero, the wear is not equal to zero. This incongruous result arises from the inapplicability to wear of Eq. (6.30) for $\sigma = 0$. For $U' = \underline{\lambda} p_r$, we have $T = T_o$, and durability does not depend upon temperature. These facts indicate the insufficiency of Eq. (6.31). As Bartenev has shown (60), another equation is appropriate for elastic-polymer durability. These and other aspects of the theory of strength are treated in detail elsewhere (60,61).

Let us examine some basic experiemtnal data that illustrate the applicability of Eq. (6.31) and some other problems of the thermo-fluctuating aspect of the theory of wear. Polymer wear on a grating has been studied (58,62). Plasticized polyvinylchloride (PVC) and polymethyl metharylate (PMMA), polyurethane, and linear polyethylene have received particular attention.

According to Eq. (6.31) there should be straight lines in a plot of log J vs 1/T for various pressures. [The actual pressure $p_r = p_n/\phi$

enters into Eq. (6.31).] Straight lines are actually obtained for various pressures, converging at a common point that matches value J_o (Fig. 6.12). According to Eq. (6.31), at this point $T \to \infty$. However

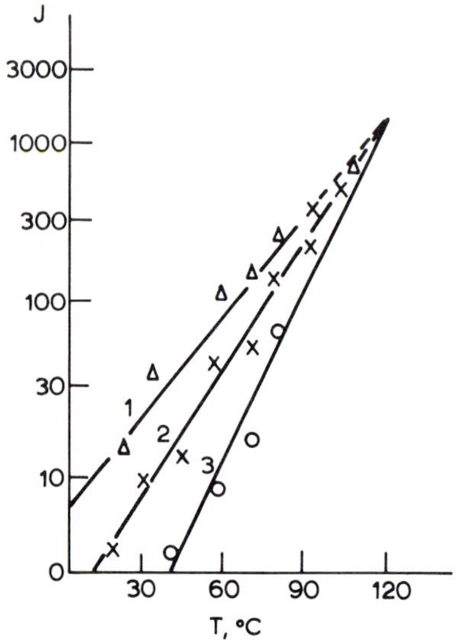

Fig. 6.12 Dependence of wear intensity J upon temperature, plotted for comparison with Eq. (6.31) for PVC + 30% plasticizer, for wear on a grating (63): (1-3) for real pressures 1.5, 1.0, and 0.5 kg/cm respectively.

one observes $T \neq \infty$; therefore, researchers have introduced an adjustment factor, and Eq. (6.31) takes on the form

$$J = J_o \exp\left(-\frac{U - \underline{\lambda} p_r}{kT}\right)\left(1 - \frac{T}{T_d}\right), \qquad (6.32)$$

where T_d is the base temperature. Therefore, at $T = T_d$, we have $J = J_o$, and Eq. (6.31) satisfies the boundary conditions. The base temperatures are close to the temperatures of decomposition for thermosetting resins or the friction temperatures for thermoplastic (63). An activation energy fairly close to those of degradation has been determined according to results of research on temperature relations. For PVC, U = 36 kcal/mole; for PMMA, U = 55 kcal/mole; and for PE(LD), U = 62 kcal/mole.

For abrasive wear (58,63), Ratner and Lure presented values $\underline{\lambda} p_r >$ kcal/mole, that is, the bond-rupture energy barrier. Hence, they concluded that abrasive wear is an energy-barrier-free phenomenon. For values $\underline{\lambda}p_r > U$, the equation has no real significance, because the exponent changes sign, and consequently the wear J is larger than the maximum value J_o, which is impossible. Generally, in our view it is not correct to speak of energy-barrier-free wear for the fluctuation theory of strength. Moreover, Ratner and Lure indicated that abrasive wear of plastics depends upon temperature (56); however, barrier-free wear should not depend upon temperature, and consequently contrary theories and experiments exist for abrasive wear. Therefore, it is not surprising that the values obtained for the activation energy (5 and 10 kcal/mole) are significantly less than the activation energy for rupture of chemical bonds (58,63).

Although the thermo-fluctuating theory of wear is used to explain molecular-fatigue wear, quantitatively the two are incompatible. Let us illustrate with only one example. According to the fatigue theory [Eq. (6.24)], wear depends upon load and the friction coefficient according to a power law; according to Eq. (6.31) the dependence is exponential--that is, even stronger; this prediction also does not agree with results obtained earlier by the authors.

The thermo-fluctuating theory of strength takes temperature as the most important factor in wear. It is true that at a constant temperature, wear, according to Kragelsky and Rybalov (51), essentially does not depend upon pressure, which contradicts Eq. (6.31). Perhaps this is related to the lack of correlation between wear on a grating and wear on smooth steel surfaces. Bilik has indicated this possibility (28). However, there is no doubt that the concepts of the fluctuation mechanism of wear are promising, and further refinement of the theory with the help of experimental results is needed.

EFFECT OF SLIDING VELOCITY. By the theory of fatigue wear [Eq. (6.24)], the sliding velocity affects wear indirectly, through a change in the polymer's elastic-strength properties and the friction coefficient. With increase in velocity, the temperature also rises in the contact area; therefore, study of the effect of velocity upon wear is possible only when the velocity is low (one can disregard the heating up of the surface) or at a constant temperature. For low sliding velocities, the friction coefficient is practically constant for plastics, but for vulcanized rubbers it rises proportionally to the logarithm of velocity (see Chap. 4). The modulus of elasticity depends weakly upon velocity, but noncontinuous expansion increases quickly with velocity. The concurrence of these factors makes wear at low sliding velocities increase only a little for vulcanized rubbers (4,7,64) and be practically constant for plastics (56,58).

For high sliding velocities, when the friction coefficient falls with a rise in velocity (37,46,49-51), wear also falls if the tempera ture is constant (Fig. 6.13). At first sight, the dependence of

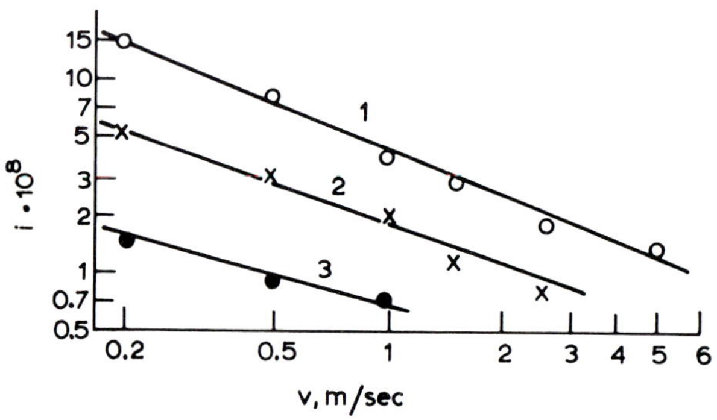

Fig. 6.13 Dependence of specific wear of a vulcanized rubber upon the sliding velocity for various temperatures (51): (1) 200°C; (2) 150°C; (3) 100°C.

wear upon the sliding velocity seems to contradict the known fact that wear is proportional to the power factor pv. However, under ordinary conditions an increase in velocity leads to a rise in temperature and growth in wear. Figure 6.13 shows results obtained for the dependence of wear upon the sliding velocity at T = const, a in this case, the wear decreases rather than increasing with increas velocity. The theory of fatigue wear can provide an explanation for this. It has been demonstrated (51) that at constant temperatur the friction coefficient decreases with increasing sliding velocity (for high velocities). See Fig. 6.14

Based on Eq. (6.24), it is possible to calculate the intensity of wear according to known relations between the friction coefficient and the change in the polymer's elastic-strength properties. As shown in Fig. 6.15, there is a good correlation between the calculat and experimental curves for wear. These results once again clearly confirm the conclusion that temperature is the most important factor determining polymer wear. From this point of view too, contrary information on the effect of the sliding velocity upon wear is also

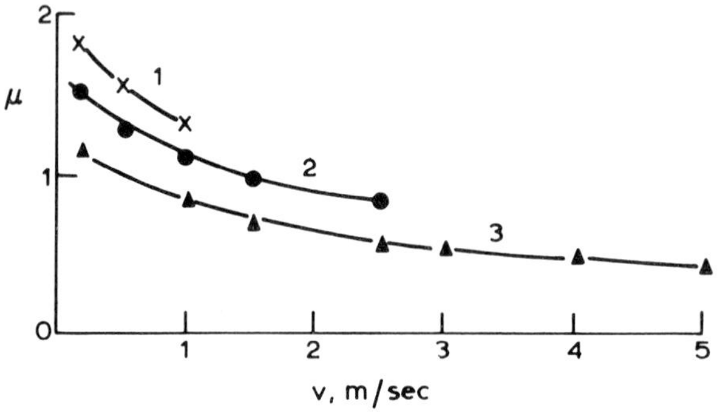

Fig. 6.14 Dependence of the friction coefficient upon high sliding velocities for a vulcanized rubber on steel for various temperatures (51): (1) 100°C; (2) 150°C; (3) 200°C.

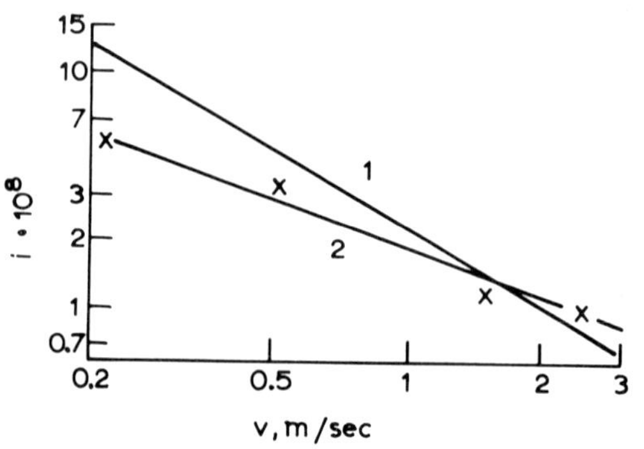

Fig. 6.15 Dependence of specific wear upon sliding velocity for a vulcanized rubber on steel (51): (1) calculated; (2) experimental (T = 150°C).

understandable, inasmuch as the temperature was not fixed. Similarly, it is necessary to examine the effect of sliding velocity upon plastic wear.

Effect of polymer physico-mechanical properties and composition

According to Eq. (6.24),

$$J = \text{const}\ E^{(t - \beta t - 1)}$$

Hence, it follows that with a rise in the modulus, wear grows because the exponent $t - \beta t - 1 > 1$. For example, for a sealing vulcanized rubber in friction against a steel surface (∇9-class roughness), $t - \beta t - 1 = 3.1$ (24). For other vulcanized rubbers and surfaces, exponents are often on the order of several units. Consequently, wear depends rather strongly upon the polymer modulus. Figure 6.16 presents data (41) that show the dependence of wear upon the modulus of a protective vulcanized rubber. Analogous results have been obtained by other researchers (32,43,65,66).

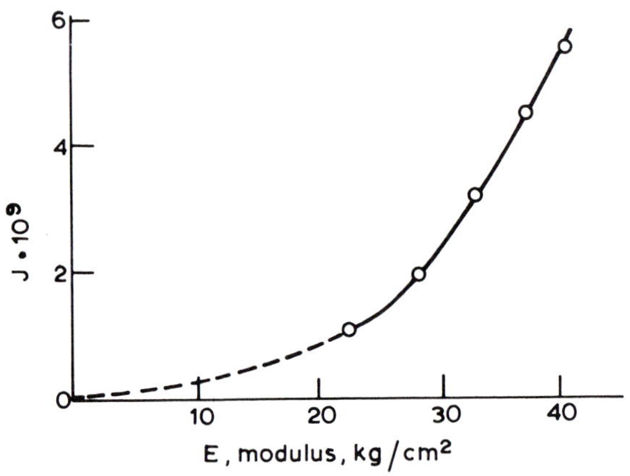

Fig. 6.16 Dependence of wear resistance on the modulus for a protective vulcanized rubber (41).

For wear of vulcanized rubber on a grating (32), the specific wear V_1 at a pressure $p_n = 1$ kg/cm^2 can be related to the physico-mechanical properties of the rubber by the following empirical equation:

$$V_1^{\zeta} = k[\mu\Delta/(\sigma_o E^{\alpha-1} K)], \quad (6.33)$$

where ζ is a constant obtained from testing, k is the proportionality coefficient; $\Delta = 100D$ (D is the hysteresis loss due to rebound), and K is the hardening coefficient.

For a series of vulcanized rubbers that differ in polymer type [natural rubber, SKB butadiene rubber, SKS-10 butadiene-styrene rubber, SKN-18, SKN-26, SKN-40, and Nairit or chloroprene rubber (equivalent to neoprene)] and in quantity and type of fillers and other ingredients, Fig. 6.17 shows the relation of wear to physico-mechanical characteristics, plotted for comparison with Eq. (6.33).

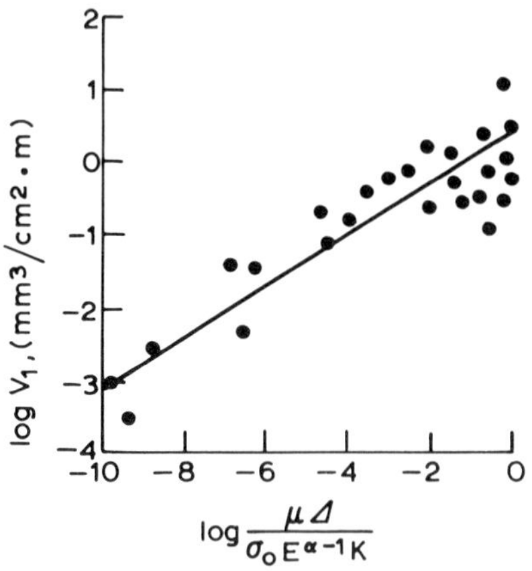

Fig. 6.17 Dependence of the logarithm of specific wear (i) upon physico-mechanical properties of a vulcanized rubber (32).

We see that the use of such empirical formulas is rather difficult, because during experimentation it is necessary to determine several constants, and moreover the quantities μ, Σ_o, E, as pointed out above, depend upon the sliding velocity and temperature. Therefore, Eq. (6.33) expresses only qualitatively the relation between wear and polymer physico-mechanical properties. The relation between Eq. (6.33) and the basic equation of the theory of fatigue wear (6.24) is not clear, nor is it clear that Eq. (6.33) reflects this mechanism of wear. From our point of view, it is doubtless preferable to use Eq. (6.24) for preliminary evaluation of the effect of various factors on the wear intensity.

According to some data (67), wear is directly proportional to the modulus and the friction coefficient, and inversely proportional to the fracture strength. Comparison of experimental data and empirical formulas that tie wear to the modulus does not allow a definitive conclusion about the effect of a polymer's modulus upon wear. This is explained by the fact that it is impossible to change a material's modulus without changing its other properties. Therefore, it is correct to use only Eq. (6.24) for an evaluation of the effect of the modulus on wear.

For semiquantitative evaluations, the equation for wear intensity can include hardness in the form

$$J = p_n/(nHB) \quad . \qquad (6.34)$$

From Eq. (6.34), we see that it is convenient to use the quantity $p_n/(HB)$ to determine wear resistance because the value of hardness alone cannot determine wear. This equation is also convenient for a calculation of the effect of pressure on wear for vulcanized rubbers of varying rigidity (68).

For plastics, according to Ratner, Eq. (6.34) cannot be accepted in principle, because with changes in hardness the ductility changes too. Generally speaking, in polymers, changing the hardness or modulus of elasticity changes other physico-mechanical characteristics and therefore, interpreting the effect of hardness with Eq. (6.34) is not always possible (55).

It is important to know how and in which direction a polymer's composition affects the wear resistance. Without setting up an experiment, it is impossible to determine this, inasmuch as the theory allows the prediction of the wear intensity only in terms of physico-mechanical and fatigue properties which are related in a complex manner. Nevertheless, enough experimental material has been gathered in this field to make some practical recommendations.

It is known that vulcanized rubbers represent a large class of rubbery materials, whose nature and properties are completely determined by the nature of the raw rubber and other ingredients (fillers). Generally, vulcanized rubbers are the end product used in industry. Let us examine the effect of the raw rubber's nature and the filler on wear of vulcanized rubbers. A change in the composition of vulcanizate leads to change in its physico-mechanical properties; therefore, for a precise examination of the effect of vulcanizate composition on wear it is necessary to calculate separately the effects of change in physico-mechanical properties and in compounding factors. Let us turn to some studies in this field (69-72).

Kuperman and Karmin (69) have shown that high wear resistance in vulcanized SKD rubbers is due to the formation in the vulcanizate of thermally and mechanically stable rubber-carbon-black structures. In stocks from SKD and carbon black of the type XAF, a stable network rubber-carbon-black structure forms. Data gathered in independent experiments on plasto-elastic properties of these stocks over a wide range of temperatures confirm this observation. Wear resistance studied on an IMI-3 machine well matches the indicators that characterize the degree of development and stability of the rubber-carbon-black structure. Vulcanized SKD rubbers also have little internal friction and high thermo-oxidative stability. This is related to the high ductility of natural rubber SKD molecules, and the absence of lateral groups and branches which usually mix in molecules of other rubbers on the surface of carbon-black particles, and is the result of the higher molecular weight and the narrow M.W. distribution of the rubber SKD. These special properties of SKD determine the physical state of the rubber on the surface layer.

As an illustration of SKD's good wear resistance, Fig. 6.18 shows

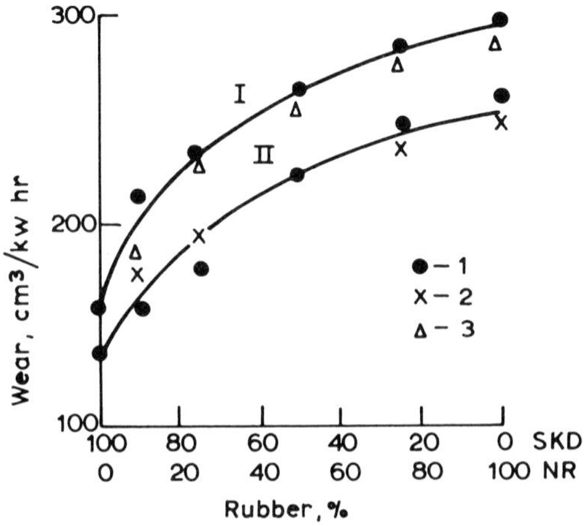

Fig. 6.18 Relation of wear intensity of vulcanizate (50 wt. % of XAF carbon black) to type of rubber at various temperatures (69): (I) 20°C; (II) 100°C; (1) SKD; (2) natural rubber; (3) SDK.

wear for vulcanized raw SKD rubbers, natural rubber, and BSK for two temperatures (20 and 100°C). As shown in figures, an increase in content of SKD in the stock leads to a decrease in intensity of the vulcanized-rubber wear. Carbon black is a basic filler of vulcanizates that raises their wear resistance (67,70-79). There are a large number of carbon blacks, which differ from one another in degree of dispersion and surface state. With increase in dispersion of the carbon black, strength properties and wear resistance improve for vulcanizates, but elasticity falls. The chemical nature of the carbon-black surface affects the formation of the rubber-carbon-black structure, the vulcanization process in rubbers, and the type of vulcanizate that resulting from the vulcanization of bonds (80). The nature of carbon-black particles also affects the fatigue properties of vulcanizates.

Other compounding factors that affect wear resistance in vulcanizates are plasticizers, antioxidants, vulcanizing agents, etc. These affect wear indirectly by means of change in the material's physico-mechanical properties (65,81,82). Research on wear of vulcanizates on a grating has shown (30,83) that the constant α, which depends upon polymer polarity, fillers, and degree of swelling, is a structure-sensitive parameter in Eq. (6.25). Increase in a rubber's polarity raises its value from 1.8 for SKN-18 to 3.84 for SKN-40. Here we observe relation to the cohesive energy, which also rises with rubber polarity. There is a rather good correlation between the experimental and theoretical values of α.

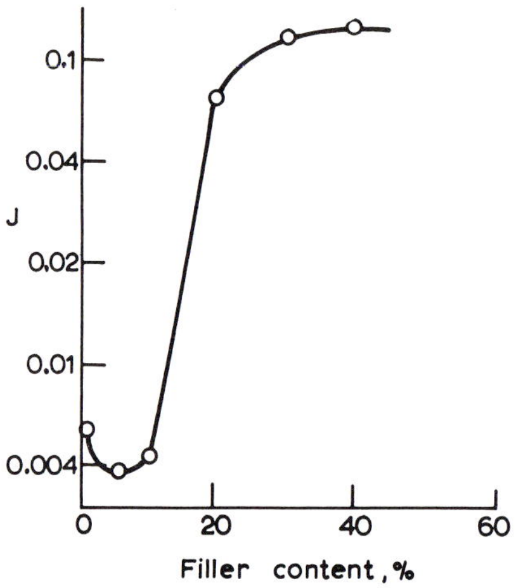

Fig. 6.19 Dependence of wear intensity upon amount of filler (talc) for Polyamid-68 on a grating (85).

Ratner and Farberova (16,59,84,85) have conducted a systematic study on the effect of composition on plastic wear resistance J. They have demonstrated that

$$J = \mu/(HB\sigma\varepsilon_f). \qquad (6.35)$$

An increase in the relative strain at rupture, ε_f, enhances the wear resistance. Thus, for example, the addition of filler to polyfluoroethylene resin and Polyamid-68 brings about an increase in wear resistance. As a rule, filling first leads to a rise in wear resistance for plastics as a result of increase in strength and rigidity. With further increase in the amount of filler, wear increases because of a reduction in strength. A plasticizer acts analogously to temperature, that is, an increase in the amount of plasticizer is equivalent to rise in temperature.

In general, it is possible to conclude that the most wear-resistant plastics have high strength and elasticity. One of the most wear-resistant plastics is Polyamid-68, with nylon 6 (polycaprolactam), Polyamide AK-7, and Caprolon V after it. Other plastics suffer much wear (Table 6.3). It is necessary to keep in mind that these results are for plastic wear on a grating and with change in surface type and mode of testing some special characteristics of wear in plastics can appear: their transfer to the surface of a solid, change in physical state, etc.

Nevertheless, thanks to the known correlation between wear on a grating and on smooth surfaces, general principles of the fatigue mechanism in polymers survive. A quantitative analysis of wear principles is difficult because the ingredients introduced into the polymer affect a material's mechanical properties in a complex manner.

Effect of surface friction. Transition to abrasive wear.

The wear intensity and wear resistance for polymers are determined not only by physico-mechanical properties, but also to a rather large degree by conditions of testing, the nature of the surface, and the environment.

For wear on a grating, the nature of the surface cannot be controlled; therefore, for example, the role of the friction coefficient, important in the theory of fatigue wear, cannot be determined. From the practical point of view, a change in the friction coefficient brings about a significant change in wear. Moreover, the nature of the hard surface on which a polymer slides determines the wear mechanism (86). But plastics that have sufficient adhesion to a hard surface during friction can be transferred onto it, sharply changing the character of wear.

For fatigue friction, wear increases with adhesion of the polymer to a hard surface. Thus, for example, vulcanized-rubber wear is greater against a polymer than when against a steel, where the rubber specimens have the same geometry (86). The friction coefficient changes similarly. However, for wear, the friction on a polymer surface is more important as a consequence of the polymer's poor thermal conductivity, which leads to a rise in the contact temperature. On smooth surfaces with low thermal conductivity, vulcanized-rubber

TABLE 6.3

Wear indices (or exponents) and elasto-strength properties of various cases of plastics

Material	Wear Index V (mm/m-cm)		Expo-nent[a] α	Friction Coefficient k	HB (kg/mm^2)	Σ (kg/cm^2)	ε_f (%)	E (10^3 kg/cm^3)
	Abrasive cloth	Grating						
Polyamides								
Polyamid--68	0.9	0.0001	1.8	0.20	12	440	160	4200
Capron	0.6	0.00015	2.0	0.13	10	580	180	8000
Polyamide AK-7	0.7	0.0003	1.6	0.15	15	470	75	3500
Caprolon B	0.8	0.0009	1.2	0.20	20	900	10	900
Polyolefines								
Low pressure polyethylene	0.5	0.0007	1.5	0.10	4.5	200	600	5400
High pressure polyethylene	0.6	0.0530	1.7	0.45	2	100	150	67
Polypropylene	2.0	0.0060	2.5	0.30	7	290	150	1000
Halogenated Polymers								
Polytetrafluoro-ethylene (PTFE)	5.0	0.0065	1.1	0.10	4	220	340	3000
Polyfluoroethylene								
Resin 40	2.0	0.0010	1.4	0.10	7	350	200	4900
Vinyl plastics	4.2	0.0050	1.3	0.20	20	480	20	1000
Phenoplastics								
Voloknite	2.9	0.0060	---	---	--	---	---	----
FKP-1	3.7	0.012	1.2	---	--	---	---	----
Other Plastics								
Organic glass ST-1	3.5	0.0050	1.1	---	--	---	---	----
Shock-resistant polystyrene	5.0	0.0600	1.3	---	12	4-0	4	----
Block polystyrene	4.5	0.400	1.15	---	20	270	2	----

[a]For wear on a grating (for wear on an abrasive paper α = 1).

wear occurs as a result of fatigue and thermo-mechanical degradation. Sometimes changes in wear of several orders of magnitude coincide with changes in the friction coefficient. In research on wear of plastics, the effect of the nature of the hard surface has also been noted (12-14,28,38), but studies to date have been mainly qualitative.

Changes in rough surfaces also show up in the intensity and mechanism of wear. It has already been noted that an increase in roughness of a hard surface during wear of rigid polymers leads to the appearance of longitudinal plowing trenches and to abrasive wear. Often a mixed type of wear is observed. It is possible to calculate the extent of either form of wear from the following considerations (25).

Fatigue wear is characterized by the size of the exponent α in Eq. (6.25). When $\alpha > 1$, the wear process is determined by the fatigue mechanism. For $\alpha = 1$, the wear is abrasive. This follows from theory and extensive experimental data (3,4,6,7,10-19,24,27, 29,30,32-37). Soft elastomers are known to wear by fatigue even on an abrasive paper. Rigid and brittle polymers wear abrasively even during friction on smooth surfaces. On a grating and on hard surfaces, vulcanized rubbers wear by the fatigue mechanism. Plastics that have developed constrained elasticity also wear in a fatigue manner, but with a definite tendency to abrasive wear. Rigid polymers basically wear abrasively on all surfaces.

The nature of wear can be determined by the pattern of wear. Fatigue and friction wear produces transverse bands; abrasive wear produce longitudinal bands. If $V = V_{=} + V$ (where $V_{=}$ and V are the wear with formation of longitudinal and transverse bands respectively), then according to the experimental data, the relation to pressure is $V \propto p^{\alpha}$. A wear pattern is characterized by the distances between friction bands, z; for this, $z \propto p^{B}$. Moreover, according to Schallamach (7) and Ratner (25), wear can be represented by $V \propto z^{3}$. For abrasive wear, there are no bands. Thus, $V = V_{=} + V$ or $V = \text{const}\ z^{3\Lambda}$, where $\Lambda = \alpha/3B$ indicates the proportion of friction wear. For $\Lambda = 0$, wear is abrasive, and for $\Lambda > 1$, wear is by friction. However, it is necessary to note that such an evaluation is only qualitative. This is also confirmed by the fact that according to the equations presented the relation

$$V = \text{const}\ z^{3\Lambda}$$

is purely an empirical equation that does not have strict physical significance.

The latter equation establishes a relation between the exponents of fatigue wear (α) and friction wear (β) in the hypothesis that $V \propto z^3$ for all materials.

6.3 ABRASIVE WEAR

Abrasive polymer wear consists in cutting (for example, in buffing, or with abrasive paper) by scratching the surface with sharp protuberances. The concept of "sharp protuberance" is rather relative, because abrasive wear also depends upon the polymer's rigidity. How the material fractures is important. In a given case, the wear is monoactive (by contrast with fatigue wear). Often abrasive wear develops during friction of rigid plastics on the surface of an abrasive cloth. Here the wear process is related to cutting and plowing of the surface by the abrasive-cloth granules. The presence of longitudinal plowing bands on a polymer surface is an external manifestation of abrasive wear.

Schallamach (7) first qualitatively studied elastomer wear, simulating the wear process by sliding a sharp or dull stylus on the elastomer's surface. He showed that the simple act of tearing of an elastomer surface is determined by overstrains that develop behind the moving stylus. These overstrains lead to the appearance of transverse bands, the so-called "Schallamach pattern". Ratner (25,35) believes that the model stylus-polymer surface describe friction wear rather than abrasive wear. It is not possible to concur in his conclusion, because abrasive wear is determined by the character of the interaction on the dividing boundary of the polymer-solid pair and depends upon the properties of the worn body. The sliding of a sharp stylus on a soft rubbery polymer leads to basically friction wear, while sliding of the same stylus on the rigid surface of a plastic sets in motion plowing* and cutting, that is, abrasive wear.

Now let us examine in detail the general principles of abrasive wear and their relation to polymer type.

---

*The process of plastic recovery is also possible.

General principles of abrasive wear

Krushchov and Babichev (90) have studied abrasive wearing of metals and have demonstrated that the wear is given by

$$J \approx k(p/HB). \tag{6.36}$$

Here k is a coefficient, p is pressure, and HB is hardness.

An analogous expression follows from Kragelsky's theory (3) [see Eq. (6.9)]. Ratner, Melnikova, and Klitenik (9,32) on the basis of more extensive experimental material demonstrated that polymer wear resistance can be expressed by the equation

$$J = J_1 p_n,$$

where J_1 is the wear for $p_n = 1\ kg/cm^2$. Comparing Eqs. (6.9), (6.22), and (6.36), it is possible to draw a general conclusion that the abrasive-wear resistance for all materials (except soft vulcanized rubbers) is proportional to pressure. This conclusion is confirmed by experimental data (Fig. 6.20). The linear dependence of wear

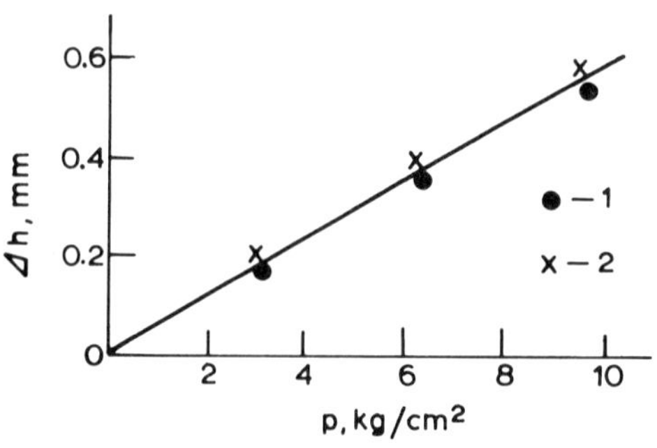

Fig. 6.20 Dependence of wear upon pressure for organic glass (PMMA) at speeds of a rotating disc (on machine X4-B) (15): (1) 15 revolutions/min; (2) 60 revolutions/min.

can be qualitatively explained by the fact that $J \propto S$ [see Eq. (6.7)]. According to Eq. (2.6) for simulated contact $S \propto p^{8/9}$. Consequently, in practice $J \propto p$, which is observed experimentally.

Often, to calculate wear resistance, a criterion on the product $p_n v$ is used in place of pressure, because it has been established experimentally that wear for p_n = const depends upon the sliding velocity. For some values of $p_n v$, wear is proportional to the load and friction track and does not depend upon the sliding velocity. This is true, for example, of PMMA on an abrasive cloth (15), as is shown in Fig. 6.21. For large values of $p_n v$, intense heat release

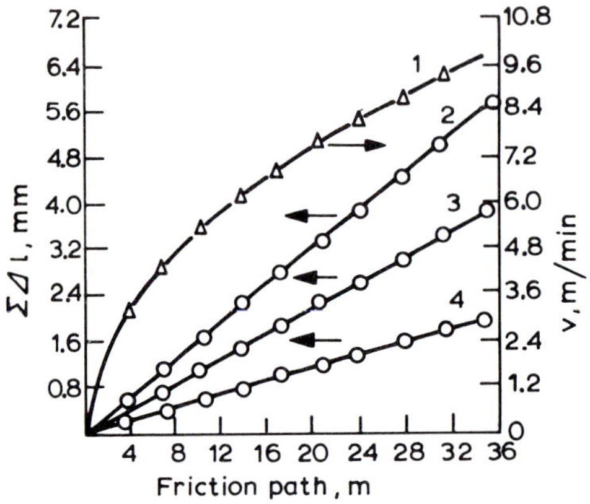

Fig. 6.21 Dependence of total wear $\Sigma\Delta\tau$ upon the friction track for polymethylmethacrylate at various friction speeds and pressures (15). (1,2) $p\underline{v}$ = 94 (kg/cm^2 (m/min); (3,4) $p\underline{v}$ = 36 (kg/cm^2) (m/min).

appears, which lowers the polymer's fracture resistance. With the attainment of a critical value of $p_n v$, the character of wear changes (as revealed by the size of particles of the severed material). The larger the value of $p_n v$, the higher the polymer's wear resistance

Generally speaking, the dependence of wear upon sliding velocity has been little studied. This is primarily because velocity affects wear mainly by changing the temperature. From the point of view of the thermo-fluctuating mechanism for abrasive wear, temperature hardly affects the wear rate. The velocity dependence of the resistance testifies either to the inadequacy of the theory or to a change in the polymer's physical state with change in temperature. The relation of vulcanized-rubber wear to temperature indicates the theory's inadequacy. For plastics, as will be shown below, the effect of temperature on wear through a change of physical state is important. Let us examine, following Ratner (56), the effect of temperature on wear of plastics.

In the range between the glass and the brittle temperatures (Fig. 6.22), hardness and strength are large, the friction coefficient μ is small. Using the qualitative relation between wear and these

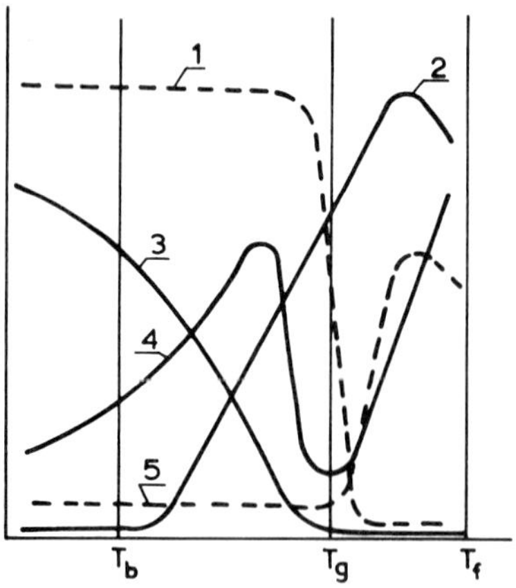

Fig. 6.22 Effect of temperature on mechanical properties and wear of amorphous polymers (16): (1) Hardness HB; (2) elongation at rupture, ε_f; (3) tensile strength σ; (4) wear rate V; (5) friction coefficient μ. T_b, brittleness temperature; T_g, glass temperature; T_f, flow temperature.

measures--see Eq. (6.35)--we learn that for an amorphous polymer, wear at temperatures up to T_b is small and increases as a result of reduction in strength (Fig. 6.22). For temperatures between T_b and T_g, the wear curve has a maximum. Here μ and HB practically do not change, but the strength falls sharply and the ductility increases. The superposition of these factors leads to the appearance of a maximum in the wear curve.

For temperatures from T_g to T_f the elongation decreases, the friction coefficient passes through the maximum, and the hardness is small. Wear increases in this range. The smallest value of the wear V corresponds to the glass temperature T_g. The amorphous part of crystalline polymers up to T_m, the melting temperature, exists in the viscoelastic state; therefore, the wear pattern is different (Fig. 6.23).

In the vicinity of T_b, wear increases because of the reduction in strength. With a rise in temperature above T_b, wear falls sharply, because in this range the polymer is characterized by increase in constrained elastic deformation with practically constant modulus of elasticity E and friction coefficient.* Actually, μ is a more complicated function of temperature (see Chaps. 2 and 4). For

*The curves for friction coefficient in Figs. 6.22 and 6.23 are qualitative only.

further changes in temperature, from T_b to T_m, the wear is minimal and practically constant.

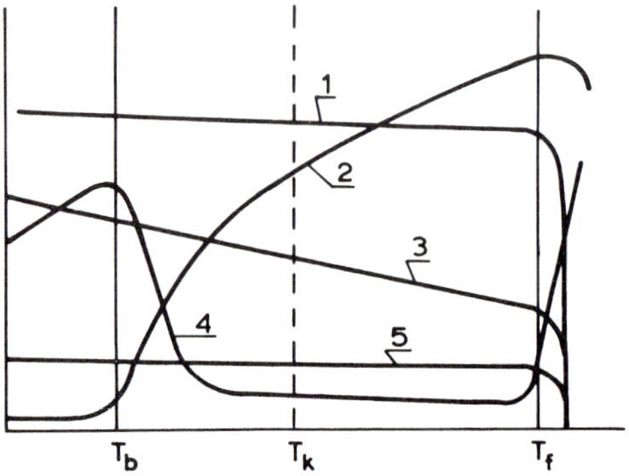

Fig. 6.23 Effect of temperature on mechanical properties and wear of crystalline polymers (16). T_k, room temperature; T_f, flow temperature, coinciding with T_m, melting temperature; other symbols are the same as in Fig. 6.22.

Above the crystalline melting temperature, wear grows as a result of a sharp drop in the modulus and tensile strength. A low value of the modulus causes a large contact area and a low tensile strength which promotes a significant increase in wear. Experimental data for PVC (vinyl plastic), PMMA, and polypropylene obtained by Ratner (25) qualitatively confirm the above expectations on the change in wear according to Eq. (6.35).

It has been demonstrated that polymer wear resistance according to the fatigue theory is inversely proportional to hardness [see Eq. (6.34)]. An analogous relation for wear resistance is also observed for abrasive wear, according to Schallamach (7); for vulcanized rubbers equations like (6.36) hold true. Consequently, most generally wear is inversely proportional to hardness (or the modulus) independent of its mechanism*--see Eqs. (6.9), (6.22) and (6.35). Babichev (15) has described the dependence of relative abrasive wear resistance R upon hardness for various plastics and metals (Fig. 6.24) and has demonstrated that

$$R = b\ HB. \quad (6.37)$$

*However, according to some data (8), abrasive wear increases with hardness.

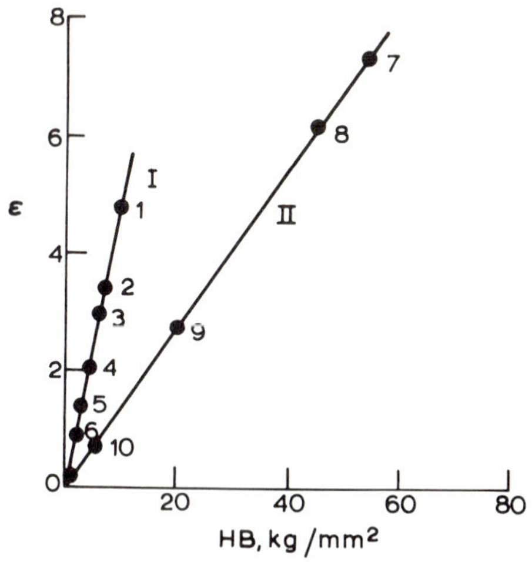

Fig. 6.24 Dependence of relative wear resistance upon hardness (HB) for (I) plastics and (II) metals: (1) L54; (2) L68; (3) nylon 6; (4) low-pressure polyethylene; (5) high-pressure polyethylene; (6) polyfluoroethylene; (7) Ag; (8) Zn; (9) Cd; (10) Pb.

It is interesting to note that the slope, characterized by the proportionality coefficient, b, is 3.2 times larger for polymers than for metals. This indicates the relatively high strength of polymers in comparison with metals at the same hardness. We note that hardness and the modulus of polymers depend upon the duration of observation (15,12,60); therefore, the size of the coefficient b is determined by the testing mode.

Effect of external factors

Wear of polymers by an abrasive cloth is a comparative method of measuring wear resistance for polymers. However, some polymeric articles work under conditions of abrasive wear by the action of a free abrasive (sand, coal or carbon, etc). It is of interest to find a relation between polymer wear by an abrasive cloth or paper and by a free abrasive. The most complete comparison has been made by Tenenbaum (18), who noted a definite correlation between the two methods of investigation. He showed that wear of polymers by sand depends upon pressure according to a power law (exponent ≥ 1).

Figure 6.25 presents some results of the study of wear of plastics by a free abrasive (sand), from which it follows that the exponents of the load are larger than one. Because wear by an abrasive paper is linearly proportional to load, correlation between wear by an

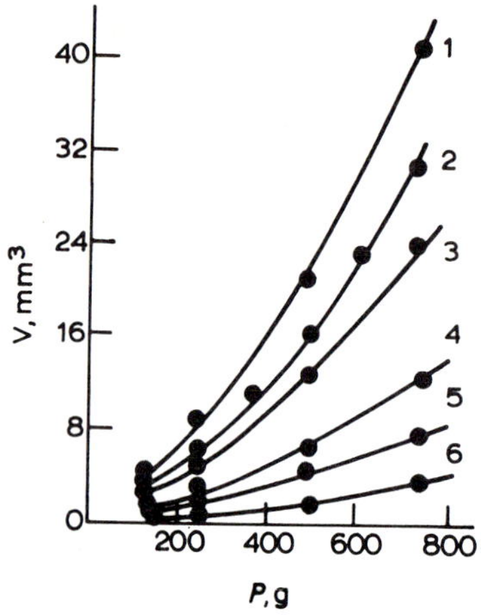

Fig. 6.25 Dependence of bulk wear of polymers by sand under normal load (18): (1) polystyrene; (2) polymethylmethacrylate; (3) high-pressure polyethylene; (4) PTFE; (5) low-pressure polyethylene; (6) Polyamid-68.

abrasive paper and by a free abrasive, and therefore the possibility of a quantitative prediction of the wear produced by a free abrasive based upon data for wear by an abrasive paper, are hampered. However Ratner (87) believes that wear by a free abrasive is proportional to pressure.

There is a more general question about the similarity among the various sizes of debris particles produced during wear by abrasives. As shown above, abrasive wear is proportional to pressure and independent of the type of material tested. One can conclude that relative wear of materials by abrasive cloths of different gages does not depend upon the size of the abrasive grain, but is determined by the polymer's properties. In practice, this indicates the equivalence of calculations of polymer wear by abrasive cloths of different coarseness.

The effect of a polymer's composition on its wear resistance can be analyzed with Eq. (6.35), which reveals the general relation between physico-mechanical properties and wear resistance of different kinds of polymers, independent of the kind of worn surface. Change in the composition of polymers, vulcanized rubbers, and plastics is attained by the introduction of various ingredients and under different conditions (temperature, pressure, and so on). More than fatigue wear, abrasive wear is characterized by the effect of elongation at rupture (59,76,84).

As a rule, wear resistance for vulcanized rubbers increases with introduction into the raw rubber of an active filler, which increases the vulcanizate's modulus and strength up to a definite amount (optimum filler). The largest effect is on strength (9,88). For example, nonactive fillers and increased modulus have little effect on wear. Active fillers bring about a significant improvement of strength properties and a corresponding increase in wear resistance (Table 6.4). It is known that up to the optimum filler the friction

TABLE 6.4
Amount of filler corresponding to optimum strength and wear resistance of vulcanizates (9)

Raw Rubber	Type of Filler	Amount of Filler Weight (Part)	
		For Σ	For J_f
SKB (Synthetic butadiene rubber)	Channel black	75	75
	Powdered silica gel	75	75
	Lampblack	100	105
	Kaolin	100	105
SKS-30M (A butadi-ene-styrene rubber)	Channel black	65	75
	Lampblack	100	100
	Lampblack	100	100
	Chimney soot	75	60
Nairit (Chloroprene rubber-equivalent neoprene)	Channel black	45	35
	Chimney soot	65	75

coefficient does not depend upon the amount of filler (see Chap. 4); therefore, according to Eq. (6.35), the wear resistance increases as a result of the increase in the polymer's strength and modulus. It is true that the elongation at rupture decreases, but this does not affect the strength and modulus. The type of carbon black is significant for wear resistance (86). Increase in the specific carbon-black surface area leads to reduction in wear. Thus, for example, for a vulcanized natural rubber a filling of channel black with specific surface area 114 m^2/g corresponds to wear resistance of J_w = 170 cm^3/kW-hr, and a filler of thermal black with specific surface area 8 m^2/g has a measure J_w = 535 cm^3/kW-hr (86). Interpretation of the effect of plasticizer on wear is more complex, because with plasticizer the friction coefficient, modulus, and elongation change. Because abrasive wear of vulcanized rubbers is more important in practical use than fatigue wear, and because the mechanical properties of vulcanized rubbers change relatively little, not much attention has been given to the effect of ingredients on wear. For plastics, the role of plasticizers and other ingredients is more significant. Plastics occur in various physical states, and their wear resistance varies within wider limits than that of vulcanizates.

Farberova (59,84) has systematically studied the effect of ingredients on plastic wear. Following Ratner and Farberovoi, we shall proceed from Eq. (6.35) in interpretating of the effect of compounding factors on plastic wear--that is, from the effect of compounding factors on physico-mechanical characteristics and strength, elongation at rupture, hardness, and the friction coefficient. As shown above, elongation at rupture has the most significant role.

The introduction of a plasticizer into a polymer increases chain mobility, and on this score, the plasticizer's action is similar to the effect of raising the temperature: the polymer's hardness and strength decrease, while elongation at rupture and wear increase. Table 6.5 shows data on the plasticizer's effect on the wear and mechanical properties of polyvinylchloride. Its effect on the wear of other polymers is analogous.

TABLE 6.5

Effect of plasticizer on the wear and mechanical properties of polyvinyl chloride (59)

Content of the Plasticizer Weight (Part)	J_1 (10^4)	HB (kg/mm^2)	Σ (kg/cm^2)	ε_f (%)
10	3.5	5.3	510	55
20	1.8	3.6	375	60
30	1.5	3.2	260	155
40	1.4	2.8	190	200
50	1.3	2.3	145	220
60	1.2	1.9	125	235
70	1.1	1.6	105	255
80	1.1	1.4	90	265
90	1.1	1.1	80	280

Filler affects plastic wear in a significantly more complex manner (59). With increase in the amount of filler, the change in wear and mechanical properties of plastics is not monotonic. As a rule, small portions of filler raises the wear resistance of plastics. With increase in content, either further increase or reduction in wear resistance can occur, depending on the filler's properties and the nature of the polymer. Figure 6.26 shows that the change in

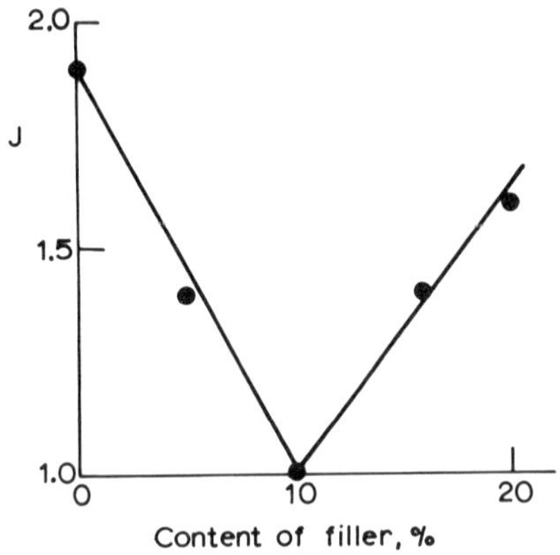

Fig. 6.26 Dependence of abrasive wear upon the amount of titanium dioxide for polypropylene (59).

wear is proportional to the amount of filler (titanium dioxide), from which it follows that the percent of wear is minimal at a filler content of 10 percent weight. Here elongation and hardness are at a maximum. Introduction of filler into a polymer to raise its wear resistance is common in practice, but there is no standard of choice for the filler and its amount. Both are determined empirically. If we consider that correlation between wear on abrasive cloths or papers with wear on smooth surfaces is not always observed, then we can comprehend the difficulty of selecting an optional amount of filler.

Improvement in the wear resistance of plastics can also be made by the blending of different polymers. Thus, for example, the introduction of natural rubber into crosslinked phenol-formaldehyde at first reduces wear, and then increases it (59). To achieve positive results, it is helpful to use the following principle: with increase in fatigue wear a plastic's general resistance to wear increases. This holds true for wear of polymers in general.

Polymer wear resistance also depends upon the structure of the polymeric substance. Here we shall not examine this problem, but note that sometimes one or another polymer component is sufficient to change the structure. This field of polymer wear has been investigated by various scientists (18,68,89-91,etc.).

In conclusion, it must be said that abrasive wear of polymers has been very inadequately studied. Researchers have had no general, consistent concepts of the nature of wear. Basically, only applied studies have appeared; wear is studied by various methods and under different conditions. Such a research approach can add nothing new to our understanding of wear. Principles that have been obtained have qualitative, descriptive character and do not reflect the wear mechanism. Regrettably, there are few studies by simulation of the initial wear behavior (of the type begun by Schallamach). There are only poorly constructed hypotheses that often are simply inconsistent Thus, for example, abrasive wear is considered a monoactive process, micro-cutting an energy-barrier-free activation process. Not one of these possible mechanisms has been studied thoroughly. To the presen there is no theory of abrasive wear of polymers.

6.4 SPECIAL TYPES OF WEAR FOR FRICTION ON SMOOTH HARD SURFACES

Wear of vulcanized rubber on smooth hard surfaces basically proceed by the fatigue mechanism (see Sec. 6.2). However, in some cases, especially at large values of the friction force, sliding velocity, and temperature, wear is related to another mechanism--the so-called wear by roll formation.

The wear mechanism in question was first studied by Reznikovsky and Brodsky (31,92) and by Rudakov and Kuvshinsky (93). The main occurrence of this mechanism on the surface of a vulcanized rubber; intense wearing is observed. The process and basic principles of wear in this case differ from both fatigue and abrasive wear.

Roll formation during friction of a vulcanized rubber against a smooth surface is illustrated in Fig. 6.27. As a result of contact

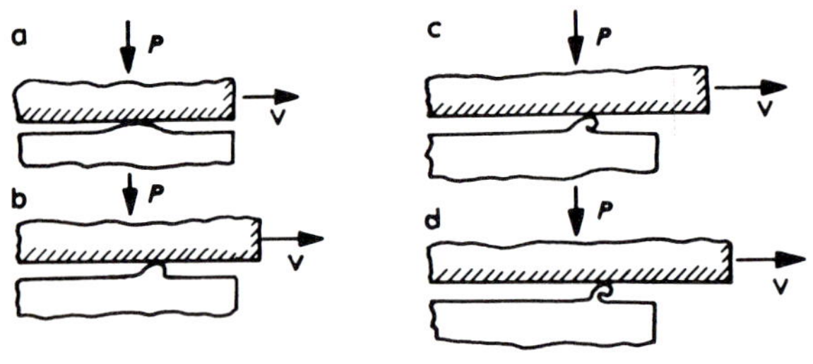

Fig. 6.27 Successive stages (a-d) of roll formation in friction of a vulcanized rubber on a smooth surface (31).

discreteness, separate rubber protuberances interact during relative displacement with a hard smooth surface. The motion of the hard surface provides a friction force which moves the contacting section of the polymer, deforming it in a rather complex way. The deformation of a protuberance is determined by the polymer's elastic properties and by the friction force. Heterogeneity of the surface and the formation of micro-defects and micro-fissures on the surface will lead on the average to leveling of protuberances into the space between friction surfaces (Fig. 6.27b). Here either the breaking off of protuberances or a change of their form and size is observed. Change in size of the protuberances is more probable; because of the buildup of irreversible deformation and surface strain as a result of repeated deformations and strains, a protuberance can turn into a "wavelet" (Fig. 6.27d). Under fixed conditions the number of wavelets increases, the defective nature of the surface increases, and wear becomes catastrophic.

The formation of wavelets can be examined quantitatively (31). Let us assume that the relative sliding velocity of a hard surface is equal to v. The absence of slipping* indicates that the velocity of forward motion of wavelet, or of the deformation strand linking a wavelet to the surface, is equal to the peripheral speed in contact with the surface.

Resistance to the movement of wavelets is determined by their rolling friction and by resistance to tearing of the strands holding them in place. But the losses during fracture, as calculated from nonelastic relaxation properties of vulcanized rubber, will also depend upon the deformation of the strand that connects a wavelet to the rubber surface. Generally, write

$$W' = W'_{tear} + W'_{ten} + W'_{roll}, \tag{6.38}$$

where W' is friction work or total work performed in moving the wavelets. The tearing work is equal (94,95) to

$$W'_{tear} = \overline{W}_o w(du/dt), \tag{6.39}$$

where $\overline{W}_o$ is the characteristic tear energy, w is the width of a shred, and du/dt is the tear velocity.

*This condition is introduced only to simplify the calculation. In the presence of slipping the general wear pattern does not change.

Because the movement of wavelets determines the tear velocity through the rate of shred deformation, to a first approximation one can assume that $x = w_o(du/dt)$. Then

$$W'_{tear} = \overline{W}_o w(X/2W_o). \qquad (6.40)$$

The tension can be expressed in terms of the specific tensile energy $\overline{W}_v$:

$$W'_{ten} = hwW_v(dv/dE) = hw\overline{W}_v(X/2W_o), \qquad (6.41)$$

where h is the thickness of a detached shred.

The roll formation energy to a first approximation is equal to

$$W'_{roll} = 0.45 \frac{P^{4/3}(1 - \overline{E})^x}{(Ezw)^{1/3}}, \qquad (6.42)$$

where $\overline{E}$ is the elasticity of vulcanized rubber (determined by Soviet State standard 259-41), and r is the radius of a wavelet. From Eqs. (6.40)(6.42), it follows that

$$W' = (\overline{W}_o wx)/(2w_o) + (hw\overline{W}_v x)/(2w_o) + 0.45 \frac{P^{4/3}(1 - \overline{E})x}{(Ezw)^{1/3}}. \qquad (6.43)$$

The specific wear [if it is assumed that $d(\Delta V)/dt = ab(v/2b_o)$] is

$$i = W'/[d(\Delta V)/dt] = W_o/h + W_v + 0.9 \frac{P^{4/3}(1 - \overline{E})2w_o}{hw(Ezw)^{1/3}}.$$

By further simplifications and calculations, it is possible to obtain (31) a final equation for the specific wear resistance:

$$i = W_o/h + (h^2W_o^2)/(2h^2E) + 0.9 \frac{P^{4/3}(1 - E)[W_o/Eh + 1]}{hw(Ezw)^{1/3}}. \qquad (6.44)$$

It follows from this equation that the specific wear resistance basically depends upon the characteristic tear energy W, the modulus E, and the elasticity $\overline{E}$ of vulcanized rubber resin. We can say that for other conditions equal, the higher the wear resistance, the larger the tear energy. Unfortunately, some researchers (31,92) do not present experimental data on the basis of Eq. (6.44); therefore their data have only qualitative significance. It is noted only that roll formation promotes an increase in the friction coefficient.

Roll formation evidently is not unique. Kuvshinsky and Rudakov (93) show that it is basically related to degradation of the polymer. The surface layer must be soft. They note (93) that it is difficult to relate polymer degradation to heat release, because calculation shows that the temperature rises negligibly on account of friction work. However, these results contradict the work of Kragelsky and Rybalov on vulcanized-rubber fatigue wear, in which the decisive role of temperature in the transition from fatigue wear to roll formation is shown.

The process of wear for vulcanized rubbers on smooth surfaces is without a doubt a fatigue process. Therefore, the surrounding environment plays an especially important role affecting the kinetics of the polymer's degradation during wear. Keeping in mind compounding factors, it is important to know what effect antioxidants have.

Oxygen has the largest effect on the wear of vulcanized rubbers. The structure of the rubber is quite significant here (92,93). Table 6.6 shows the wear intensity for various vulcanized rubbers on rough metal surfaces in air and in argon. The ratio of wear in air (J_W^{air}) to wear in argon (J_W^{Ar}) characterizes the effect of oxygen on wear of vulcanized rubbers obtained from various raw rubbers.

The raw rubbers are set out in Table 6.6 in decreasing order of the environment's effect on wear. It is evident that oxygen significantly affects raw-rubber wear. This is related to the effect of oxygen on fatigue of the corresponding vulcanized rubbers (60). Oxygen least affects fatigue of vulcanized butyl and urethane rubbers, because these rubbers have few double bonds.

Unfortunately, studies on the effect of the environment and type of raw rubber on wear are only qualitative. The detailed mechanism and kinetics of degradation are not known, which hampers the development of new wear-resistant vulcanized rubbers. However, the fatigue mechanism of wear permits an approach to a choice of the type of raw rubber and prescription of a vulcanized rubber, based on the better-studied aging and fatigue processes in the dynamic mode (60).

Vinogradov and others (96-101) have conducted special research on the effects of oxygen on plastic wear. They have shown that unlike vulcanized rubber wear, a decrease in the intensity of oxidation processes does not affect wear of soft plastics. They have noted that atmospheric oxygen affects the formation of metal-oxide films.

TABLE 6.6

Wear of various vulcanized rubbers in air and in argon (92)

Raw Rubber	Wear Intensity (cm^3/kW-hr)		Air J_W / Arg J_W
Natural rubber	163	60	2.72
Butadiene-styrene (SKS-30-1)	23	15	2.53
Sodium butadiene	204	88	2.32
SKS-30M(a butadiene-styrene rubber)	121	55	2.19
SKS-25 MPV-5 (a butadiene-styrene rubber)	16	9	1.78
Nairit (chloroprene rubber, equi	111	66	1.70
SKN-26 (butadiene-acrylonitrile rubber)	8	5	1.60
Butyl rubber	120	119	1.01
Urethane (SKU-2)	78	71	1.10

During friction a transfer of the metal to plastic is observed, and thus the metal actually wears.* Reduction in the concentration of oxygen leads to an increase in the friction coefficient and metal wear. These experiments have already shown that wear of the metal depends more upon the properties of plastics than upon those of the metal. Very complex physico-chemical processes arise on the contact boundary. In this respect, neither wear by an abrasive cloth or paper nor wear on a grating can be compared with wear on smooth hard surfaces. Special consideration should be given to the thermal degradation of plastics under heavy friction modes and its effect on wear (10). Rigid polymers act to chemically modify metal surfaces at elevated temperatures and also bring about an increase in surface wear (96).

In conclusion, we should note that wear of vulcanized rubbers and plastics on hard smooth surfaces has several special characteristics. In practice, this is the most important type of polymer wear. But, especially in theoretical design, this type of wear has been all too inadequately studied; therefore, at the present time it is impossible to make adequate general conclusions about wear.

*The experiments were conducted on machine MT-6; see Chap. 7.

6.5 PRACTICAL RECOMMENDATIONS AND PROSPECTS FOR USE OF POLYMERS AT FRICTIONAL CONTACTS

From the theoretical analysis and experimental material presented in this chapter, we can reach definite practical conclusions. Engineers and designers who develop and adapt products made from polymers have an interest in the conditions required for raising wear resistance. We cannot offer general prescriptions in this connection. However, this does not mean that there is no possibility of practical recommendations, because there are specific cases. Let us examine an approach to the choice of wear mode and polymer composition.

The mode and type of wear ought to be selected in regard to conditions of use, special configuration characteristics, and special demands on the polymer's structure. The polymer composition ought to satisfy production demands and requisite wear resistance. As a rule, the type of wear of the articles is difficult to determine without special analysis. Nevertheless, there are morphological and structural tests that can determine the basic wear type: abrasive, fatigue, or roll formation.

Morphological tests include the dimensions and form of wear products and the pattern of the worn surface. By examining the wear pattern, it is possible to distinguish fatigue and abrasive wear. Transverse bars characterize fatigue wear, and longitudinal bars, abrasive. For abrasive wear, wear products are measured on the basis of the decrease of some dimensions of the polymer bulk. For fatigue wear, the wear products are fine particles and the wear surface is slightly rough.

Structural tests allow a determination of changes in the surface's physico-chemical structure, the role of the destruction process, the effect of the environment, the change in the polymer's molecular weight, and other similar changes. The wear mode ought to be chosen by a calculation of wear type and conditions of use. It has been shown that change in the type of wear mode leads to change in the type of wear. The parameters that characterize the wear mode are pressure, temperature, and sliding velocity. Often it is not necessary to study the behavior of the polymer article as a function of all parameters. Thus, for example, for abrasive wear within a definite range of velocities, linear wear does not depend upon velocity, and its relation to pressure is linear. For fatigue wear, temperature is a basic parameter.

Articles usually perform under conditions of constant deformation or friction force (moment). Wear resistance for various polymers is not identical under these modes; furthermore, effects on wear sometimes are conflicting. Thus, for example, wear of a vulcanized rubber decreases with increase of filler content under a fixed load and increases under a fixed friction force (102).

We think that to compute a material's wear resistance, irrespective of the testing mode and conditions of practical use, employing the standard method STATE 426-66, is completely unsatisfactory and sometimes can lead to error. STATE 426-66 enables the computation of wear resistance against an abrasive cloth or paper for a given mode of testing. Polymer materials can be ranked according to the degree of rise in wear resistance by this method. However, it absolutely does not follow from this that wear resistance for these materials during actual use will change in analogous fashion. This set of circumstances very much hampers selection of wear-resistant materials and leads to the necessity of processing instruments and equipment that simulate conditions of actual use. The conclusion to be drawn from this is that standard methods of testing and a general theory of wear are required.

Nevertheless, some recommendations can be given on the basis of an analysis of wear research to date. Below are general recommendations to guide both concepts of polymer wear and more specific recommendations on basic types of manufactured articles and their processing.

According to Kragelsky (3), for the selection of wear-resistant materials it is useful to employ two principles: the presence of a positive gradient of mechanical properties and minimal interpenetration of the interacting surface. The presence of a positive gradient of mechanical properties inhibits deep impression and leads to a reduction in wear intensity. Minimal interpenetration of the interacting surfaces requires rigidity and smoothness of the hard surface. This condition contributes to the reduction of abrasive wear.

Reduction in wear intensity requires a decrease in the friction coefficient. Methods of decreasing it are examined in Chaps. 3 and 4 above. The friction coefficient determines not only the material's general wear resistance, but the kind of wear. The temperature mode, which plays an important role in the wear process, depends upon the magnitude of the friction coefficient.

Fatigue wear is the most beneficial type and guarantees the highest wear resistance for polymer products. Moreover, for fatigue wear, polymer fatigue resistance, which has been well studied both experimentally and theoretically, is a fixed factor. Wear by both the abrasive and fatigue mechanisms depends upon a polymer's physico-mechanical characteristics, among which the basic factors are the modulus E or hardness HB, the strength Σ_o, and the rupture strength ε_f. Qualitatively, wear is proportional to $\mu/(HB\ \Sigma_o\ \varepsilon_f)$. We noted above the special role of constrained elasticity in plastic wear.

Some specific recommendations on selection of wear-resistant polymers can be made in relation to the type of wear, the structure of the product, and the mode of its operation. The weak effect of abrasive particles on wear and friction is a special characteristic of polymer wear and tear. In purely abrasive wear, polymer wear resistance is significantly lower than that of hard metals and alloys. Therefore, polymer products are most widely used under conditions of sliding on smooth, hard (metal) surfaces. This does not pertain to a special group of products--rubber tires, which are used for both abrasive and fatigue wear and tear, for reasons other than achieving long life.

The five most common groups of products that work under dry friction are tires, seals, rolling bearings, coatings, and film coverings. (These products can work with lubrication; however, we shall not consider this question here.) A series of studies (103-112) have been devoted to automobile-tire wear and have shown that tire wear resistance depends upon the vulcanized rubbers' fatigue properties, the structure of the tire, and the mode of operation. Tire tread wear depends, for example, upon tread-element slip and the character of the active forces. The temperature in the contact area between the tire and the road has great significance (107).

Modern flexible packings (113-115) should guarantee a tight seal at elevated temperatures, pressures, and sliding velocities. Wear resistance for packings is determined at a base temperature (37,46, 49,50,116-119). The wear is by fatigue. The optimal roughness, guaranteeing the best fatigue life for the packing, is $\nabla 9$. Because temperature is the determining factor in the contact zone, it is necessary to improve heat removal and decrease the friction coefficient.

Rolling bearings of rigid polymers and their composits (120-125) have attained wide acceptance, particularly those based on PTFE, polyamides, laminated fabrics, and graphite composites. The wear resistance of rolling bearings depends upon contact temperature, which itself depends upon polymer type and conditions of use: sliding velocity, pressure, and lubricant, and also upon structural parameters: diameter, wall thickness, character of the metal and plastic surfaces, and size of clearance.

Improvement in the fatigue life of rolling bearings can be achieved by the introduction of plastic coatings that have low friction coefficients (PTFE). Furthermore, the introduction of filler of high thermal conductivity into the plastic, together with the choice of optimal thickness of the antifriction coating and other condition- that allow reduction in the friction coefficient, enhance bearing life (see Chap. 4). Materials based on polymer composites combined with metals and graphites are especially promising. Often it is useful to increase the polymer's modulus.

Among the basic deficiencies of plastic rolling bearings, in comparison with metals, is their weak supporting power and poor thermal conductivity. For this reason, thin polymer coatings in metal bearings, which have good supporting power and better thermal conductivity, have gained wide distribution. These two characteristics depend upon thickness. For example, for polyamide coatings the optimal thickness is 0.3 mm (91). Generally, the optimal coating thickness depends upon the operation mode and the type of polymer from which the coating is produced. The introduction of coatings in industry is limited by their weak adhesion to metal.

Thus, practical recommendations on the choice of polymer are based on conditions of operation and polymer properties. There is no doubt that the most promising methods of creating wear-resistant polymers are related to obtaining low friction coefficients and good thermal conductivity for required structural demands.

REFERENCES

1 I.V. Kragelsky, Izv. Vyssh. Uchebn. Zaved. Fiz., (No. 5)(1959) 119-127.
2 I.V. Kragelsky, Kauch. i Rezina, (No. 11)(1959)20-26.
3 I.V. Kragelsky, Treniye i Iznos (Friction and Wear), Mashgiz, 1968.
4 I.V. Kragelsky and Ye.F. Nepomyaschii (Eds.), Friktsionnyi Iznos Rezin (Friction Wear of Polymers), Khimiya, 1964, pp. 5-9.
5 G.M. Bartenev and V.V. Lavrentev, AN SSSR OTN, (No. 9)(1958)126.
6 M.M. Reznikovsky, Kauch. i Rezina, (No. 5)(1960)34.
7 A. Schallamach, Wear, 1, No. 5, (1957)385.
8 V.F. Yevstratov, M.M. Reznikovsky et al. (Eds.), Friktsionnyi Iznos Rezin (Friction Wear of Polymers), Khimiya, 1964, pp. 56-75.
9 S.B. Ratner and M.V. Melnikova, Kauch. i Rezina, (No. 8)(1958)15.
10 S.B. Ratner, G.S. Kliterik and Ye.G. Lure (Eds.), Teoriya Treniya i Iznosa (Theory of Friction and Wear), Nauka, Moscow, 1965, pp. 156-159.
11 M. Kerrige and J. Lancaster, Proc. Roy. Soc. (London), A 236(1956) 250.
12 K. Tanaka, J. Phys. Soc. Jpn., 16(1961)2003.
13 R. Matveevsky, Wear, 4(1961)300.
14 J.F. Archard and W. Hirst, Proc. Roy. Soc. (London), A 236(1956) 397.
15 M.A. Babichev (Ed.), Plastmassy kak Antifriktsionnyye materialy (Plastics as Anti-Friction Materials), AN SSSR, Moscow, 1961, pp. 80-85.
16 S.B. Ratner and I.I. Farberova, Plastmassy, (No. 9)(1960).
17 Ye.F. Nepomnyashchy, Dissertation abstract, IMASh AN SSSR, 1964.
18 Tenenbaum, J. Appl. Phys., 32(1961)1420.
19 E. Powell and S. Gough, Rubber World, 132(No. 2)(1955).
20 G.M. Bartenev and F.A. Galil-Orly, Dokl. Akad. Nauk SSSR, 100(No. 3) (1955)477.
21 M.M. Saberin, Tr. Tsentr. Nauchno-Issled. Inst. Tekhnol. Mashinostr., Book 2, Mashriz, 1946.
22 M.M. Reznikovsky and K.N. Lazareva, Kauch. Rezina, (No. 3)(1963)17.
23 N.B. Demkin, Fakticheskaya Ploshchad' Kasaniya Tverdykh Poverkhnostei (The Actual Contact Area of Hard Surfaces), AN SSSR, Moscow, 1962.
24 I.V. Kragelsky, Ye.F. Neponipyashchii and G.M. Kharach, Ustalostnyi Mekhanism Iznosa i Kratkaya Metodika Analiticheskoi Otsenki Velichiny Iznosa Poverkhnostei Treniya pri Skol'Zhenii (The Fatigue Wear Mechanisms and a Short Method of Analytical Evaluation of Friction Surface Wear Magnitude During Sliding), Nauka, Moscow, 1967.
25 S.B. Ratner, Dokl, Akad. Nauk SSSR, 144(No. 2)(1962)327.
26 A. Schallamach, J. Polym. Sci., 12(No. 5)(1952)385.
27 A. Schallamach, Rubber Chem. Technol., 26(No. 1)(1953)230.
28 M.M. Bilik (Ed.), Sukhoye Treniye (Dry Friction), AN Latv. SSSR, 1961, pp. 185-197.
29 S.B. Ratner, Dokl. Akad. Nauk SSSR, 150(No. 4)(1963).
30 S.B. Ratner and G.S. Klitenik, Kauch. Rezina, (No. 3)(1960).
31 M.M. Reznikovsky and G.I. Brodsky, Kauch. Rezina, (No. 7)(1961)18.
32 G.S. Klitenik and S.B. Ratner, Friktsionnyi Iznos Rezin (Friction Wear of Polymers), Khimiya, 1964, pp. 77-87.

33 J. Burns and E. Storye, Ind. Eng. Chem., 44(No. 4)(1952)825.
34 S.B. Ratner and G.S. Klitenik, Zavod. Lab., (No. 11)(1959).
35 S.B. Ratner, Friktsionnyi Iznos Rezin (Friction Wear of Polymers), Nauka, Moscow, 1964, pp. 32-44.
36 S.B. Ratner and E.G. Lure, Mekh. Polim., (No. 6)(1966)867.
37 S.L. Rybalov, Dissertation abstract, NIFKhl im Karpova, 1966.
38 V.P. Mitrovich, in Plastmassy kak Antifriktsionnyye Materialy (Plastics as Anti-friction Materials), AN SSSR, Moscow, 1961, pp. 43-52.
39 L.Yu. Pruzhanskii, in Plastmassy kak Antifriktsionnyye Materialy (Plastics as Anti-friction Materials), AN SSSR, Moscow, 1961, pp. 74-79.
40 G.M. Bartenev and V.V. Lavrentev, Dokl. Akad. Nauk SSSR, 141(1961) 334.
41 I.V. Kragelsky and E.F. Nepomnyashchi, Wear, 8(1965)303-319.
42 A.V. Chinchinadze and G.I. Troyanovskaya, Peredovoi Nauchno-tekhnicheskii i Proizvodstvennyi Opyt (Advanced Scientific-technical and Industrial Testing), tema 20, No. M 57-127/6 izd VINITI, 1957.
43 A.P. Rudakov and Ye.V. Kuvshinsky, Friktsionnyi Iznos Rezin (Friction Wear of Polymers), Khimiya, 1964, pp. 46-55.
44 A. Cameron, J. Inst. Pet., 40(1954)367.
45 N.L. Sakhnovsky, L.A. Smirnova and V.F. Yevstratov, in Friktsionny Iznos Rezin (Friction and Wear of Polymers), Khimiya, 1964, pp. 141-161.
46 S.L. Rybalov, Ye.V. Mal'chikova and B.S. Tsybuk, Friktsionnyi Iznos Rezin (Friction Wear of Polymers), Khimiya, 1964, pp. 192-199.
47 K.A. Grosch and A. Schallamach, Wear, 4(No. 5)(1961)356.
48 M.M. Bilik (Ed.), Iznos i Treniye Metallov i Plastmass (Friction and Wear of Metals and Plastics), Nauka, Moscow, 1964, pp. 87-100.
49 I.V. Kragelsky and S.L. Rybalov, Dokl. Akad. Nauk SSSR, 164(No. 5) (1965)1035.
50 S.L. Rybalov and I.V. Kragelsky, Mekh. Polim., (No. 6)(1965)120.
51 S.L. Rybalov and I.V. Kragelsky, Rezina Konstruktsionnyi Materialy Sovremennogo Mashinostroyeniya (Structural Material for Contempora Mechanical Engineering), Khiniya, 1967, pp. 286-295.
52 A.V. Chichinadze, Raschyet i Issledovaniye Vneshnego Treniya pri Tormozhenii (Calculation and Research on External Friction during Frictional Action), Nauka, 1967.
53 A.M. Snisarenko and Z.N. Tarasova, Friktsionnyi Iznos Rezin (Friction Wear of Polymers), Khimiya, 1964, pp. 88-94.
54 A. Schallamach, Wear, (No. 3)(1960); (No. 5)(1963).
55 S.B. Ratner, Dissertation Abstract, NIFKhl, im. Karpova, 1963.
56 S.B. Ratner, Dokl. Akad. Nauk SSSR, 150(No. 5)(1963)1084.
57 S.B. Ratner and Ye.G. Lure, Vysokomol. Soedin., 8(No. 1)(1966)88; Dokl. Akad. Nauk SSSR, 166(No. 4)(1966)909.
58 Ye.G. Lure, Dissertation abstract, NIFKhl im. Karpova, 1966.
59 I.I. Farberova, Dissertation abstract, NIFKhl im. Karpova, 1965.
60 G.M. Bartenev and Yu.S. Zuyev, Prochnois'i Razrusheniye Vysoko-elasticheskikh Materialov (Strength and Failure of Viscoelastic Materials), Khimiya, 1964, Trans. F.F. and P. Joray, Pergamon Press, 1968.
61 V.Ye. Gul, Prochnost'Polimerov (Strength of Polymer), Khimiya, 1964.
62 S.B. Ratner and Ye.G. Lure, Mekh. Polim., (No. 6)(1966)867.
63 S.B. Ratner and Ye.G. Lure, Dokl. Akad. Nauk SSSR, 166(No. 4) (1966)909-912.
64 V. Vogt, Ind. Eng. Chem., 20(No. 2)(1928).
65 Anon., Engineering, 204(No. 5285)(1967)185.

66 A.E. Juve, Rubber Age, 78(No. 6)(1956)911.
67 R. Zapp, Rubber World, 133(No. 1)(1953)59.
68 C.M. White and D.F. Denny, The Healing Mechanism of Flexible Packing, Great Britain, Ministry of Supply, Sci. & Tech., Memo No. 3-47, London, 1948.
69 F.Ye. Kuperman and B.K. Karmin, Symposium Friktsionnyi Iznos Rezin (Friction Wear of Polymers), Khimiya, 1964.
70 G.N. Buiko, L.A. Smirnova, N.A. Pruzhanskaya and A.I. Karmanova, Symposium Friksionnyi Iznos Rezin, Khimiya, 1964, p. 162.
71 V.F. Yestratov and G.N. Buiko, Ibid., p. 141.
72 D. Bianca, Can. Min. Metall. Bull., 60(No. 665)(1967)1017.
73 P.S. Hammer and W.T. Ceoner, Rubber Age, 89(No. 6)(1961)963.
74 T.D. Bolt, E.M. Dannenber et al., Rubber Plast. Age, 41(No. 12) (1960).
75 J. Drogin, Rubber Age, 88(No. 3)(1960)517.
76 V.A. Garten, K. Eppiger and D.E. Weiss, Rubber Chem. Technol., 29(No. 4)(1956)1434.
77 M.E. Samuels, Rubber Age, 86(No. 4)(1960).
78 V.A. Lepetov, Rezinovyye Tekhnicheskiye Izdeliya (Polymeric Technological Products), Khimiya, 1972.
79 F.F. Koshelev and N.S. Klimov, Obshenaya Tehhnologiya Rezin (General Processing of Polymers), 1958.
80 B.A. Dogadkin, M.S. Feldshtein and Z.V. Skorodumova, Kolloidn. Zh., 23(No. 6)(1961)679.
81 G. Ebert and V. Weidner, Rubber Chem. Technol., 28(No. 4)(1955) 1082.
82 N.L. Sakhnovsky, Dissertation abstract, NIIShP, 1964.
83 G.S. Klitenik, Dissertation abstract, MITKhT, 1964.
84 I.I. Farberova and S.B. Ratner, Plast. Massy, (No. 1)(1967)64.
85 I.I. Farberova and S.B. Ratner, Plast. Massy, (No. 4)(1967)68.
86 G.I. Brodsky and M.M. Reznikovsky, Friktsionnyi Iznos Rezin, Khimiya, 1964, pp. 95-106.
87 S.B. Ratner, Dokl. Akad. Nauk SSSR, 144(No. 2)(1962)327.
88 A. Schallamach, Wear, 1(No. 5)(1957)384.
89 M.M. Bilik, Pary Treniya Metall-Plastmassa v Machinakh i Mekhanizmakh (Metal-plastic Friction Pairs in Machines and Mechanisms), Mashinostroyeniye, 1966.
90 M.M. Krushchov and M.A. Babichev, Issledovaniya Iznashivaniya Metallov (Studies on Metal Wear and Tear), AN SSSR, Moscow, 1960.
91 V.A. Bely and R.A. Rutto (Eds.), Teoriya Smazochnogo Deistviya i Novyye Smazochnyye Materialy (Theory of Lubricating Action and New Lubricating Materials), Nauka, Moscow, 1965.
92 M.M. Reznikovsky and G.I. Brodsky, Friktsionnyi Iznos Rezin, Khimiya, 1964.
93 Ye.V. Kuvshinskii and A.P. Rudakov, Friktsionnyi Iznos Rezin, Khimiya, 1964, pp. 46-55.
94 A.G. Thomas, J. Polym. Sci., 18(No. 86)(1955)177.
95 G.V. Vinogradov and M.D. Bezborodko, Wear, 5(No. 6)(1962)467.
96 G.V. Vinogradov and M.D. Bezborodko, Wear, 5(No. 6)(1962)467.
97 V.A. Mustafaev, G.V. Vinogradov and Yu.Ya. Podol'skii, Iznos i Treniye Metallov i Plastmass (Friction and Wear of Metals and Plastics), Nauka, 1964.
98 V.K. Pirankov and A.J. Yelkin, Kauch. Rezina, (No. 8)(1972)29.
99 G.V. Vinogradov, Lyan Go-lin and N.G. Pavlovskaya, Neftekhimiya, (No. 2)(1961)275.
100 V.A. Mustafaev and G.V. Vinogradov, Teoriya Smazochnogo Deistviya i Novyye Materialy (Theory of Lubrication Action and New Materials), Nauka, 1964, p. 118.
101 V.A. Bely and S.V. Cherbakov, in Teoriya Treniya i Iznosa (Theory of Friction and Wear), Nauka, 1965, p. 298.

102 N.L. Sakhnovsky, S.A. Ivanova, S.B. Ratner et al., Kauch. Rezina, (No. 10)(1958).
103 A. Schallamach and D. Turner, Wear, 3(No. 1)(1960)1.
104 V.I. Novokolsky, Ye.F. Nepomnyashchy and S.P. Zakharov, Friktsionnyi Iznos Rezin, Khimiya, 1964, p. 200.
105 Ye.F. Nepomnyashchy, Rezina-Konstruktsionnyi Material Sovremennog Mashinostroeniya (Structural Material for Contemporary Mechanical Engineering), Khimiya, 1967.
106 V. Gough, Wear, 2(No. 2)(1958)107.
107 K.A. Grosh and A. Schallamach, Wear, 4(No. 5)(1961)356.
108 G.I. Brodsky, M.M. Reznikovsky and N.N. Sizikov, Rezina Konstruktsionnyi Material Sovremennogo Machinostroeniya, Khimiya, 1967, p. 118.
109 W. Viehmavn, Kautsch. Gummi, 8(1955)227.
110 A.E. Juve, Rubber Age, 78(No. 6)(1956)911.
111 D. James, Rubber Plast. Age, 43(No. 2)(1962)138.
112 V.I. Knoroz, Rabota Avtomobil'noi Shiny (Automobile Tire Service) av Totransizdat, 1960.
113 G.V. Makarov, Uplotnitel'nyye Ustroistva (Principles of Packing), Mashinostroeniya, 1965.
114 A.I. Golubev, Sovremennye Uplotneniya Vrashchayushchikhsya Valov (Contemporary Rotating Axle Compressions), Masheiz, 1963.
115 Ya.Yu. Shats, Uplotneniya Podshipnikovykh Uzlov (Bearing Joint Compression), Masheiz, 1963.
116 V.T. Vlasenko, Voprosy Nadezhnosti Gidravlicheskikh Sistyem (Reliability PRoblems in Hydraulic Systems), Issue 3, 1963.
117 V.T. Vlasenko, Izd. Vuzov Aviazioonaja Technika, No. 3(1963).
118 N.N. Smykov, Bul. Omskogo shinnogo zavoda, No. 1(1956).
119 Anon, Uplotneniya (Compression), Transl. from English, Mashinostroeniye, 1964.
120 A.P. Semenov, Plastmassy Kak Antifriktsionnyye Materialy, AN SSSR, Moscow, 1961, p. 103.
121 Anon, Tezisy Coveshchaniya po Primeneniyu Plastmass dlya Podshipnikov Skol'zheniya (Conference on Use of Plastics for Rolling Bearings-Theses), Nauka, 1967.
122 A.D. Kuritsyna, Nestn. Machinostr., (No. 4)(1967)40.
123 Anon., Machine Design, 30(No. 13)(1967).
124 Anon, Engineering, 185(No. 4814)(1958).
125 Anon, Proceedings of the Conference on Lubrication and Wear, London, October, 1957.

Chapter 7

INSTRUMENTS AND METHODS FOR FRICTION AND WEAR RESEARCH

7.1 General Information

In accordance with a general classification scheme proposed by Mooney (1), polymer research methods can be divided into three groups: scientific, official standard, and composite. According to Kragelsky (2), friction testing is diveded into laboratory tests on small specimens, laboratory tests on full-scale specimens, and testing in actual use.

Only the first group of tests can have specific significance, because characteristics measured in these have a fixed physical value and usually are obtained with a complete knowledge of external parameters. Official standard tests are intended for calculating actual use properties of manufactured articles. Composite (intermediate-type) testing cannot have scientific significance and should be viewed at best as preliminary.

The present chapter is concerned with scientific tests. Detailed information on other kinds of testing can be found in other references (3-16). The static and kinetic friction forces can be measured at a constant sliding velocity (v = const) or at a constant tangential force (F_f = const). As Roth, Driscoll, and Holt (17) have shown, with steady sliding, these modes are equivalent. However, the mode v = const has practical advantages, because in this case it is easy to attain motion at a given speed. Moreover, if $F_f(v)$ has a maximum, then for reduction of friction force with increase in sliding velocity the mode F = const cannot be attained, because the sliding process is unsteady (v = const).

The mode v = const is used to study friction properties of a different kind of compressed component that operates at a constant pressure. In this mode, the friction force is determined at a constant load or compression deformation ε = const; as a result of relaxation processes that occur from time to time in the polymer, the contact stress changes.

Choice of the friction-process parameters and external conditions is not less important for determining the friction force. The normal pressure p, sliding velocity v, and temperature T are the basic parameters. An uncontrolled value of each friction parameter can lead to an inaccurate interpretation of results and change in friction conditions.

For other conditions equal, the friction force is determined by the composition, geometry, and state of the surfaces. The effect of the composition of surfaces has been examined in the preceding chapters. The surface geometry affects the friction force in two ways. Under static conditions the size, character, and dimensions of the asperities determine the actual contact area, that is, the initial friction force. Under kinetic conditions the relationship between the surfaces in contact is a significant factor. Chichinadze first focused attention on this relationship (18) and introduced an indicative measure--the so-called "contact period coefficient" K_{BZ} (which can be taken to be the ratio of slider diameter to distance of sliding without repeated passes) (see also Chap. 2). Cooling is related to the coefficient K_{BZ} under kinetic conditions, and consequently, so is the friction-surface temperature.

Roughness of friction surfaces depends upon the conditions of obtaining polymer specimens. Because most specimens are prepared by molding, the polymer surface's macro-geometry is conditioned by the mold surface's macro-geometry. The macro-geometry of polymer specimens prepared by mechanical processing is determined by operating conditions. Surface smoothness depends upon the relationship between height and length of asperities. For long asperities the surface is smooth, regardless of the absolute height of the asperities. If the polymer surface is obtained by formation on extremely smooth hard surfaces through solution coating, fusion, and other similar methods, then the character of the supermolecular structures is important (see Chap. 1). Therefore, the minimal dimensions of micro-asperities are determined by the dimensions of the supermolecular formations. Electron-microscop studies of the surfaces also allow a judgment of the nature, size, and distribution of the friction-surface micro-discontinuities.

A layer of adsorbed gases and liquids covers the polymer surface. Adsorption of solvents by the polymer's surface can lead to change not only in the surface's physical state, but also in the polymer's bulk properties (19,20). Oxidizing processes due to the action of ozone or oxygen exert a large effect on the structure of the polymer surface (21,22). As a rule, these processes lead to degradation of the polymer, and to changes in its mechanical properties with time. Rubberlike polymers (elastomers) are more subject to oxidizing processes than plastics.

Among many methods of cleaning and preparing friction surfaces, we prefer Akhmatova's (5), in which surfaces are cleaned by ethyl alcohol and activated carbon. As demonstrated in practice, treatment by activated carbon followed by ethyl alcohol cleansing is satisfactory basic care. After cleansing, polymer materials must be preserved in a desiccator shielded from heat and light.

In those cases when the polymer surface roughness for one reason or another is difficult to measure or greatly differs from what is expected, in order to obtain comparable friction surface roughness we usually polish the polymer surface with an abrasive cloth or paper of a fixed gage. The dimensions and distribution character of roughness are determined on a profilometer (23,24).

Among the kinematic methods of measuring F and μ, we distinguish forward, rotary, and oscillatory motion. The most common methods are those that employ rotary motion of a hard surface against the polymer specimen. We usually conduct testing of rigid polymers and plastics on instruments intended for solids (metals). Some hard elastomers can also be tested on these instruments.

Kragelsky (2) presents in detail the classification of methods and machines by the kinematic principle. Figure 7.1 shows the kinematic schemes most often encountered in machines and equipment, differentiated according to their contact-period coefficient (K_{BZ} varies from 1 to 0.01).

7.2 Methods for Measuring External Parameters of Friction

In equipment and machines designed for the study of polymer friction properties, various methods find use in determining the friction force, sliding velocity, and temperature. The accuracy of the measured values depends upon the method chosen.

Measuring mechanisms can be divided into two kinds: direct operation and with transducers. Direct-operation mechanisms are simpler, and machines based on them permit a visual calculation of the measured dimension (25). This simplifies the experimental method and makes possible a restriction to standard measured patterns. Mechanisms with transducers are more complex and demand supplementary equipment, but have several advantages, the most important of which are parameter change in the dynamic mode with automatic recording, the possibility of simultaneous measurement of several values, and freedom from measurement error (26). Given these advantages, measuring mechanisms with transducers are more often used to study polymer friction properties than direct operation.

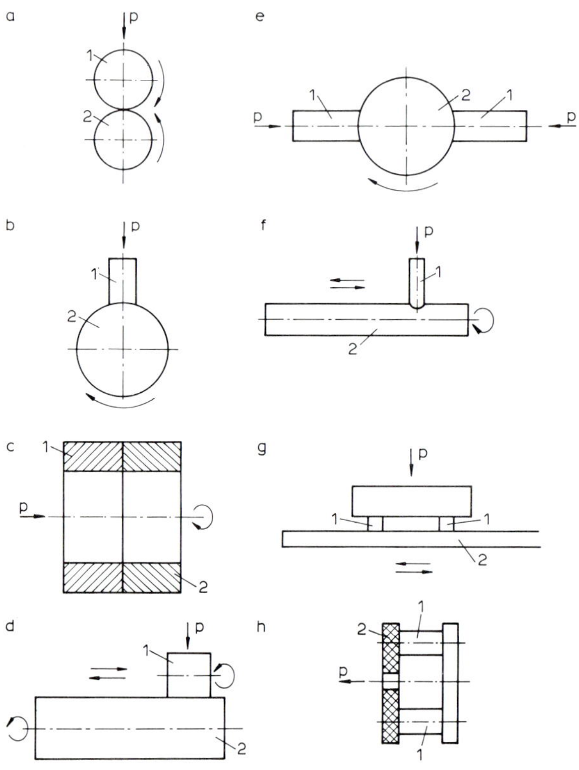

Fig. 7.1 Kinematic drawings of friction pairs: a-g, cylinder on cylinder; d, cylinder on cylinder front; e, cylinder faces on plate; (1) specimen; (2) counter surface.

Measurement of friction force

On direct-operation machines and instruments, the most prevalent methods of measuring the friction force are the dynamometric (spring) and the pendulum. For well-defined polymer relaxation processes, measurements produced by these methods have serious shortcomings. Thus, for example, in studying rubbery polymers it has been demonstrated (27) that the measure of the friction forces depends upon the dynamometer's stiffness coefficient K (Fig. 7.2). As its stiffness increases, its effect on the initial friction force decreases. The duration of the tangential force is a significant reason for this decrease. With an increase in the dynamometer's stiffness (at a constant traction velocity), the duration of the force decreases and leads to a weaker relation of F_f to the force-measuring instrument's stiffness coefficient K. Besides these reasons, the nonlinearity of the dynamometer scale has great significance and also affects the initial friction force. In particular, pendulum dynamometers are supplied with nonlinear scales (28).

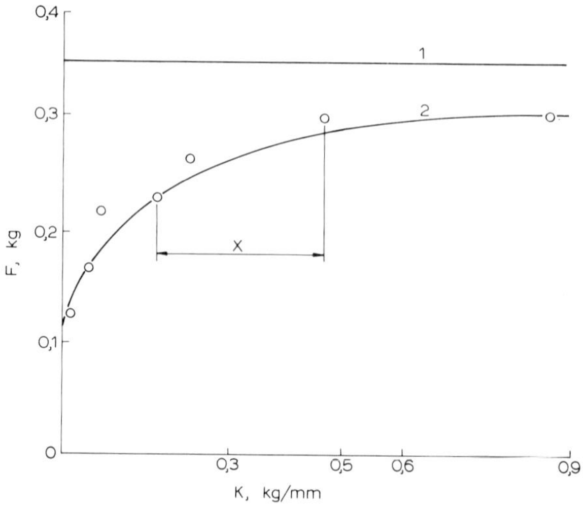

Fig. 7.2 Dependence of the friction force F upon dynamometer stiffness coefficient K (of vulcanized SKN-26, butadiene-acrylonitrile rubber, on steel, traction velocity 0.08 mm/sec) (27): (1) kinetic friction; (2) initial friction; (X) measurement range by a pendulum tribometer.

In modern machines and instruments, dynamometers of high stiffness with transducers are used--tensometer force measurers. They use transducers cemented on elastic gage elements (usually on a steel cantilever arm or ring) (25). Wire transducers and tensometers based on them have significant advantages by comparison with direct-operation of measuring machines: little visual persistence, the possibility of automatic recording of readings, high sensitivity, the possibility of cyclic study, interval measure, and so forth.

The dependence of readings upon temperature is a shortcoming of transducers because heating up produces change in the converter's resistance comparable with change in resistance due to deformation.

The use of a bridge circuit with the transducer eliminates the initial heating-up effect. The measurement error for transducers is ≈1.5% (25). The relative sensitivity θ for transducers can be calculated by the relationship:

$$\theta = [(\Delta R/R)/(\Delta L/L)] ,$$

where L is the length of a transducer wire and R its resistance.

The most common system is that in which the wire transducer is engaged in one bridge arm, and an inoperative transducer (similar to the working one) is engaged in an adjacent arm. As a result the temperature error decreases substantially. The two transducers are glued on similar metals, for example a measuring (tensometer) steel ring, or on the arm (Fig. 7.3). As a rule, an EMP-0.9 machine is connected to the bridge circuit as shown in Fig. 7.3.

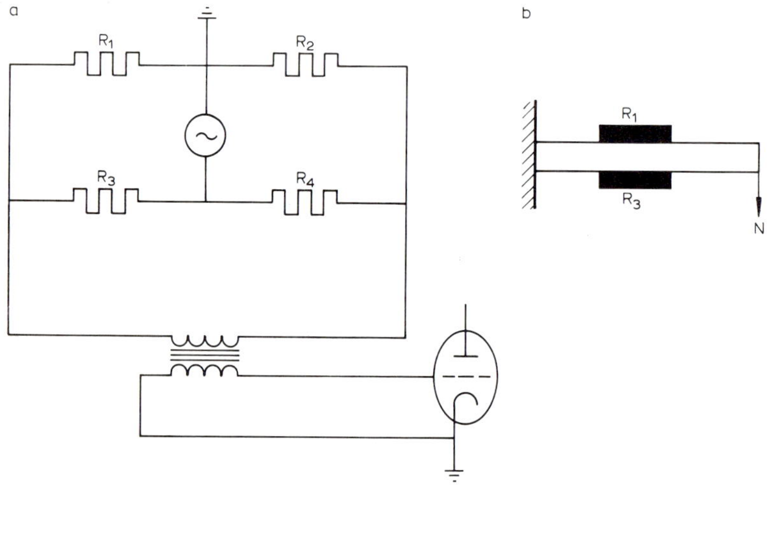

Fig. 7.3 A tensometer bridge circuit for measurement of the friction force: (a) electric circuit with tensometers R_1-R_4; (b) arm circuit with compensating transducers R_1 and R_3.

Such a circuit performs static and dynamic measurements (in accordance with the type of applicable automatic recorder). To determine the friction moment with the help of tensometric systems, hydraulic manometric measurements are used. Figure 7.4 shows one of the possible manometric circuits.

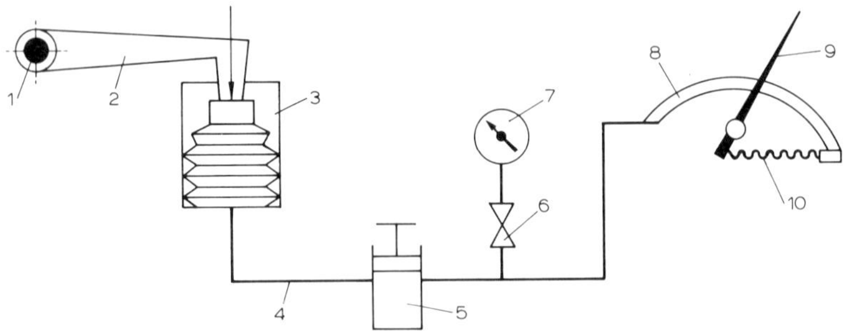

Fig. 7.4 Manometric circuit for measuring the friction moment (29). (1) roller; (2) arm; (3) dynamometer; (4) tubing; (5) piston compensator; (6) valve; (7) pressure gage; (8) Burdon tube; (9) automatic recorder; (10) spring.

Measurement of temperature

To study polymer friction properties, it is necessary to measure the friction-surface temperature. Under static conditions, with adequate insulation it is possible to confine temperature measurements to the bulk of the bodies, which does not present great difficulties. It is another matter to measure the surface temperature in motion. In practice, this is the more important case. Many Soviet (2,3,12,15,16,18,26,29-44) and non-Soviet (45,49) studies give attention to methods of calculating and measuring friction-surface temperatures.

The calculation and experimental determination of the friction temperature are complex theoretical and experimental tasks. There are many methods of determining surface temperature: by thermal emission, by change in structure and in the properties of the various layers, by thermometric methods, and so on.

Thermometric methods are not widely used. The most important are manufactured thermocouples and thermometric resistances, as well as sliding and combined thermocouples. Most common are methods of measuring temperature by using manufactured thermocouples (couples such as chromel-alumel, or chromium-nickel-alloy-copper-constantan, nickel-copper-alloy, etc.), resistance thermometers, and thermistors (temperature-sensitive resistors). The merits of these methods are their good sensitivity and the possibility of determining the temperature gradient. The thermocouple method is preferable for determining the temperature gradient and is applicable to any material.

The natural thermocouple method is used to measure the temperature of dissimilar metal friction pairs and does not apply to polymer-metal and polymer-polymer pairs. It is necessary to keep in mind that thermocouple methods allow the calculation of mean surface temperatures only. Temperature heterogeneity arising as a result of discrete contact, cannot be studied by these methods. However, by using very small "localized" thermocouples, the task of determining the temperature field is significantly simplified (29).

For use of thermocouples, their linkage capacity and geometric form are quite significant. The most common thermocouples are spherical and lamellar ones (Fig. 7.5). Lamellar thermocouples, welded by shock-contact welding, are convenient for measuring the temperature of metal surfaces. Such thermocouples have little inertia and adhere well to a surface (29).

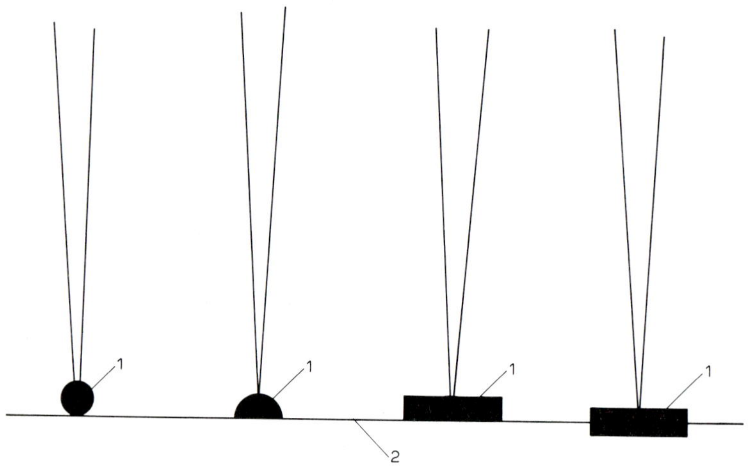

Fig. 7.5 Types of sensitive thermocouples most prevalent in practice (29): (1) thermocouples; (2) friction plane.

Measurement of sliding velocity

Usually, the sliding velocity is a known quantity and is given in terms of the reduction-gear ratio and revolution rate of a rotor. However, in several cases the sliding velocity depends upon the friction force and must be ascertained.

Standard tachometers (mechanical or electrical), and stroboscopic and photoelectric systems are used to measure the sliding velocity (25). Such measurements are made by the instrument designs and equipment described below.

Measurement of load

Like sliding velocity, load is measured by standard machines. If, during testing, the load changes, this can be detected by measuring the friction force. In some cases, the change in load is determined by the instrument's structure and is given by a formula for its change over time (50).

7.3 General-Purpose Instruments and Machines

Many instruments and machines have a place in the study of polymer wear and friction properties. In fact, almost every laboratory and researcher has an instrument or machine of its own design. Let us examine only the most typical and widely used instruments and machines in the U.S.S.R.

Reciprocal-motion instruments

These instruments are designed to study static (initial) friction and kinetic friction at low sliding velocities. Temperature and pressure can be changed within rather wide limits.

Ratner and Konenkov's Pendulum Tribometer. The construction of a pendulum tribometer (MT) was developed at NII (Scientific Research Institute) of the Rubber Industry (51). Later its structure was improved (28), to have a wide range of loads and an attached apparatus for graphical recording of the friction force. The tribometer serves to measure the initial and kinetic friction forces at sliding velocities of 0.1-50 cm/min, temperatures of 20-150°C, and specific loads of 0.5-100 kg/cm^2. This instrument allows friction-force determination for rigid vulcanized rubbers and plastics on hard surfaces. Figure 7.6 shows the tribometer.

Specimens (5) of the material studied are placed on the hard block (6) on which sliding occurs, and are loaded by 4. Tangential force is produced by the vertical travel of the connecting rod (7) of the pendulum (1). The size of the tangential force upon the specimen is determined by the pendulum's deviation angle from the vertical axis. The pendulum, connecting rod, and specimen are linked by a flexible, strong filament. During the specimen's sliding, this filament moves across the cylinder plate 3 and produces rotary motion of the automatic drum recorder.

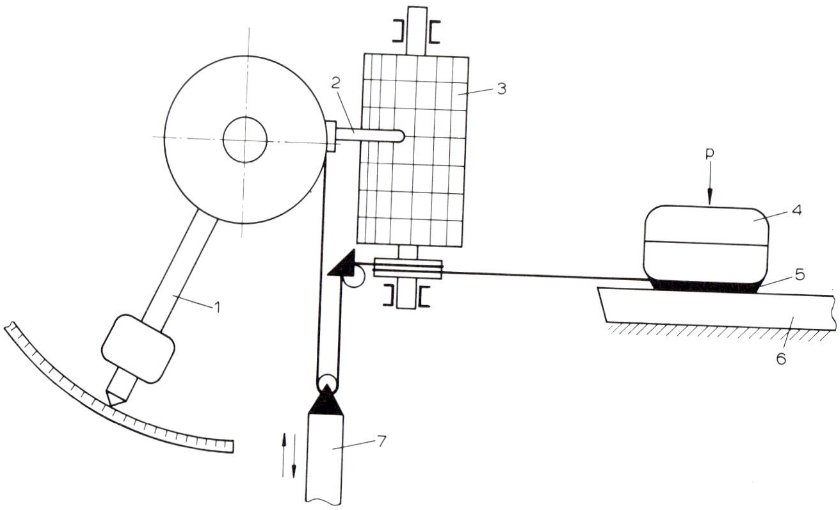

Fig. 7.6 A pendulum tribometer (28): (1) pendulum with bob; (2) automatic recorder; (3) recording drum; (4) load; (5) specimen; (6) hard block; (7) connecting rod.

Three cylindrical specimens 4.5 x 6 mm with general nominal contact area of 0.38 cm^2 are clamped in a collar. Large-diameter specimens can also be used. Specimens are fastened in sockets, protruding from them by 0.5 mm. The use of three specimens secured in an isosceles-triangle band allows the attainment of a stable band position with normal load on the friction surfaces and uniform stress for each specimen. The specimens together with the band are so oriented that one side of the triangle is perpendicula to the direction of the motion. For other specimen positions (especially with large loads) a moving pin in front of the specimen and breaking of the uniform distribution of normal loads are observed. Moreover, this leads to change in the normal and actual contact areas.

The instrument has electric booster heating to determine the friction force at elevated temperatures. The spread of experimental data does not exceed 3-5%.

Bartentev, Lavrentev, and Yelkin's Tribometer, Type TPL-2. The TPL-2 tribometer (52,53), developed in the Polymer Physics

Laboratory at V. I. Lenin Moscow Pedagogical Institute, is designed to determine the friction force of polymer films and various polymers in a range of sliding velocities from 10^{-5} to 10 cm/min and pressures from 0.2 to 20 kg/cm^2 (Fig. 7.7). This instrument is also used for bulk measurements of polymer friction coefficients under factory conditions.

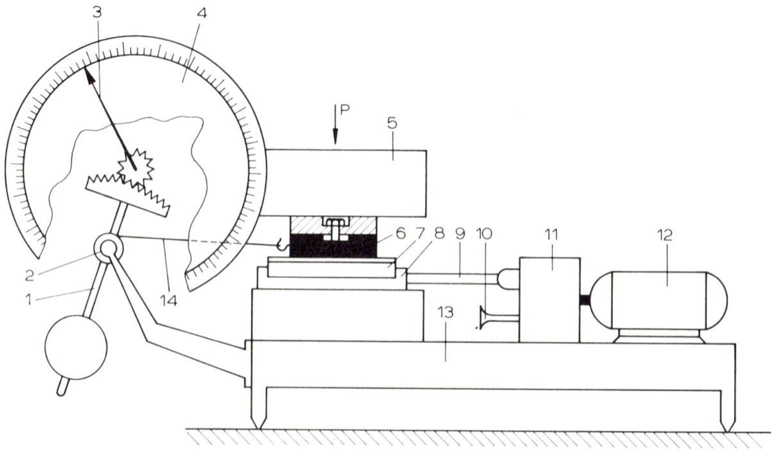

Fig. 7.7 Illustration of TPL-2 tribometer (52): (1) pendulum with bob; (2) pendulum axis; (3) arrow indicator; (4) dial; (5) load; (6) specimen; (7) slide rails; (8) sliding block; (9) propeller; (10) joint pin; (11) reduction gear; (12) motor; (13) steel base; (14) filament.

A sliding block (8) with ball bearings is set on a steel base (13). The sliding block has a groove for a friction platform arrangement (steel, polymer, etc.) taking the form of slide rails. The slide rails (7) are of unequal area. For a constant normal load, produced by the load (5), this allows one to obtain different pressures. The slide rails and the sliding block are moved with constant speed by the micrometric propeller (9). The transfer velocity is regulated by the motor's (12) reduction gear (11). (Several gears can be used.) Gear shifting can be accomplished by the joint pin (10). On the slide rails (7) rests a carrier with the specimen (6), which is connected to the pendulum (1) by a sturdy filament (14) and serves to measure the friction force. The friction force also can be determined by the tensometric method.

For a constant load p, friction-coefficient values can be calibrated on the scale (4). Measurement for p = const most often must be conducted in case of bulk determination of the friction coefficient for computation of relative friction properties of polymer materials. For plastics, the scale is calibrated in friction-coefficient values from 0 to 1.

Scale graduations are derived from the relation

$$\mu = (Q/P)(\ell/r)\cos\alpha ,$$

where Q is the pendulum weight, P the normal load, ℓ the pendulum length, r the radius of the roller on which the tangential force (friction force) presses, and α the pendulum's angle of deflection. The quantity $Q\ell/Pr$ is constant for a given machine.

TMP-3 Tribometer. This machine was also developed in the Polymer Physics Laboratory at V. I. Lenin Moscow Pedagogical Institute. It allows the study of friction of solids and various polymers in the initial and steady states of sliding under atmospheric conditions for temperatures from -100 to +250°C, normal loads from 0.1 to 1 kg, and sliding velocities from 0.01 to 10 mm/min. Figure 7.8 is a diagram of this tribometer.

On the frame (1), assembled in a boxlike steel structure, rests a heat-insulating chamber (2). A thermoblock (3) is set within the chamber and consists of a copper bar with a system of channels for the circulation of a temperature-regulating material. In the lower part of the bar is set an electric heating element (14), which automatically maintains a fixed temperature in the chamber.

The slider (4) is set up in the thermoblock, and on the slider is fastened an interchangeable base, relative to which polymer friction is studied. On the base rests the carriage (8) with polymer specimens (6) of different forms (plates, films, cylinders, etc.) attached to it.

Before testing, various loads are applied to the carriage with the specimens. The loading point consists of a magnetized P-shaped clamp (7), which is strengthened by means of a pin and a groove, and a shaft system (13) connected by a clamp with a hinged suspended disc (16) and (15). Loads (12) are fastened on the disc. With a turn of the arm the disc, rising, unloads the specimens. Here the clamp (7) moves upwards, bus bars come out of the groove, and simultaneously the carriage with the specimen is freed from the load and dynamometer. (At this point it is necessary to apply a small force to the carriage with the specimens in order to overcome the attraction of the magnetized clamp.) This greatly facilitates the experimenter's work, because while conducting the experiment one must remove the carriage from the chamber to change specimens, polish them, and so forth. Moreover, for repeated experiments the samples are brought into an initial position relative to the base without the normal load. In this case, by turning the lever the magnetized clamp is raised together with the carriage on the base.

Pulling force on the slider is transmitted from the motor through the reduction gear, the gear system, and the micrometric screw. When the motor is switched on, the carriage with specimens is carried away (as result of rising friction) from the base and across the traction system linked with clamp, and tightens the tensometric dynamometer, secured on the stand. The tensometric dynamometer is a steel ring with tensometers glued on it.

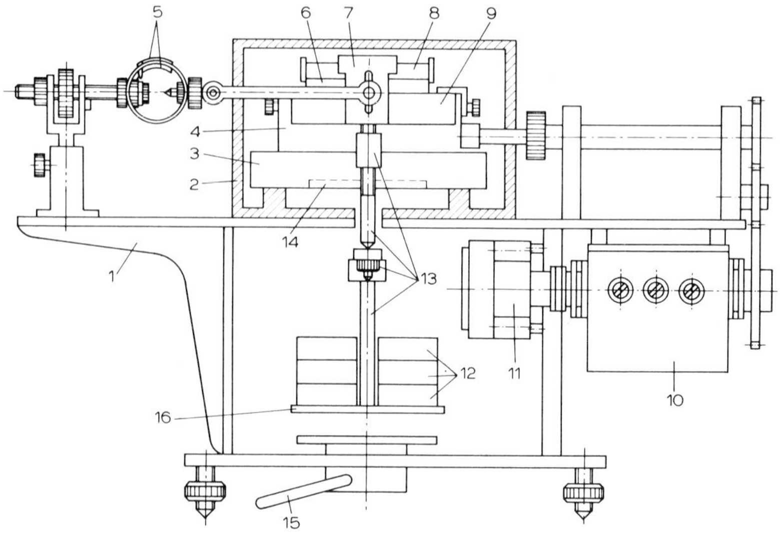

Fig. 7.8 Diagram of TMP-3 tribometer: (1) frame; (2) heat-insulating chamber; (3) thermoblock; (4) slider; (5) ring with tensometers; (6) specimen; (7) clamp; (8) carriage; (9) base; (10) reduction gear; (11) motor; (12) load; (13) shaft system; (14) heating element; (15) lever; (16) disc.

Tribometer for Measuring Oriented Polymer Friction Force. It is known that polymers in the oriented state have anistropic mechanical properties. As has been demonstrated (54), the friction coefficient of oriented polymers also depends upon the direction of the polymer's orientation. A simple instrument used to study such polymers is shown in Fig. 7.9. This instrument, in addition to creating the tangential force, can provide a measurement of the friction coefficient for normal loads of 0.1-1 kg, sliding velocities of 0.01-1 mm/min, and temperatures of 20-100°C.

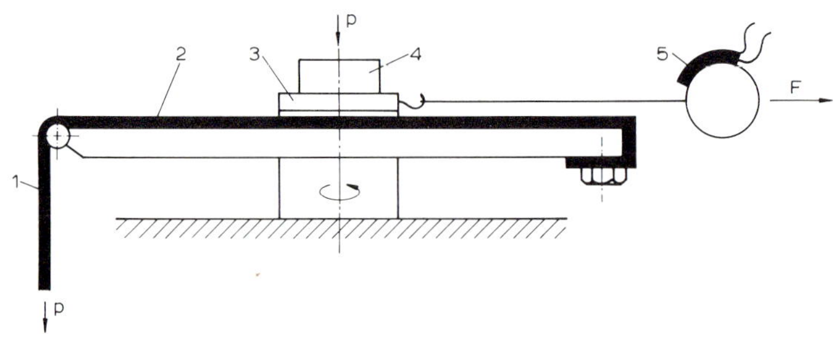

Fig. 7.9 A tribometer used to measure oriented-polymer friction forces (54): (1) specimen; (2) mobile stage; (3) carriage (slide rails, spheres); (4) load; (5) tensometric ring.

Lavrentev's Recording Tribometer, Type RTL. The RTL tribometer (Fig. 7.10) serves to determine the friction force of various polymers at normal loads from 1 to 150 kg and sliding velocities from 10^{-5} to 10 cm/min. This tribometer also provides for measurement of the friction forces at elevated temperatures (to 150°C). Determination of the friction force can be conducted on various hard backings, steel, aluminum, plastic, etc.

The polymers to be tested, in the form of three cylindrical specimens in diameter 4 mm and height 6 mm, are fastened in tracks of the mobile carriage (8), which lies loosely on the steel plate (6). The carriage is connected to the reduction gear by a flexible filament (2), and the reduction gear is driven by the motor (1), which aids in providing the required tangential force and sliding velocity of the specimen.

Normal load is created by the lever (7), one end of which is fastened on the lever support and the other is loose. A load is applied on the lever's loose arm. The normal stress on the specimen depends upon the relative orientation of the trolley and the lever.

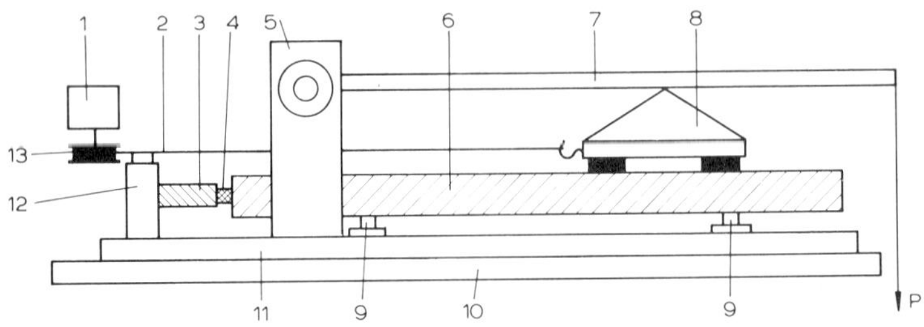

Fig. 7.10 RTL tribometer (50): (1) motor; (2) filament; (3) tensometric dynamometer; (4) insulated packing; (5) lever support; (6) steel plate; (7) lever; (8) carriage with specimens; (9) rollers; (10) table; (11) block; (12) stand; (13) reduction gear.

The friction force that arises from the motion of the specimens shifts the bulky steel plate (6). The plate lies loosely on the rollers (9) and pushes against the steel ring (3), on which are glued the transducers, which are incorporated into the bridge network. The friction force is determined by the aid of the transducers, distributed on the ring (3). Under tangential force action the carriage, loaded with specimens, is moved at a constant speed along the lever (7), and gradually passes through all values of the normal force. Here the friction force F_f is recorded by the automatic recorder. Knowing the speed of the specimen's passage and the recorded force, and having determined the constant correlation between them, it is possible to read the experimental curve of friction force vs time as a graph of friction force vs normal load. It is actually a plot of F_f vs $1/P$, because the normal load changes in inverse proportion to time. This instrument is usually used for studying the dependence of the friction force upon normal load.

Istomin and Kuritsyn's Machines. Machines and instruments with plane-hemispherical-indenter friction pairs have come into wide use. Istomin and Kuritsyn's machine belongs to this category. It allows the study of plastic friction for loads from 0.1 to 3 kg and sliding velocities from 0.2 to 3 mm/sec. This machine's operating principle is shown in Fig. 7.11.

The specimen to be tested is fastened on the carriage (10) and is transferred in a horizontal direction with smoothly variable speed within the limits 0.2-3 mm/sec. The required constant speed is attained by a motor of type MSH-2, a reduction gear, and a special starter-regulating rheostat.

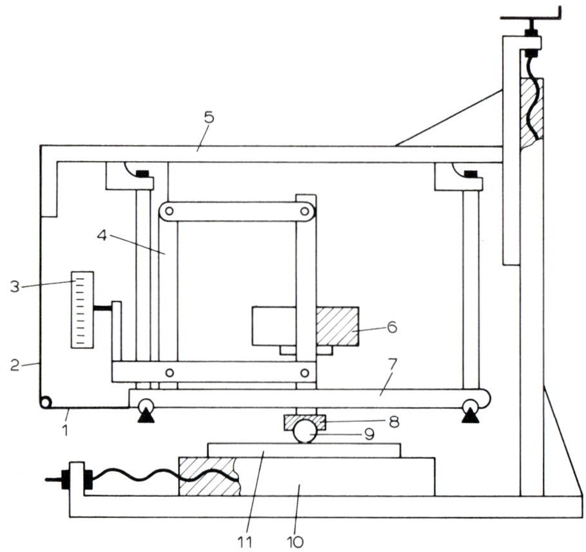

Fig. 7.11 Istomin and Kurityn's machine (55). (1) Connecting rod; (2) small beam with transducer; (3) counterweight; (4) four-link connector; (5) cantilever; (6) load (interchangeable gears); (7) gib; (8) holder; (9) sliding block; (10) carriage; (11) specimen.

A slider (9), is fastened on a holder (8), which is located on one of the four links (4), of the hinged linkage. Normal load is created by the interchangeable gears (6). The counterweight (3), serves to balance the slider's mobile suspension system without gears.

The measured friction force that arises between the specimen and the slider is transferred to the board and further across the flexible bond is taken up by the elastic cantilever arm (2), to which are attached the transducers.

For testing at elevated temperatures, between the specimen and the carriage a little plate with an electric heater is set up, and the sliders are fastened onto the holders, which are heated.

MPT-1 Machine. The MPT-1 machine (56) for testing materials under friction is designed for mass use and has a wide range of loads (to 20 kg), sliding velocities (to 0.01 m/sec), and temperatures (from -60 to +150°C). Moreover, it can be used to study friction auto-oscillation. Figure 7.12 shows this friction machine.

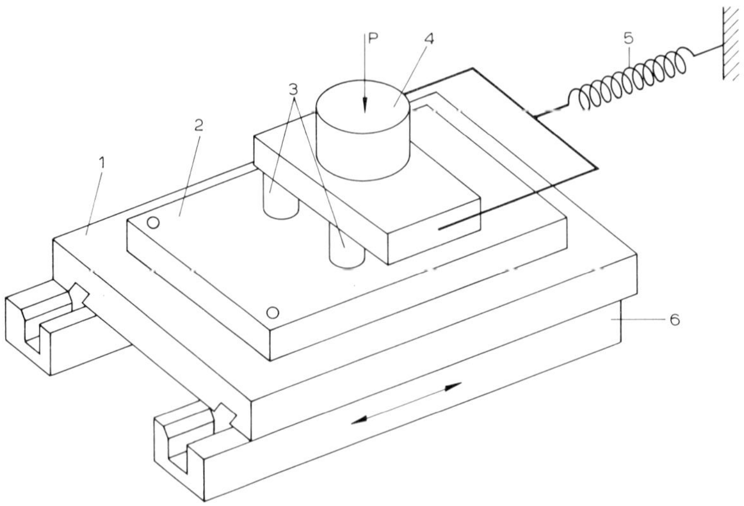

Fig. 7.12 Assembly of MPT-1 machine: (1) slider; (2) lower specimen (friction surface); (3) higher specimens; (4) load; (5) spring; (6) base.

The lower specimen (2), in the form of a plate, is fastened to the slider, which is in reciprocal motion. The three higher specimens (3), have a rectangular or circular cross-section, are pressed against the lower specimen by the interchangeable loads (4), and are prevented from shifting by the spring (5). Forces that arise in the friction zone deform the spring. The deformation causes a change in resistance of the transducer attached to it. A signal from the transducer is transmitted to the recorder. The slider is set in motion by the d.c. motor, the two-speed transmission, and the screw couple. Regulation of the slider's speed is managed by the magnetic amplifier. The regulation of the slider's speed is within 1% at the maximum speed of 0.01 m/sec.

Measurement of friction forces is performed by strain gages, of temperature by the thermocouple, and of the slider's speed by the tachometer generator. A record of friction forces and friction auto-oscillations is made on an MPO-2 oscillograph. This instrument has a chamber for conducting tests at high and low temperatures. The capability of testing in an oil environment is provided (see Table 7.1).

TABLE 7.1
BASIC CHARACTERISTICS OF GENERAL PURPOSE TRIBOMETERS

PURPOSE OF MACHINE	DESIGNATION		MAXIMUM LOAD Kg	MAXIMUM SPEED OF RELATIVE MOTION /sec	PRECISION OF MEASURE-MENT %		NUMBER OF SPECIMENS TESTED	THERMO-CrY6-CHAMBER	CHAMBER FOR TESTING IN ENVI-RONMENTS	MFG	REMARKS
	OLD	NEW			LOADS	FRICTION FORCES					
Machine to test mat'ls for friction & wear	SMT_S-1	T-1186-1 (T-60017.9)	600	7.9	± 3	± 5	1-2	No	Yes	ZIP (Ivanovo)	
Machine to calculate polymer antifriction Mat'ls	MAST-1	T-21-51-1	11	0.0004	± 3	± 5	1 or 3	Yes	Yes	ZIP (Ivanovo)	
Four Ball friction machine	4ShM-3.2	T-2184-2 (T1000/0.68)	1,000	0.68	± 1	—	1 or 3	Yes	Yes	ZIM (ARMA-VIR)	Semi-automatic machine control
Disc Machine for testing materials for friction	MDP-1	T-2177-1 (T-450/47)	450	47	± 2.5	± 5	3	No	Yes	ZIP (Ivanovo)	
Machine for determining materials heat resistance	MFT-1	T-2176-1 (T300/8.8)	300	8.8	± 2.5	± 5	1	No	Yes	ZIP (Ivanovo)	Testing method by RTM 6-60
Machine for testing plastics for friction	MPT-1	T-2552-1	20	0.01	—	± 3.5	3	Yes	Yes	ZIP (Ivanovo)	
Machine for testing rubber brake materials	MIR-1	T-1340-1 (T-5/21)	5	11	—	—	1	Yes	No	Metallist (Leningrad)	
Machine for testing rubber wearing	MI-2	T-2144-1 (T-2.6/0.32)	2.6	0.32	—	—	2	No	No	Metallist (Leningrad)	Testing according to Soviet State Standard 426-57

Rotary-Motion Instruments. The instruments and machines described above, used mainly for research, are characterized by the specimen's reciprocal motion and thus have low sliding velocities (not above 10 cm/sec). Now, let us examine a group of instruments and machines that are designed to measure friction force in a wide range of sliding velocities. These instruments are characterized by rotary motion of the specimen or friction surface (in constant mutual contact). The most common are disc instruments (57-59).

Bisson and Johnson's Machine. This machine allows the study of friction in a wide range of sliding velocities (from 10^{-2} to 3000 m/min), temperatures (from 20 to 500°C), and loads (from 20 to 500 g). Figure 7.13 shows this machine. The specimen, fastened on a ring (6), is pressed against rotating disc (8), by the load (3). With the aid of the motor, the specimen can move radially. The friction force is measured by the tensometric method. The merit of these instruments is smooth variation of sliding velocity in a rather wide range. Their chief shortcoming is the difficulty of eliminating disc vibration.

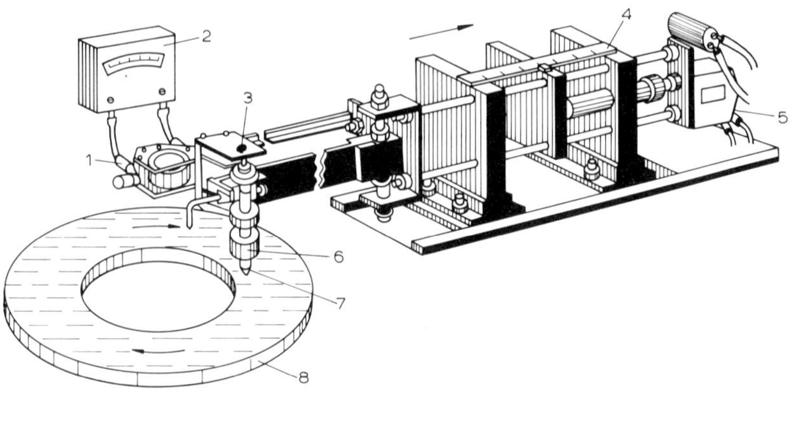

Fig. 7.13 Bisson and Johnson's machine (58): (1) dynamometer-ring strain-gage assembly; (2) friction-force indicator; (3) load; (4) radial-position scale; (5) rider-assembly motor; (6) rider holder; (7) specimen; (8) rotating disc.

Krushchov's Machine, Type X-4B. The X-4B machine (Fig. 7.14) is used to test materials for abrasive wear and friction. The material to be tested, in the form of a cylindrical specimen, can be moved radially on the disc. This allows the measurement of wear on a new surface. One half of the abrasive disc is designed for the testing standard and the other for the material to be studied. The ratio between the reference's wear and the specimen's wear characterizes the relative wear resistance.

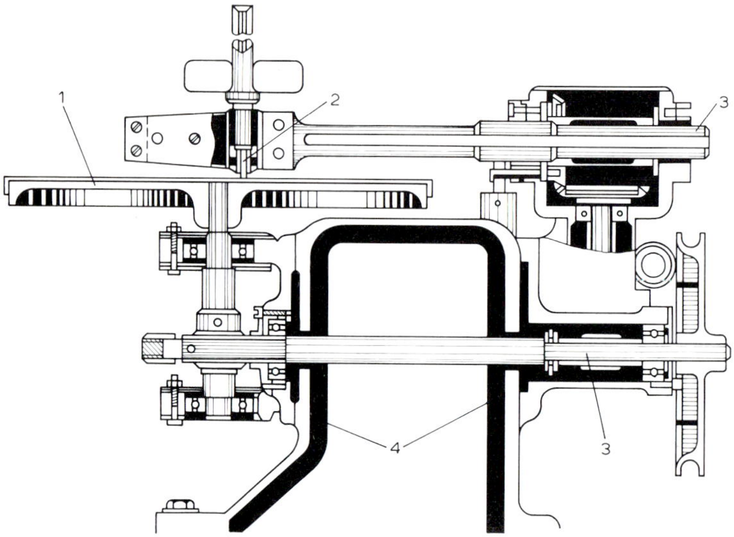

Fig. 7.14 Diagram of X-4B machine (59): (1) disc; (2) specimen; (3) driving gear; (4) mount.

MDP-1 Machine. This machine is designed to determine friction coefficients and the wear intensity of metallic and plastic materials (60). Three finger-shaped specimens of diameter 5-10 mm, fixed at an angle of 120° one to another and fastened on tong clamps, vibrate about the rotating disc. The turning of the finger-shaped specimens impedes the spring. The pressing of the specimen to the disc is accomplished pneumatically. Measurement of the total linear wear and friction coefficient is performed by inductive monitors. Temperature in the contact zone is measured by a thermocouple. Smooth regulation of speed from 9 to 900 revolutions/min is made possible by the generator-motor system. The speed is constant, independent of the oscillations of the load (which may vary from 10 to 450 kg). Table 7.1 shows the machine's basic parts.

MFT-1 Machine. This machine is designed to determine frictional-heat thermal resistance on plastics according to Soviet State Standard 10851-64. The frictional thermal resistance of rubbing materials maintains stability and intensity of wear for a wide range of temperatures that are produced during friction. The MFT-1 machine (Fig. 7.15) is designed on the basis of the 1-47-K-54 (2,62).

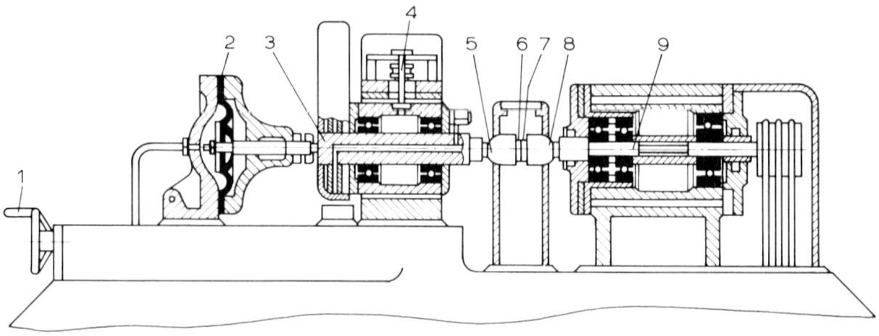

Fig. 7.15 MFT-1 machine (61): (1) crank handle of the measuring-cap apparatus; (2) membrane mechanism; (3) stationary roller; (4) pivot with friction-moment inductive monitors; (5,8) caps for fastening specimens; (6) specimen; (7) rotating specimen; (9) axle.

The friction assembly (see also Fig. 7.1d), consisting of rotating (7), and stationary (6), specimens (external diameter 28 mm, internal diameter 20 mm, height 15 mm) and caps (5-8), for fastening them, corresponds to RTM 6-60. The cylinder's driving gear 9 is put in motion through a two-speed gear box with gear ratios 1:1 and 1:25 and an inclined alternating transmission from a d.c. motor. The rate of cylinder rotation can change smoothly from 10 to 6000 revolutions/min. The electric circuit maintains speed constancy with load oscillations.

The spindle revolves in roller bearings, fixed on the head. On its housing channels are cast, allowing the circulation of cooling liquid. Owing to this cooling, the experiment can be prolonged. On the end of the spindle there is a thread. A cap (8), is screwed to this thread to hold the specimen (7). A specimen (6), is fastened to the cap (5), which is screwed on the end of a stationary roller (3). The roller is fixed on the free bearings on the cap body and is supported by the elastic element's turning. Depending upon the moment to be measured, a spring can be substituted for the required rigid element, selected from a set accompanying the machine.

Pressing of the specimens is accomplished by the pneumatic membrane mechanism (2). Regulation of pressure is performed by the reduction valve, and the reading appears on the pressure gage. On the machine is provided a device of inductive type, which allows change in position of the measuring knob according to the wear of the specimens, and thus a measurement of the total wear of specimens rubbing together. According to the specimen wear, the roller (3), together with the housing, is transferred to the right under the action of the coupling rod and carrier along the joint pin (4), which is connected to the inductive monitors. Using the lever (1), the measuring cap, together with the membrane housing, can be adjusted in the required directions. Liquid, where necessary, moves through a passage drilled into the stationary roller with the aid of a funnel in the specimen's cavity. A special chamber permits testing in various gas environments. All the assemblies are fastened to the bulk cast mount. The MFT-1 machine allows visual observation of the specimen friction temperature (measured by a thermocouple setup on the stationary specimen by RTM 6-60), the friction force, the wear speed of spindle rotation, and the total number of revolutions.

The friction moment and specimen wear are measured by the inductive monitors, which are switched onto the differential system through step-up transformers. Two specimen-wear limits are provided for: the first from 0 to 1 mm, the second from 0 to 8 mm. For measurement of the specimen friction temperature, six thermocouple are provided, adjustable on the stationary specimen at the required distance from the friction surface. The specimen's rotation speed is measured by a tachometer, to which impulses are given by the electro-mechanical monitor, but the disc rotation speed is measured by a voltmeter graduated in revolutions/min and connected to the technogenerator clamps. This machine also has a signal system for the operation of one or another assembly.

MIR-1 Machine. This machine is designed to study rubber wear resistance in a rolling mode with slipping on a renewable specimen surface. A sketch of the friction machine assembly is given in Fig. 7.16. The specimen (3), in the form of a ring, is pressed against a roller (1) with an abrasive cloth on it. With rotation of the running screw (6), the carriage (2), is placed in contact with the generating cylinder; here also rotation of the specimen (3) is created.

Testing on the MIR-1 machine can be conducted in three modes: (1) fixed sliding and fixed friction force; (2) fixed sliding and a fixed normal load; (3) a fixed friction force and a fixed normal load. A detailed account of testing in these modes is given elsewhere (63).

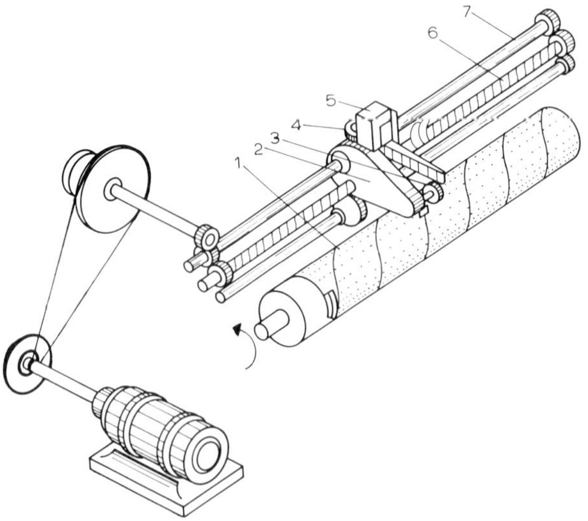

Fig. 7.16 Assembly of MIR-1 friction machine (63): (1) roller; (2) carriage; (3) specimen; (4) level; (5) load; (6) running screw; (7) rod.

MAST-1 machine for determination of polymer antifriction properties. This machine is designed for testing the friction of materials with and without lubrication at normal and elevated temperatures. It can determine metal and nonmetal friction coefficients, and also the critical temperatures of lubricating materials and the wear intensity. The MAST-1 machine (Fig. 7.17) utilizes a friction pattern of four balls (1) against a small friction ball or a flat specimen in the form of a circular band (11). The upper specimen (3), in the form of a small steel ball in diameter 8 or 12 mm, is fastened to a removable holder, a small conical tail, which is joined to the spindle (2), directly or with the aid of a belt transmission. The spindle is rotated by a motor at 1 revolution/min. The three lower specimens pressed into the ball ring 4 (scheme I) or circular specimen (scheme II) are fastened on the bottom to the electric heater of the small vessel, which can be filled with lubricating material in the amount of 25 grams. The small vessel with the lower specimens is forced into contact with the yielding balls and can be turned by the friction force under the load of the higher rotating ball. By its turning, it tightens a replaceable spring, whose angle of twist is proportional to the friction-force moment. Force that rotates the small vessel with the lower specimens during friction is recorded by the needle (with a pen) on the recording drum (6). The load is applied with the aid of a lever system with an arm ratio of 1:10.

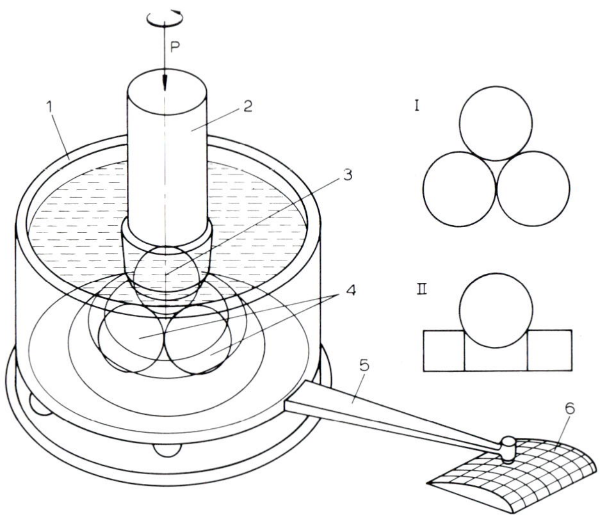

Fig. 7.17 MAST-1 machine (64): (I) friction assembly of four balls; (II) ball casing; (1) metal vessel with lubricating material; (2) spindle; (3,4) specimens; (5) recording needle; (6) recording drum.

Besides those described above, MI-2, SUM-1, MPI-1, 4ShM, and other machines are commonly used. Brief descriptions of some of these are given in Table 7.1.

7.4 Special-Purpose Instruments and Machines

Below we shall examine instruments that are designed for special purposes or have special structure characteristics. To this group belong, for example, instruments for determining the actual contact area, the friction force for high pressures, and sliding velocities, as well as vacuum tribometers.

Instruments for measuring the area of real contact

According to Demkin (24), there are four methods for determining the real contact area: (1) based on electrical conductivity in the contact zone; (2) based on the effect of mass transfer; (3) based on the measurement-surface bonding; (4) based on an optical principle. These methods are not direct, and give only a qualitative or semiquantitative determination of the real contact area. Demkin has examined in detail (24) the merits and inadequacies of various methods and has come to the conclusion that the most acceptable is Mekhau's optical method.

The electrical methods have serious shortcomings, and therefore, they have not been well accepted. The methods based on effects of mass transfer (e.g. tracer techniques, luminescent dyes, etc.) have special properties and are used in advanced work. Their principal deficiency consists in the impossibility of determining the contact area of moving surfaces. Methods of determining the contact area by surface bonding are examined in detail elsewhere (2,23,24). They are rarely used for polymer friction. The most common are optical methods.

Mekhau's optical method has been examined in studies by Kragelsky and Demkin. Its basic merit is the possibility of determining the static and kinetic contact areas S_r. Use of this method is somewhat limited by the necessity that one of the friction pairs be transparent and sufficiently flat. A glass prism surface usually serves this purpose.

Shvetsova (65) has proposed a transparent-model optical method for determining S_r, but this suggestion has not gained wide acceptance. In the Soviet Union, Mekhau's optical method has been applied in the construction of instruments by Demkin and Lankov (66), by Chestnov (67), and in the most complete form by Demkin and Lazarev (68) and also by Lavrentev (69).

Demkin's machine. Demkin's machine (24) is based on the interference of the total internal reflection of an incoming light beam (Fig. 7.18) on the glass-air boundary. The interference of the total internal reflection occurs as a result of light beam's bending in the environment (II), when it penetrates half a wavelength, and its absorption at places where the specimen makes real contact with the prism surface.

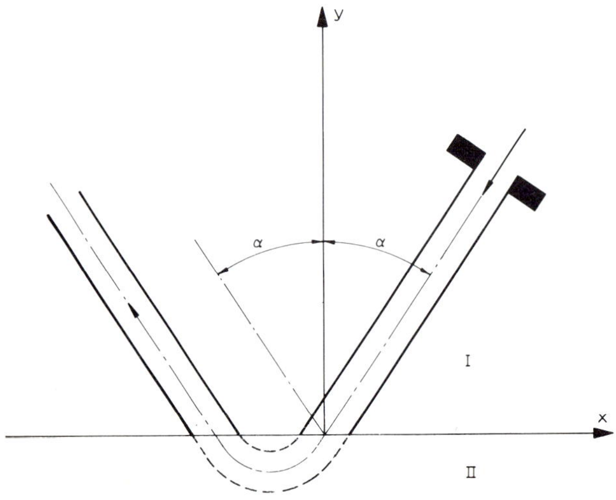

Fig. 7.18 Path of light beam in the optical tribometer with full internal reflection: (I) glass; (II) air.

The optical design of the instrument is based on the balance principle and includes the light source, the prism, two barrels, and two photocells. The recording galvanometer assembly can be defined by electrical and optical methods. Later, Lazarev and Demkin (68) perfected the machine, allowing measurement in the dynamic mode by a prism with a mobile disc specimen of the material to be tested. With motion of the specimen, the machine records some mean value for the real contact area studied, because the tested specimen (disc) slides over the same section. The machine does not allow comparative study of the change in contact area of a selected specimen section with change in sliding velocity.

A simpler and more reliable system for studying the sliding contact area with simultaneous measurement of the friction force has been proposed by one of the present authors (69). This scheme is based on the TOL-2 optical machine (70).

TOL-2 optical equipment. This apparatus is designed for simultaneous study of the actual contact area S and the friction force F for polymers. It consists (Fig. 7.19) of two basic assemblies: an optical one, designed for measuring the real contact area, and a mechanical one (tribometer), used to measure the friction force.

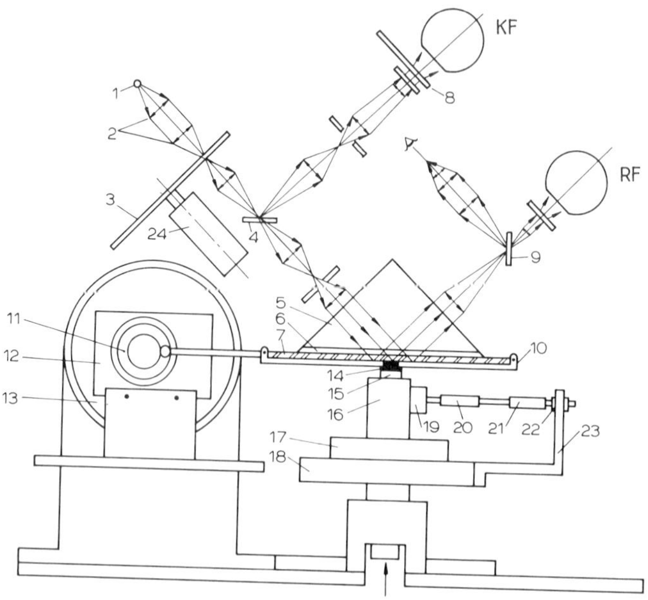

Fig. 7.19 TOL-2 optical apparatus (70): (1) light source; (2) lens system; (3,8) diaphragms; (4,9) semitransparent mirrors; (5) prism; (6) oil film; (7) mobile glass; (10) frame; (11) cam; (12) reduction gear; (13) motor; (14) specimen; (15) holder; (16) heater; (17) mobile plate; (18) loading table; (19) heat insulator; (20,22) connecting rods; (21) steel ring; (23) cantilever; (24) motor; KF, control photocell; RF, working photocell.

A light beam from a stabilized light source (1) strikes the semitransparent mirror (4) through the lens system (2) and the diaphragm (3), and the mirror divides the light beam into two. One of these beams passes through the left barrel's optical system, the total internal reflection prism (5), a thin film of cedar oil (6), then a mobile glass plate (7), and, reflecting off its lower face, strikes the working photocell RF through the optical system of the right barrel. The other beam strikes the control (compensating) photocell KF through the lens system and the diaphragm (8).

In the light field, the plastic specimen (14), contacts the glass* (7), causing interference with total internal reflection.

The essential difference between this optical system and earlier well-known methods consists in the fact that the lower prism face (5),

*One of the hard surfaces must be transparent: glass, organic glass, etc.

with the aid of the cedar-oil film (6), optically contacts the similarly smooth glass (7) that is set in the frame (10), and can be moved together with it. Because the refractive indices of the glass prism, cedar oil film (6), and mobile glass plate (7) are nearly the same (1.5163), the light beam passing through this optically uniform system does not suffer significant change. The semitransparent mirror (9), allows visual observation or photographing of the actual contact area during the experiment.

The reciprocal motion of the frame (10) is accomplished by the cam (11), which is connected to a motor (13), through a reduction gear (12). The specimen's sliding velocity on glass and the frequency of the frame's reciprocal motion can vary within limits from 3.24 to 0.000443 m/sec and from 25 to 0.0032 sec^{-1}, respectively.

The specimen (14), is clamped to the draw-in chuck (15) set on the heater (16), which is made fast to the mobile plate (17), through the heat-insulating packing (freely moving on pellets), with centering to the stationary mirrors of the loading table (18). For measuring the friction force, the steel ring (21) with transducers is fastened by the connecting rod (20) to the top part of the heater (16) through the heat insulator, and by another connecting rod to the stationary cantilever. With the aid of the lever system, the mobile table (18), together with the heater (16) and the specimen (14), can be raised to bring the specimen in contact with the glass (7). By this system, a specific normal load is created on the specimen within limits from 0.4 to 240 kg/cm^2.

Sliding of the specimen's contacting surface is brought about not on the prism, but on the interstitial glass because of the glass's motion. The specimen is stationary relative to the prism. This allows observation of changes in the real contact area and the friction force on the very same selected part of the material.

Recording and measuring of the parameters studied is accomplished by two electronic recorders: the change in the real contact area by an EPP-09 potentiometer, and the change in the friction force by an EMP-109 bridge.

Instruments for measuring polymer friction force at high pressure

The range of high pressures has not only theoretical, but also great practical significance. Modern polymer packing components operate under pressures to 10^4 kg/cm^2.

For polymers, the modulus is several orders of magnitude lower than for metals. If the pressure p is related to the modulus E, then the ratio p/E can characterize the range of pressures. In the instruments described below, the range of high pressures means the range 2.10^2-2.10^3 kg/cm^2. For $p/E > 1$, a polymer is strongly deformed; therefore, the usual research methods are inapplicable. In principle, there are two possibilities for measuring the friction force at high pressures: (1) the use of thin-film specimens, and (2) extreme compression.

The thin-film method serves to limit the thickness in comparison with asperity height and surface undulations. With little thickness, there is direct contact with the hard surfaces between which the film is placed. For larger thicknesses, under pressure and with the passage of time, there occurs a decrease in the film's thickness as a result of rheological processes that occur within the polymer. Consequently, the film method requires a very careful empirical selection of the optimum film thickness.

The extreme-compression method is free of the thin-film methods' shortcomings. It is mainly applicable to elastomer friction. However, it is possible to combine these two methods. A viscous elastic polymer specimen consists of a coating of the substance studied. This enables us to conclude that pressure is transmitted evenly (the thorough-compression method), and that the friction force of a film on a polymer is higher than that of a film on a metallic surface. This method is discussed in detail in Chap. 4.

Let us examine three instruments that are designed to measure polymer friction forces at high pressures:

Deryagin and Toporov's instruments. The conical instrument (Fig. 7.20) consists of a metallic cone-shaped plug (8), which is inserted in a conical opening in a thick-walled sleeve (9) made of organic polymer. The taper of the plug and the sleeve is the same. The specimen (10), in the form of a ring, is placed between the plug and the sleeve. Normal load is created by the lever. The pressure can reach 10^3 kg/cm^2 and is calculated by the formula

$$p = [(N_{BT})/(2\pi r_K z_K \sin \alpha)] ,$$

where N_{BT} is the load on the sleeve, r_K is the ring radius, z_K is the ring thickness, and α is the taper angle. The measurement of α and r is performed visually through the transparent sleeve wall.

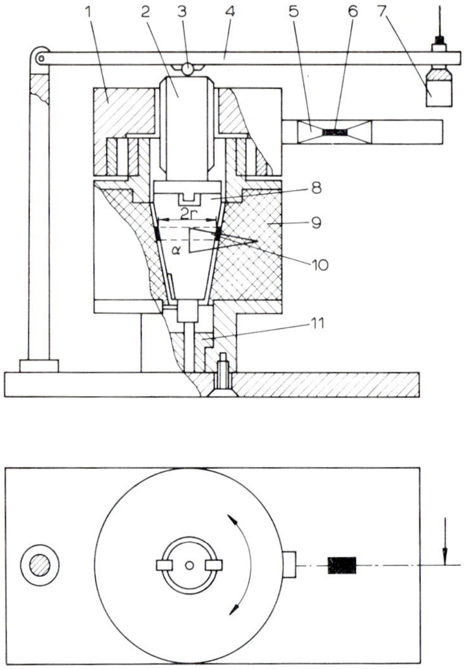

Fig. 7.20 Deryagin and Toporov's conical instrument (71): (1) block; (2) bar; (3) small ball; (4) arm; (5) spring; (6) strain gages; (7) weight; (8) cone-shaped plug; (9) sleeve; (10) specimen; (11) fixed sleeve.

Determination of the friction force is accomplished from the plug rotation path as recorded by a resistance tensometer incorporated in the bridge system. During rotation of the plug (8), rubbery materials usually break down. In this way, the initial friction force is determined.

A uniaxial-compression instrument (Fig. 7.21) consists of a screw press and two dynamometric rings (1,7) with attached strain gages (2,9). Ring (1) serves to measure the normal load, and ring (7) to measure the friction force. There are two systems of testing. The system shown in Fig. 7.22, is used for metallic gages; the one shown in Fig. 7.22, for polymers. Tangential force is created with a power screw propeller (10) by rotation of its disc axis (14) (see Fig. 7.21).

Instrument for thorough compression (Lavrentev's design). This instrument allows the determination of static and kinetic friction forces of rubbery polymers in a range of pressures from

20 to 1500 kg/cm^2 with sliding velocities from 10^{-5} to 10^2 mm/min. Figure 7.23 shows this instrument.

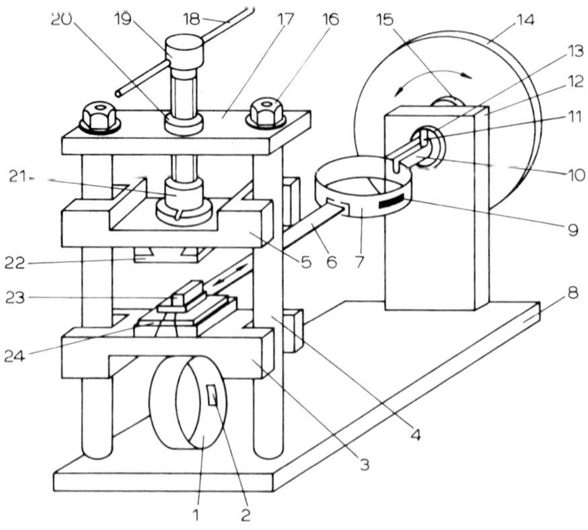

Fig. 7.21 The uniaxial compression instrument (71): (1,7) rings; (2,9) strain gages; (3) lower support; (4,12) pillars; (5) upper support; (6) connecting rod; (8) base; (10,16) screws; (11) joint pin; (13) nut; (14) disc; (15) roller; (17) fixed (center) plate; (18,19) loading screw; (20) washer; (21) stopper screw; (22) upper part of the friction pair; (23) specimen; (24) specimen holder.

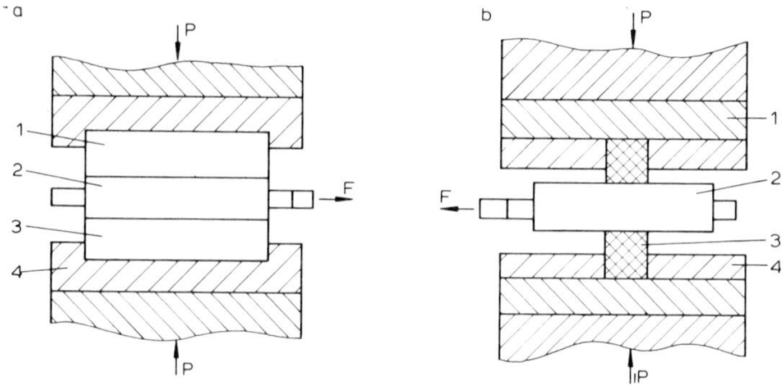

Fig. 7.22 Testing patterns on the uniaxial compression instrument (71): (a) using metallic gages; (b) using cylindrical specimens. (1) Upper stationary specimen; (2) free specimen; (3) lower stationary specimen; (4) holder.

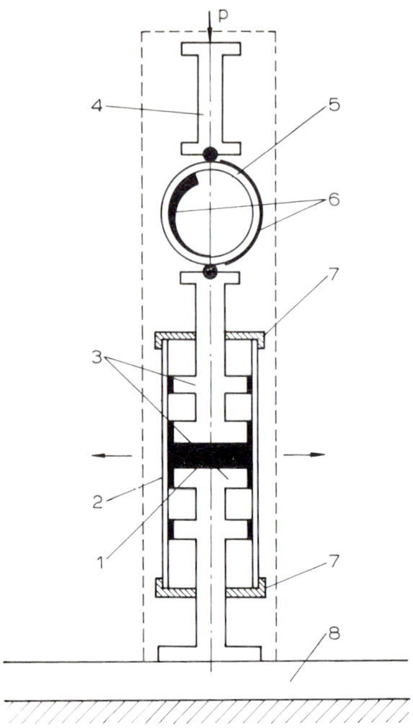

Fig. 7.23 Diagram of compression instrument (72). (1) Specimen; (2) cylinder; (3) pistons; (4) connecting rod; (5) steel ring; (6) tensometric ring with monitors; (7) flanges; (8) table.

A specimen (1) of the material is injected into the metallic or polymer cylinder (2) and condenses on the two pistons (3). One of the pistons rests upon the stationary table surface (8), and the other upon the steel tensometric ring (5), to which the normal compression load N is applied through the connecting rod (4).

As a consequence of the easy deformability of rubbery materials (72), the polymer in practice behaves under load as a liquid, obeying Pascal's law (Poisson's ratio for vulcanized rubbers ≈ 0.5). The load creates a pressure $p = p/S_{\pi}$, where S_{π} is the piston base area.

The cylinder and pistons are manufactured with high precision, and there is little clearance between them when in operation. In order to avoid misalignments under load, the pistons are supplied with extra guiding rings. The piston's friction force on the cylinder wall during idle run does not exceed 0.5 kg and does not depend upon the normal load.

The friction force of the polymer specimen on the cylinder wall is determined by the tensometric method and is equal to the force required for moving the cylinder relative to stationary pistons (3) and the specimen (1). Normal pressure is created by the screw, which presses on the piston through the coupling rod (4) and the ring (5). The friction force is measured during the relative movement of the cylindrical piston (3) by the usual tensometric system.

Vacuum tribometers

As a rule, vacuum tribometers are designed to determine the friction force at low temperatures, inasmuch as the vacuum allows the avoidance of condensation and the freezing out of moisture. Moreover, these instruments can be used to study external friction on extremely clean sliding surfaces. Vacuum tribometers assure testing in a medium vacuum in small ranges of pressures and velocities. An important shortcoming of these instruments is that replacing specimens demands nonsolder sealing of the chamber works.

Bartenev and Yelkin's tribometer. This instrument is used to measure polymer friction coefficients at temperatures from -150 to +150°C, pressures from 0.1 to 10 kg/cm^2, and sliding velocities from 10^{-5} to 10 mm/min in a vacuum of 10^{-3} mm Hg. Figure 7.24 shows this tribometer.

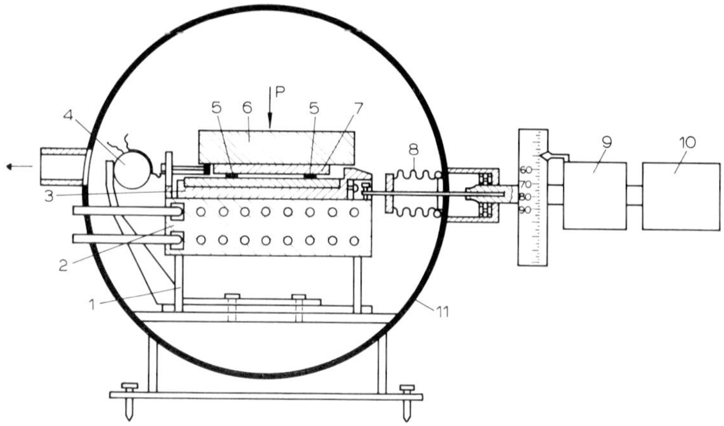

Fig. 7.24 The vacuum tribometer (74): (1) porcelain pillars; (2) thermal element; (3) slider; (4) proof-dynamometer ring with inductive transducer; (5) specimens; (6) load; (7) carriage with specimens; (8) bellows; (9) multi-stage reduction gear; (10) motor; (11) vacuum chamber.

Specimens (5) of the material to be studied, in the form of three cylinders, are placed in the cradles of the sliding carriage (7), and during interaction with the load (6) make contact with the surface of the slider. Tangential force is created by a screw that is set in motion by the reduction gear (9) and by the motor (10). The friction force is fixed by the tensometric ring (4).

Tr-6 and Tr-7 tribometers. These instruments, developed by Korepova and Mustafaev (76), are designed to study friction and wear of plastics for a wide range of sliding velocities and temperatures in gaseous environments and in a vacuum. Table 7.2 outlines their technical characteristics.

The Tr-6 tribometer is shown schematically in Fig. 7.25. The friction assembly, consisting of three lower balls (2) and one upper ball (3), is set in the vacuum chamber. Normal load is created by the hydraulic system (9). The upper ball is secured to the roller end (7), driving it in rotation. The friction force is measured by the tensometric dynamometer. Rotation is produced by the hydraulic driving gear and is measured by a tachometer.

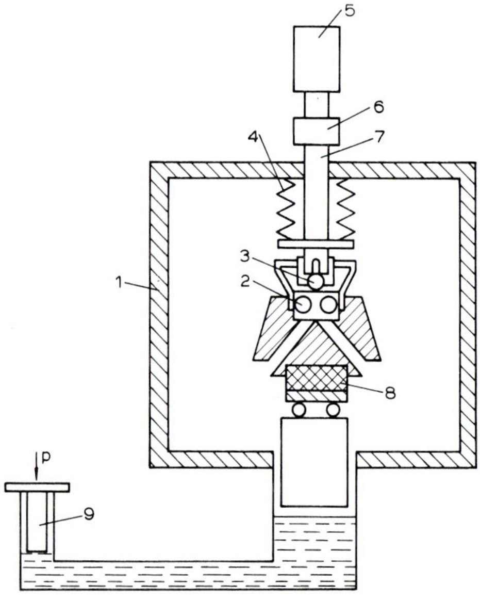

Fig. 7.25 Tr-6 tribometer (76): (1) vacuum chamber; (2) lower specimen (ball); (3) upper specimen (ball); (4) bellows; (5) motor; (6) reduction gear; (7) axle; (8) insulating packing with fixed collar; (9) hydraulic system.

TABLE 7.2

Characteristics of Tr-6 and Tr-7 Vacuum Tribometers

Tribometer	Friction pattern	Compressive forces, (kg)	Specific loads (kg/cm^2)	Sliding velocities (cm/sec)	Temperatures (°C)	Pressures in working chamber (mm Hg)
Tr-6	Four balls	1-800	6.7×10^3 - 62×10^3	0.2×10^{-8} - 50	20-600	10^{-5}-760
	Cylinder-plane (at ends)	1-800	1.1-800	0.4×10^8 - 100	200-600	10^{-5}-760
Tr-7	Hemisphere-plane	10^{-2}-1.0	----	0.5×10^{-4}-	20-250	10^{-6}-760
	Cylinder-end-plane	10^{-2}-1.0	2×10^2 - 2.0	0.5×10^{-4} - 1.7	20-350	10^{-6}-760

The Tr-7 tribometer is shown in Fig. 7.26. The specimen (13) of the material, in the form of a hemisphere or a cylinder, is brought into contact with the moving disc (14). The resulting friction force is stopped by the dynamometric spring (11) with transducers attached to it. Normal load is created by the lever (9), to which the load is transferred. Simultaneously with the disc's rotation, the specimen (13) can be moved from the center to the periphery. In this way, the friction track is a spiral. Rotation of the disc (6) is provided by the motor and the reduction gear. Low temperatures are maintained by a cooling jacket through which the coolant (methanol) is pumped, and high temperatures are supplied by the radiant heaters. Temperature regulation in this instrument is provided by an EPV-2 potentiometer. Thermocouplcs serve as temperature monitors.

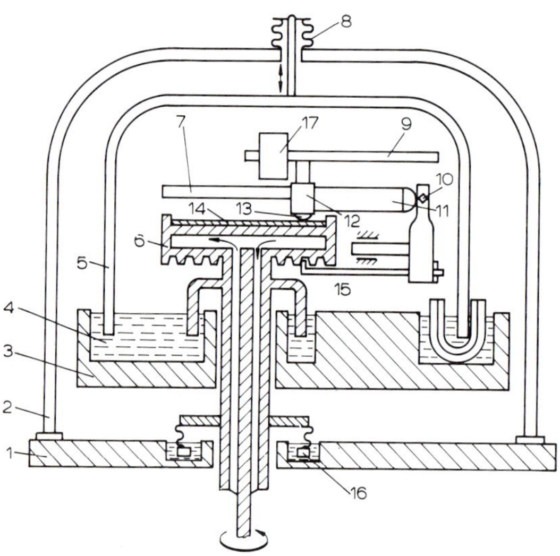

Fig. 7.26 Tr-7 tribometer (76): (1) base; (2) housing; (3) plate (or tile); (4) liquid; (5) cap; (6) disc; (7) radiant heater; (8) bellows; (9) lever; (10) tilting bearing; (11) dynamometric spring; (12) holder; (13) specimen; (14) friction disc; (15) connecting rod; (16,17) loads.

Matveevskii's four-ball machine, type KT-4. This is a modification of the same scientist's KT-3 machine (77,78). Basically, ball friction machines are designed to study lubrication properties. The KT-4 machine (Fig. 7.27) is composed of a vacuum cover (10) in which the friction assembly (5,6) is placed with recording equipment. The friction force arising from the upper ball's rotation is recorded by the elastic element (14) of the tensometric system. Normal load is created by the lever (8), one end of which acts on the three ball bearings through the spring (15) and the valve (1). Heating of the friction assembly to 70°C and recording of the temperature are provided for by the EPD-12 potentiometer. The rarefaction in the chamber can reach 5 x 10^{-5} mm Hg.

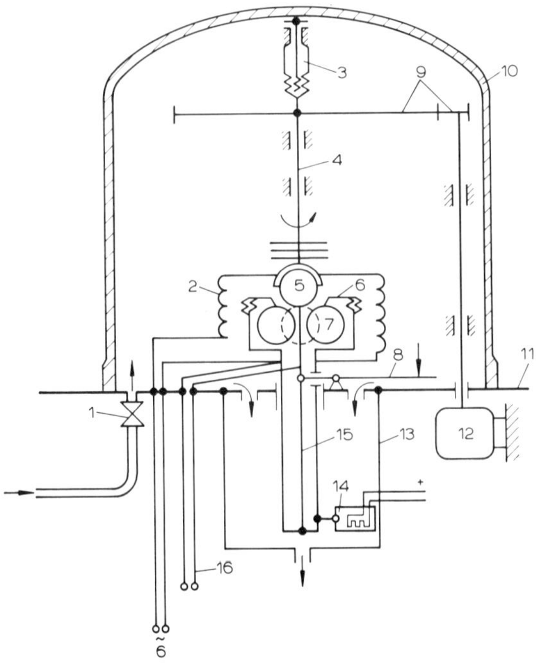

Fig. 7.27 Matveevskii's four-ball machine, type KT-4 (78): (1) valve; (2) vacuum chamber; (3) sliding mechanism; (4) spindle; (5) upper ball; (6) cup; (7) lower balls; (8) lever; (9) gear and piston; (10) vacuum cover; (11) mount; (12) motor; (13) heater; (14) spring with low modulus; (15) flat spring; (16) thermocouple.

Furthermore, for operation at a high vacuum Gib's instrument is provided, which is a modernization of the 1-47 instrument, (2). It allows testing friction properties in a vacuum of 10^{-9} mm Hg, at pressures to 5 kg/cm^2 and sliding velocities to 5 m/sec.

A simple and compact vacuum tribometer designed to study static friction in a vacuum to 10^{-3} mm Hg was built by Toporov and Lazarev (80). Many general-purpose tribometers have vacuum chambers or can be easily modified to have them (81-83).

Instruments for special uses

To this group belong instruments and machines designed for specialized studies and also special testing-mode instruments.

Tolstoy's tribometer. For the sliding friction, normal friction-surface displacements occur, which cause friction auto-oscillations and cause the friction force to depend upon sliding velocity (84-89). General-purpose instruments do not permit determination of the effect of normal displacements on the friction force. To do this, Tolstoy (86) constructed a specialized tribometer, shown in Fig. 7.28.

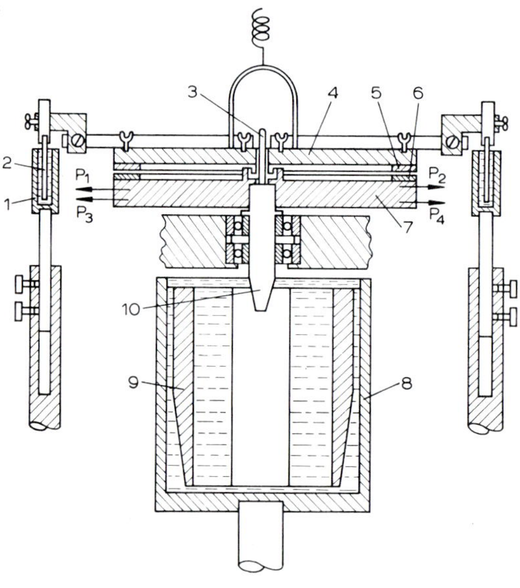

Fig. 7.28 Diagram of Tolstoy's tribometer (84): (1) vessel with damping liquid; (2) plate; (3) pivot; (4) upper disc; (5) specimen; (6) counter body; (7) lower disc; (8) vessel; (9) cylinder; (10) shaft.

This instrument has two metal discs (4,7), between which are friction pairs (5,6). The lower disc (7) rotates to loads P_1, P_2 or P_3, P_4. Stabilization of sliding velocity is attained by internal friction of the cylinder (10) on a viscous liquid in the vessel (8). A shaft (10) unites the cylinder (9) with the lower disc (7). Measurement of the friction force is accomplished by a tensometric apparatus that is connected with the lower disc.

A pivot (3), serves to measure normal displacements. Determination of normal displacements is accomplished by interferometric methods with an accuracy of 0.003 MKM. Normal oscillation of the upper disc with the specimen (4) is damped by a system consisting of a vessel with viscous liquid and with a plate immersed in it (1,2).

The sliding velocity and magnitude of damping are regulated by the viscosity of the damping liquid. This instrument permits measurement of the friction force for low sliding velocities from 10^{-3} to 10^{-9} cm/sec at room temperature.

Deryagin's instrument. This instrument makes use of the crossed filaments, which have received rather wide acceptance for the study of fiber friction and adhesion (71). The instrument's operation principle can be seen in the diagram in Fig. 7.29. It measures the force required to rub two elastic (quartz) filaments during tangential or normal motion of a sliding filament. The force of breaking and shear is determined by deformation of the mobile filament. The extent of deformation is measured visually through a microscope. The crossed filament is conveniently used for simultaneous determination of both friction and adhesion.

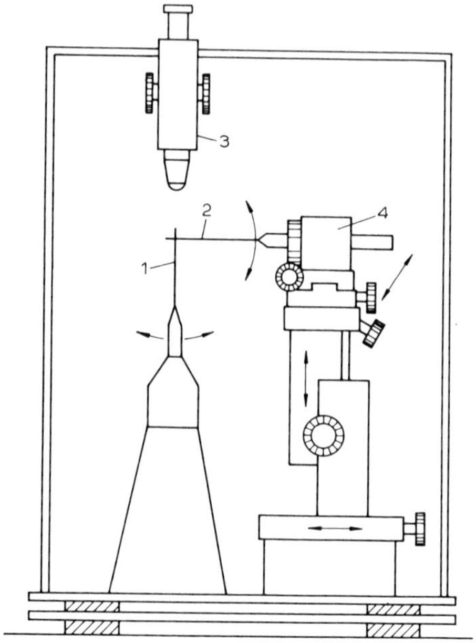

Fig. 7.29 An instrument to measure friction forces and adherence by the crossed-filament method (71): (1,2) quartz filament; (3) microscope; (4) device for moving the filaments.

Equipment for measuring friction force at high sliding velocities. Fairly high sliding velocities (to 50 m/sec) can be obtained on disc instruments. However, for higher velocities other methods must be used. For velocities on the order of 100 m/sec, an ultracentrifuge is used (90,91).

As an example, Fig. 7.30 shows the friction apparatus assembled on a UT_s-P-A ultracentrifuge by the Microtechna Company (Czechoslovakia). Specimens (4) of the material to be studied are fastened by clamps on the solenoid's core. The solenoid serves to produce a normal load between the specimens (4) and the ball (8), which rotates in magnetic field of the untwisting coil (5). The sliding velocity is determined according to the angular velocity of the rotating ball, measured by a photocell, which is supplied with light by a rotary chopper. Determination of the number of revolutions is based on a comparison of rotation with frequency.

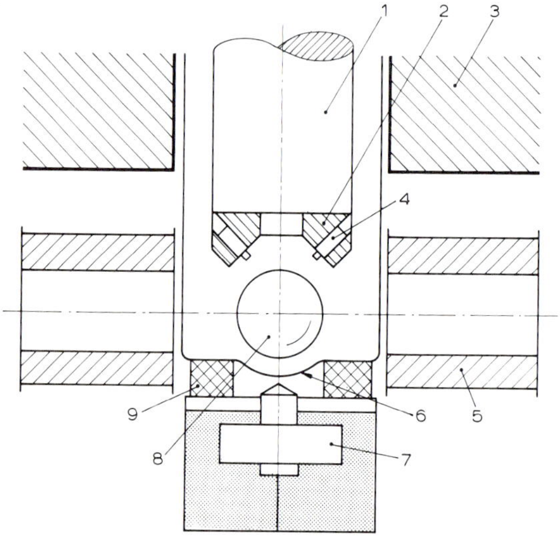

Fig. 7.30 High-velocity friction assembly (90): (1) core; (2) clamp; (3) solenoid; (4) specimen; (5) untwisting coil; (6) surface that limits the vacuum field; (7) damper; (8) ball; (9) monitor for the ball's position.

It follows that for this instrument, in the study of highly elastic materials in the friction mode with constant deformation, we have ε = const (92). Bartenev, Lavrentev, and Yelkin's instrument (92) serves to study the relation of the elastic friction force to the nominal contact area for p = const. The IM-58 inertia machine is designed to study friction and wear of friction pairs used in braking systems and is described in detail elsewhere (29). Rybalov's interesting instrument, created on the basis of 1-47-K-54 machines, serves to study polymer wear at constant temperature.

REFERENCES

1 M. Mooney, Sympsoium on Consistence, June 20, 1937, pp. 9-12.
2 I.V. Kragelsky, Treniye i Iznos (Friction and Wear), Mashriz, 1968.
3 M.M. Krushchov and M.A. Babichev, Issledovanize Iznashivaniya Metallov (Studies on Metal Wear), AN SSSR, Moscow, 1960.
4 V.D. Kuznetsov, Fizika Tvyordogo Tela (Solid Physics), Vol. 4, Tomsky, 1947.
5 A.S. Akhmatov, Molekulyarnaya Fizika Granichnogo Treniya (Boundary Friction Molecular Physics), Fizmatgiz, 1963.
6 D.N. Garkumov, Povysheniye Iznosostoikosti Detalei Mashin (Advancement in Wear Resistant Machine Components), Mashriz, 1960.
7 M.M. Krushchov (Ed.), Metody Ispytaniya na Iznashivaniye (Methods of Research on Wear), Nauka, 1967.
8 M.M. Krushchov (Ed.), Plastmassy kak Antifriktsionnye Materialy (Plastics as Anti-friction Materials), Nauka, 1967.
9 Anon., Collection, Novyye Raboty po Treniyu i Iznosu (New Studies on Friction and Wear), IL, 1959.
10 V.F. Yevstratov (Ed.), Friktsionnyi Izn Rezin (Friction Wear of Polymers), Khimiya, 1964, p. 56.
11 Anon., Collection, Teoriya Treniya i Iznosa (Theory of Friction and Wear), Nauka, 1965.
12 F.P. Bowden and D. Tabor, The Friction and Lubrication of Solids, Clarendon Press, Oxford, Part I, 1950; Part II, 1964.
13 I.V. Kragelsky and U.S. Shchedrov, Razvitiye Nauki o Trenii (Development of Friction Science), AN SSSR, Moscow, 1956.
14 B.I. Kostetsky, Soprotivleniye Iznashivaniyu Detalei Mashin (Machine Components Wear Resistance), Mashgiz, 1959.
15 N.G. Golego, Skvatvaniye v Machinakh i Metody Yego Ustraneniya (Hardening in Machine and Its Elimination), Tekhnika, Kiev, 1965.
16 Anon., Collection, Sukhoye Treniye (Dry Friction), AN Latv. SSR, 1962.
17 E.L. Roth, R.L. Driscol and W.L. Holt, Rubber Chem. Technol., 16(1943)155.
18 A.V. Chichinadze, Dissertation abstract, IMASh AN SSSR, 1954.
19 A.D. Kuritsyna and N.P. Isotomin, Mechinovedeniye, 1(1965)104.
20 B. Golding, Polymers and Resins, Their Chemistry and Chemical Engineering, Van Nostrand-Reinhold, New York, 1959.
21 G.M. Bartenev and Yu.S. Zuyev, Prochnost' i Razrusheniye Tverdykh Tel (Strength and Failure of Solids), Khimiya, 1964.
22 V.A. Mustafaev, G.V. Vinogradov and Yu.Ya. Podolsky, in Treniye i Iznos v Mashinakh (Friction and Wear in Machines), Vol. 9, Nauka, Moscow, 1964, pp. 111-127.
23 P.Ye. D'yachenko, N.N. Tolkacheva, G.A. Andreev and T.M. Karpova, Ploshchad' Falsticheskogo Kontakta Sopryazheronykh (The Actual Contact Area of Connected Surfaces), AN SSSR, Moscow, 1963.
24 N.B. Demkin, Fakticheskaya Ploshchad' Kontakta (The Actual Contact Area), AN SSSR, Moscow, 1962.
25 T. Turichin, Elektricheskiye Izmereniya Neelektricheskikh Velishin (Electric Measurement of Non-electric Dimensions), Energoizdat, 1959.

26 M.F. Malikov, Osnovy Metrologii (Formulations of Metrology), Moskua, 1949.
27 G.M. Bartenev and V.V. Lavrentev, Vysokomol Soedin., 2(1960)238.
28 S.B. Ratner and V.V. Lavrentev, Dokl. Akad. Nauk. SSSR, 108(No. 3)(1956)461.
29 A.V. Chichinadze, Rashet i Issledovaniye Vneshnogo Treniya pri Tormozhenii (Calculation and Research of External Friction During Braking), Nauka, Moscow, 1967.
30 A.V. Chichinadze and G.I. Troyanovskaya, in Peredovoi Nauchnotekniicheskii i Proizvodstvennyi Opyt (Advanced Scientific-Technical Industrial Experience), tema 20, No. M-57-127/6 Izd., VINITI, M., 1957.
31 G.I. Troyanovskaya, Dissertation abstract, IMASh, AN SSSR, 1956.
32 Yu.I. Kogan, Dissertation abstract, Tomsk, ros. unit, 1964.
33 S.A. Sukhov, Trudy 3 oi Vsesoyuznoi Conferentsii po Treniyu i Iznosu v Maschinakh, 2, Izd. AN SSSR, 1960.
34 L.M. Pyzhevich, Rashet Friktsionnykh Topmozov (Calculation of Friction Brakes), Mashinostroyeniyye, 1964.
35 I.V. Kragelsky, T.Ye. Chupilko and A.V. Chinchinadze, Protsessy Treniya v Tormozakh Aviakoles (Friction Processes in Airplane Wheel Brakes), AN SSSR, Moscow, 1955.
36 B.T. Kelgin and B.I. Krapov, Zaved. Lab., (No. 8)(1963).
37 V.M. Goryunob, in Novoye v Teorii Treniya (Recent Advances in Friction Theory), Nauka, 1966.
38 M.P. Levitskii, Zh. Tekh. Fiz., 9(1949)1010-1014.
39 Collection, Metody Izmereniya Temperatury (Methods of Measuring Temperature), IL, 1954.
40 V.S. Shchedrov, in Treniye i Iznos v Machinakh, Vol. 10, AN SSSR, Moscow, 1955.
41 L.M. Pyzhevich, in Teoriya Treniya i Iznosa, Nauka, 1968.
42 A.I. Belonsov, in Teoriya Treniya i Iznosa, Nauka, 1965.
43 A.V. Chichinadze, in Treniye Tverdykh Tel (Friction of Hard Solids), Nauka, 1964.
44 V.A. Kudinov, in Sukhoye Treniye, AN Latv. SSR, 1961.
45 J.F. Archard, Wear, 2(No. 6)(1959)438-456.
46 H. Blok, Inst. Mech. Eng., (No. 2)(1937)14-20.
47 A. Holm, J. Appl. Phys., 19(No. 4)(1948)361-366.
48 F.P. Bowden and K.E. Ridler, Proc. Roy. Soc., 154(1936)640.
49 A. Kovacs and A. Mesler, Proc. Roy. Soc., 35(No. 4)(1964).
50 V. Lavrentev, Plaste Kautsch., 21(1963)72.
51 S.B. Ratner and V.D. Sokolskaya, Sh. Khim. Prom., (No. 1)(1956)27.
52 G.M. Bartenev, V.V. Lavrentev and A.I. Yelkin, AVT. svid, 171646.
53 G.M. Bartenev, V.V. Lavrentev and A.I. Yelkin, in Teoriya Treniya i Iznosa, AN SSSR, Moscow, 1965, p. 290.
54 V.V. Lavrentev, in Teoriya Treniya i Iznosa, AN SSSR, Moscow, 1965, p. 17.
55 N.P. Istomin and A.D. Kuritsyna, Mashinovedeniye, (No. 1)(1965)104.
56 M.M. Reznikovsky and A.I. Lukomskaya, Mekhovicheskiye Ispytaniya Kauchuka i Rezina, Khimiya, 1968.
57 N.A. Anichikov and L.V. Elin, Zavod. Lab., (No. 3)(1940).
58 Collection, Novyye Raboty po Treniyu i Iznosu (New Studies on Friction and Wear), IL, 1959, p. 228.
59 M.M. Krushchov and M.A. Babichev, in Treniye i Iznos v Mashinakh, Vol. 9, Izd. AN SSSR, Moscow, 1954.
60 I.V. Kragelsky and F.P. Nepomnyashchy, Wear, 8(No. 4)(1965) 303-319.
61 G.Ye. Lazarev, Dissertation abstract, IMASh AN SSSR, 1963.
62 G.Ye. Lazarev, Friktsionnaya Teplostoikost; Materialov (Friction Resistance to Heat of Materials), Nauka, Moscow, 1969.

63 M.M. Reznikovsky, E.I. Goloskov, B.N. Atlas, Z.V. Shcherbach, G.I. Brodsky and S.B. Merezhannyi, in Friktsionnyi Iznos Rezin (Crosslinked Rubber Friction Wear), Khimiya, 1964.
64 B.L. Smushkevich, Mekh. Polim., (No. 1)(1966)151.
65 Ye.M. Shvetsova, Treniye i Iznos v Mashinakh, Vol. 7, AN SSSR, 1953, pp. 12-33.
66 N.B. Demkin and A.A. Lankov, Zavod. Lab., (No. 6)(1965)739.
67 L. Chestnov, Stanki i Instrum., (No. 9)(1954).
68 G.Ye. Lazarev and N.B. Demkin, Trudy i Konferentsii Molodykh Nauchnykh Rabotnikov (Proceedings of the Young Scientific Workers Conference), TSNIITMASh, 1961.
69 V.V. Lavrentev, Dissertation abstract, 193129, Bull. Izobr., (No. 6)(1967)89.
70 V.V. Lavrentev, G.Ye. Lazarev and A.I. Serebrenikov, in Proizvodstvo i Pererabotka Plastmass (Plastics Production and Reprocessing), or NIIPM, (No. 8)(1967)39.
71 B.V. Deryagin and Yu.P. Toporov, Tribory i Tekhnika Eksperimenta, (No. 6)(1960)132; Kolloidn. Zh., 23, No. 1, (1961)118-121.
72 V.V. Lavrentev, Kauch. Rezina., (No. 9)(1961)33.
73 N.G. Kolyadina, Dissertation abstract, ITI im. Lensobeta, 1963; G.M. Bartenev and N.G. Holyadina, Kauch. Rezina., (No. 5)(1962) 29.
74 G.M. Bartenev and A.I. Yelkin, Vysokomol. Soedin., 7(No. 6) (1965)992.
75 A.I. Yelkin, in Iznos i Treniye Metallov i Plastmass (Wear and Friction in Metals and Plastics), Nauka, 1964, p. 101.
76 I.V. Korepova and V.A. Mustafaev, in Teoriya Treniye i Iznosa, Nauka, 1965.
77 R.M. Matveevsky, in Plastmassy Kak Antifriktsionnyye Materialy, AN SSSR, Moscow, 1961.
78 R.M. Matveesky, in Iznos i Treniye Metallov i Plastmass, Nauka, 1964.
79 V.V. Grib, in Teoriya Treniya i Iznosa, Nauka, 1965, p. 281.
80 Yu. P. Toporov and V.P. Lazarev, Pribory i Tekhnika Exkperimenta, (No. 3)(1959)123.
81 O.V. Lozovskaya and R.M. Matveevsky, in Teoriya Treniya i Iznosa, Nauka, 1965, p. 312.
82 T.V. Larin and B.M. Astashkevich, Zavod. Lab., 26(No. 2)(1960).
83 B.V. Deryagin, V.P. Lazarev and Yu.P. Toporov, Zavod. Lab., 26 (No. 8)(1960)1060.
84 D.M. Tolstoy and R.L. Kaplan, in Teoriya Treniya i Iznosa, Nauka, 1965, p. 44.
85 A.S. Akhmatov, Trudy Konferentsii po Treniyu i Ixnosy, T.2, AN SSSR, Moscow, 1960.
86 D.M. Tolstoy, Stanki Instrum., (No. 1)(1961).
87 V.A. Kudinov, Issledovaniye Kolebany Metallorezhushchikh Stankov pri Rezanii Metallov (Research on Oscillation of Metal Cutting Machine Tools During Metal Cutting), AN SSSR, Moscow, 1958.
88 B.V. Deryagin, V.E. Push and D.M. Tolstoy, Trudy III Vsyesoyuznoi Konferentsii po Treniyu i Iznosu v Mashinakh (Proceedings of All-Soviet Conference on Machine Friction and Wear) Vol. 2, AN SSSR, Moscow, 1960, p. 132.
89 B.V. Deryagin, V.E. Push and D.M. Tolstoy, Zh. Tekh. Fiz., 26(No. 6)(1956)132; Trudy III Kontferentsii po Treniyu i Iznosu, T.II, AN SSSR, Moscow, 1960, pp. 132-152.
90 A.V. Semenov, in Teoriya Treniya i Iznosa, Nauka, 1965.
91 F.F. Bowden and E. Freitag, Nature, 176(1955)944.
92 G.M. Bartenev, V.V. Lavrentev and A. I. Yelkin, in Teoriya Treniya i Iznosa, Nauka, 1965.
93 S.L. Rybalov, in Teoriya Treniya i Iznosa, Nauka, 1965, pp. 302-306

AUTHOR INDEX

We ask the reader's indulgence with regard to the spelling of many of the authors' names in the list. As a result of transliteration from the Russian, there may be inaccuracies and inconsistencies.

SUBJECT INDEX